Transformation of Human Epithelial Cells

T0259031

Molecular and Oncogenetic Mechanisms

Editors

George E. Milo, M.S., Ph.D.
Professor, Department of Medical Biochemistry
Director of Carcinogenesis and Molecular Toxicology
Comprehensive Cancer Center
The Ohio State University
Columbus, Ohio

Bruce C. Casto, M.S., Sc.D.
Director of Research
Environmental Health and Research Testing, Inc.
Research Triangle Park, North Carolina

Charles F. Shuler, D.M.D., Ph.D.
Assistant Professor, Center for Craniofacial Molecular Biology
University of Southern California
Los Angeles, California

CRC Press
Taylor & Francis Group
Boca Raton London New York

CRC Press is an imprint of the
Taylor & Francis Group, an **informa** business

PREFACE

Several years ago, in the 1960s and 1970s, there were but a few human cell lines available to study human cell carcinogenesis. At that time the "Hayflick" hypothesis suggested that human cells cultured *in vitro* have a finite life-span of approximately 60 PDs; little was known of the "real" relationship between the limited life-span of the human fibroblast *in vitro* and aging *in vivo*. This area of research piqued the interest of the scientific community because the expression of cancer was thought to be an escape of the neoplastic phenotype from the limited proliferative potential, i.e., the finite life-span. However, over the years many human cancers were determined to be of an epithelial origin, and it was not possible to isolate, except on a rare occurrence, stable tumor phenotypes *in vitro* that exhibited infinite life-spans. Consistently heterogenous human tumor phenotypes either ceased to proliferate or terminally differentiated *in vitro*. Rarely did we observe in a routine fashion an escape from a limited life-span to a phenotype with an unrestricted proliferative potential. With the advent of collagen-coated plastic substratum, feeder layers, "quasi"-chemically defined growth media containing the likes of pituitary extracts, and growth factors, epithelial cells from different human tissues could be isolated and cultured for limited defined periods *in vitro*. Still the concept of the limited life-span of normal cells *in vitro* persisted and the cancer cell was thought to be an escape from a limited life-span. This loss of growth control and extension of life-span are discussed in Chapter 5 by Johng S. Rhim, one of the pioneers in the field of SV-40T-induced immortalization of human cells. We have included also a chapter on another developing field that will dramatically impact the field of growth control and life-span. Chapter 2 examines how autocrine and paracrine factors elicit proliferative growth responses in normal and transformed cells. In the 1970s, Heppner recognized along with others that epithelial tumors were heterogenous in cellular composition. Since that time we have learned how to identify different phenotypes from the epithelial tumors. Because of the recent achievements by Bert Vogelstein in the late 1980s of identifying, in colon carcinoma tumors, cells that occupy different stages of progression and the discovery by Patricia Steggs and Lance Liotta in the late 1980s of the molecular events that were associated with expression of the metastatic stage, we can now follow the events of metabolism of putative human carcinogens, DNA-adduct binding, early, middle and late stages of progression of initiated cells, anchorage-independent growth, identification of expression of altered phenotypes, and identification of expression of tumorigenic and metastatic phenotypes. If we agree that tumors are clonal in origin, then we need to understand that tumor heterogeneity may be a function of the composition of mixed phenotypes. Some of these questions comprise the reasons this book and the contributors were assembled to address the stages from metabolism to metastasis of epithelial cells associated with epithelial tumors.

In Chapter 1, James R. Smith addresses the role immortalization plays in cancer and how resistant human cells are to spontaneous immortalization. There has been an attempt to link the loss of expression of a finite life-span to the change in genetic program of either the 1st or the 4th chromosome. A defect in these chromosomes putatively can give rise to the expression of a cellular phenotype that exhibits immortalization. However, at present (in these authors' opinion) the question of quiescence vs. senescence in human cells has not been answered by these experiments. The loss of senescent control can lead to immortalization; the loss of control of expression of quiescence cannot. However, it is recognized that the transient down-regulation of the cyclin gene or the permanent interruption of this gene function may play a pivotal role in the two processes. The next several chapters discuss the stages of multi-stage carcinogenesis in different epithelial cell systems. Moreover, correlative stages of progression in different carcinogen-transformed epithelial cell systems are compared and contrasted. The last few chapters — 10, 11, and 12 — discuss molecular controlling mechanisms that are involved in the control of expression of stages of progression, e.g., the role of oncogenes and their interactiveness with the suppressor genes. In particular, the role of *ras* gene mutations and suppressor gene interaction with these activated on-cogenes in tumorigenic cells in controlling the progression of initiated cells into tumor cells is presented. The promotion stage in human cell carcino-genesis is a silent stage and has not been observed experimentally, and many attempts to discover this stage have led to failure.

In Chapter 13 we compare and contrast the stages from metabolism, DNA-adduct formation, anchorage-independent tumor growth, and regression be-tween epithelial cells and fibroblast (see also *Transformation of Human Dip-loid Fibroblast: Molecular and Genetic Mechanisms,* Milo, G. E. and Casto, B. C., Eds., CRC Press, 1990). Lastly, an invitation was extended to Juergen R. Vielkind to contribute a guest chapter discussion on the nature of growth control in osteichthyes. Our reasons for including this chapter are to develop and understand the biological significances of the conservation of suppressor gene function phylogenetically from lower animals up to humans and to understand the role of some of these genes in more primitive animal systems.

It was our intention to present to the scientific community a forum for focusing on the utility of how human epithelial cells as systems can be used to examine how environmental xenobiotics can be metabolized to reactive metabolites that can react with critical sites on the genomic DNA that lead to the expression of an early stage of a transformed phenotype. Later in the book we have focused on how current dogma in suppressor gene-oncogene interaction is insufficient to totally explain the existence of many diverse phenotypes in a heterogenous spontaneous epithelial cell tumor. Moreover, the concept of plasticity — the transient expression of a tumorigenic phen-otype — is not explained only by the presence of mutations in critical sites of activated oncogenes or mutations in suppressor genes. This treatise is an attempt to collate many of the scientific results, significant scientific concepts, and laboratory efforts from active investigators in this field of environmentally induced human epithelial cell cancer.

George E. Milo
Bruce C. Casto
Charles F. Shuler

THE EDITORS

George E. Milo, B.A., M.S., Ph.D., is Professor of Medical Biochemistry in the College of Medicine and Program Director of Molecular Environmental Health in the Comprehensive Cancer Center at The Ohio State University in Columbus.

Dr. Milo has actively pursued research in the discipline of human cell carcinogenesis since 1970. He published the first report on the transformation of human fibroblast cells in 1978 in *Nature* and the first report on the transformation of human epithelial cells in *Cancer Research* in 1981. Again in 1990, along with Dr. Charles Shuler, he published a new concept on the isolation and identification of a plastic anchorage-independent growth of a nontumorigenic phenotype that can be transiently converted to a tumorigenic and metastatic phenotype. This article was published in the *Proceedings of the National Academy of Science (U.S.A.)*. His National Institutes of Health (NIH)-supported postdoctoral training at the Roswell Park Memorial Institute in Cancer Research (Buffalo, NY) served him well as a stepping stone to begin his career in human cell carcinogenesis.

Dr. Milo is a member of the Society for Toxicology — Molecular Toxicology Division, the American Association for Cancer Research, the American Society for Biochemistry and Molecular Biology, the American Society for Cell Biology, and the International Society for the Study of Xenobiotics.

He has published in excess of 130 publications in the discipline of human cell carcinogenesis. He has written several chapters for different books on the subject and has served as an *ad hoc* reviewer for the National Institutes of Health — National Cancer Institute (NIH-NCI) in the discipline, as a reviewer and Chairperson for the U.S. Environmental Protection Agency Extramural Health Effects Research Review Panel, as a Chairperson for the U.S. Environmental Protection Agency "Health" Research Centers program, and as a reviewer on the NIH-NCI Parent Preclinical Pharmacology Program Panel. He has received many grants from the National Institutes of Health — National Cancer Institute and National Institute of Environmental Health Science — and from the U.S. Environmental Protection Agency — Health Effects Research. He also edited, along with Dr. Bruce C. Casto, *Transformation of Human Fibroblasts: Molecular and Genetic Mechanisms*, published by CRC Press in 1990.

His present area of interest is to investigate how exposure to environmental xenobiotics alters human gene function.

Bruce C. Casto, M.S., Sc.D., is Director of Research for Environmental Health and Research Testing, Inc. in Research Triangle Park, North Carolina. Dr. Casto received training in microbiology and virology at The Ohio State University and the University of Pittsburgh. During this time, research was conducted in the areas of viral receptors, oncolytic viruses, and adeno-associated viruses. While at the University of Pittsburgh, Dr. Casto discovered the defective nature of AAV-1 and its dependence on adenovirus for replication. He was an assistant member at the Institute for Biomedical Research — American Medical Association, professor of microbiology at Rush Medical School, senior scientist at BioLabs, Inc., and research director for health effects at Northrop Environmental Sciences. Dr. Casto's major area of research is chemical carcinogenesis, especially the enhancement of viral transformation by chemical carcinogens and the chemical transformation of mammalian cells *in vitro*.

Charles F. Shuler, D.M.D., Ph.D., is Assistant Professor in the Center for Craniofacial Molecular Biology at the University of Southern California School of Dentistry.

Dr. Shuler received his dental degree from the Harvard University School of Dental Medicine, his Oral Pathology training at the University of Minnesota, and his Ph.D. in Experimental Pathology from the University of Chicago.

Dr. Shuler is a member of the American Society for Cell Biology, the American Association for Dental Research, the American Association for Advancement of Science, and the American Academy of Oral Pathology. He has served as an Associate Editor of the *Journal of Oral Pathology*.

His current areas of research interest include mechanisms of epithelial differentiation, especially during the development of the secondary palate *in utero*, and human cell transformation and tumorigenesis.

CONTRIBUTORS

William M. Baird, Ph.D.
Glenn L. Jenkins Professor of
  Medicinal Chemistry
and Purdue Cancer Center
Purdue University
West Lafayette, Indiana

Linda L. Barrett, M.S.
School of Medicine
East Carolina University
Greenville, North Carolina

Michael J. Birrer, M.D., Ph.D.
NCI-Navy Medical Oncology
  Branch
Clinical Oncology Program
Division of Cancer Treatment
National Cancer Institute
Bethesda, Maryland

Tammela Butler, B.S.
Department of Toxicology
University of North Carolina
Chapel Hill, North Carolina

Charleata A. Carter, Ph.D.
Experimental Carcinogenesis and
  Mutagenesis Branch
National Institutes of
  Environmental Health Services
Research Triangle Park, North
  Carolina

Bruce C. Casto, M.S., Sc.D.
Environmental Health & Research
  Testing, Inc.
Research Triangle Park, North
  Carolina

Dharam P. Chopra, Ph.D.
Institute of Chemical Toxicology
Wayne State University
Detroit, Michigan

Frank C. Cuttitta, Ph.D.
Biomarker and Prevention
  Research Branch
NCI-Navy Medical Oncology
  Branch
Biotherapy Section
National Naval Medical Center
National Cancer Institute
and Department of Medicine
Uniformed Services University of
  the Health Sciences
Bethesda, Maryland

Curtis C. Harris, M.D.
Laboratory of Human
  Carcinogenesis
National Cancer Institute
Bethesda, Maryland

**David G. Kaufman, M.D.,
Ph.D.**
Department of Pathology
University of North Carolina
  School of Medicine
Chapel Hill, North Carolina

Hudson H. S. Lau, Ph.D.
Department of Medicinal
  Chemistry Pharmacognosy
Purdue University
West Lafayette, Indiana

Caroline H. Laundon, Ph.D.
GeneCare
Chapel Hill, North Carolina

Teresa A. Lehman, Ph.D.
Laboratory of Human
  Carcinogenesis
National Cancer Institute
Bethesda, Maryland

George E. Milo, Ph.D.
Department of Medical
 Biochemistry
and Department of Molecular
 Environmental Health of the
 Comprehensive Cancer Center
The Ohio State University
Columbus, Ohio

Zenya Naito, M.D., Ph.D.
Medical Technology
Yokosuka National Hospital
Yokosuka, Kanagawa, Japan

Johng Sik Rhim, M.D.
Department of Radiation Medicine
Georgetown University School of
 Medicine
Washington, D.C.
and Laboratory of Cellular and
 Molecular Biology
National Cancer Institute
Bethesda, Maryland

Clifford A. Rinehart, Ph.D.
Department of Pathology
University of North Carolina
Chapel Hill, North Carolina

**Charles F. Shuler, D.M.D.,
 Ph.D.**
Center for Craniofacial Molecular
 Biology
School of Dentistry
University of Southern California
Los Angeles, California

Jill Siegfried, Ph.D.
Department of Pharmacology
University of Pittsburgh
Pittsburgh, Pennsylvania

James R. Smith, Ph.D.
Roy M. & Phyllis Gough
 Huffington Center on Aging
and Division of Molecular
 Virology
Baylor College of Medicine
Houston, Texas

Martha R. Stampfer, Ph.D.
Department of Cell and Molecular
 Biology
Lawrence Berkeley Laboratory
University of California
Berkeley, California

Gary M. Stoner, Ph.D.
Experimental Pathology
Department of Pathology
Medical College of Ohio
Toledo, Ohio

Juergen R. Vielkind, Ph.D.
Department of Cancer
 Endocrinology
British Columbia Cancer Agency
and Department of Pathology
University of British Columbia
Vancouver, British Columbia,
 Canada

Li Hui Xu, M.D.
Department of Pathology
University of North Carolina
Chapel Hill, North Carolina

Paul Yaswen
Department of Cell and Molecular
 Biology
Lawrence Berkeley Laboratory
University of California
Berkeley, California

TABLE OF CONTENTS

Chapter 1

IN VITRO CELLULAR AGING AND IMMORTALIZATION

James R. Smith

TABLE OF CONTENTS

I. INTRODUCTION

Carcinogenesis has become widely accepted as a multistep process[2] (see References 1 and 2 for recent reviews). A number of events have to occur in order for cells to become cancerous. In many cases, if not all, cellular immortalization is one of these steps and is an obligatory process.[3] Normal human diploid cells go through various numbers of population doublings (PDs), depending on the age of the donor and the origin of the tissue from which the cells are derived.[4-6] In most cases, cells from adult donors go through fewer PDs than cells from young or embryonic donors.[6] The number of PDs that a culture can go through when derived from adult tissue is typically 20 to 30.[5] Cancers generally are of clonal origin, and for a single cell to produce a tumor 1 g in size requires approximately 30 cell divisions. Primary tumors are not the major cause of problems in carcinogenesis because of the possibility of surgical removal of the primary tumor and, hence, the threat to the individual by that tumor. Indeed, metastasis is the crucial step that causes carcinogenesis to be a life-threatening phenomenon. Metastases are also generally of clonal origin and require a cell that has already gone through a number of doublings in the primary tumor to undergo further doublings as a metastatic growth in order to be significant. Cell growth, tumor regression, and cell death are all normal parts of the processes of carcinogenesis. Therefore, the number of PDs that cells have to go through in order to become life-threatening may be more than 100 to 200. This range is clearly greater than normal cells are able to go through, as evidenced by experiences with human fibroblasts in tissue culture.[7] Other cells in the body may normally be able to go through more doublings *in vivo*. However, at this time, the doubling limit for most epithelial cells *in situ* is unknown. Therefore, it is reasonable to assume that as part of the multistep process of carcinogenesis, cellular immortalization is required for tumor progression and metastasis. The spontaneous immortalization of human cells in culture has never been observed. However, this can be contrasted with the situation that we see in rodent cells, particularly mouse and rat cells, in which spontaneous immortalization is the rule rather than the exception.[8-10] When considering whether immortalization may be necessary for tumor formation and metastasis, it is interesting to compare the rates of tumor formation in rodents with those in humans. A mouse weighs on the order of 10 g while humans weigh on the order of 100 kg, and the mouse's life-span is approximately 1/30 that of a human, yet mice very often have tumors during their $2^1/_2$- to 3-year life-span. Therefore, on a per cell unit time basis, the rate of tumor formation in mice is 10^5 to 10^6 times the rate of tumor formation in humans. It seems likely that this incredibly higher rate of tumor formation seen in mice compared to humans is due to the much higher incidence of spontaneous immortalization of mouse cells compared with human cells. Therefore, the study of cellular immortalization and of mechanisms that limit the proliferative potential of normal human cells in culture is of paramount importance in understanding the mechanisms of carcinogenesis in humans.

II. *IN VITRO* CELLULAR AGING IS DOMINANT IN SOMATIC CELL HYBRIDS

A. LIMITED *IN VITRO* LIFE-SPAN OF NORMAL CELLS

Swim and Parker were the first to show that human fibroblasts derived from biopsies had a limited proliferative potential in culture.[11] Hayflick and Moorehead in 1961 showed that these cells were karyotypically normal and that normal cells derived from a large number of different individuals all had a finite proliferative potential.[4] They also showed that a major characteristic of cells that were able to divide indefinitely, i.e., transformed immortal cells, was an abnormal karyotype. In 1965, Hayflick proposed that the limited *in vitro* proliferative potential of normal human fibroblasts in culture was a manifestation of aging at the cellular level.[5] More recently, it has been proposed by O'Brien et al.[3] that limited proliferative potential of normal cells *in vitro* and also *in vivo* is a powerful tumor suppressor mechanism. The observation of limited proliferative potential of normal cells in culture has been repeated in hundreds of labs and thousands of cultures over the last 30 years.[7] The proliferative potential of the cells depends on the age of the donor,[6] species of the donor,[12] and the site of biopsy.[6] Typically, human embryonic cells will undergo 50 to 80 PDs before growth cessation, although it has been reported that some cells are capable of going through approximately 100 PDs before proliferation stops.[13] Cells from other species go through fewer PDs than those from humans, the exception being the Galapagos turtle.[14] The number of PDs that the cells are able to undergo is correlated with the maximum life-span of the species.

Cells spontaneously immortalize at various rates, depending on the species of origin of the cells. Human cells and chick cells have never been observed to immortalize spontaneously, while rodent cells routinely immortalize in culture and species such as bovine immortalize spontaneously only rarely.[15] The mechanisms that lead to limited *in vitro* proliferative potential of normal cells in culture is not understood. A number of investigations have been carried out over the past 30 years to measure various biochemical, metabolic, and structural parameters of these cells as they age in culture, and with very few exceptions, which will be discussed later, no changes have been observed that could account for the irreversible division cessation. Likewise, the process by which cells escape the finite proliferative potential and become able to divide without limit (immortalization) is not understood. In order to try to understand the mechanisms operating in these processes, we and others have undertaken a series of experiments discussed below.

B. HYBRIDS BETWEEN NORMAL AND IMMORTAL CELLS

The early work of Littlefield suggested that the limited proliferative potential (the senescence phenotype) might be dominant in somatic cell hybrids between senescent cells and young proliferating cells.[16] However, the evidence was not conclusive and the prevailing belief at that time was that cellular immortalization was due to dominant changes in the cellular genome. Many different ideas have been presented to try to explain the limited proliferation of normal cells in culture. These can be broken into two main categories. One category proposes that cells stop dividing because they accumulate damage of various sorts, e.g., somatic cell mutations or errors in protein synthesis, so that the error burden becomes so large that the cells are no longer able to divide. The other category proposes some sort of genetic program that limits the *in vitro* life-span of cells in culture. We thought that we might be able to differentiate between these two broad categories of hypothesis by fusing normal cells with immortal cells and determining whether the hybrids resulting from that fusion had a limited *in vitro* life-span or were immortal. If normal cells stopped dividing because they had accumulated a large amount of damage, then one could argue that cells that are immortal have escaped from limited proliferative potential because either they don't accumulate damage at the same rate or they have evolved a mechanism to better cope with the damage. Therefore, one might expect in hybrids that the phenotype of cellular immortality would be dominant.

In the first set of fusions, we fused an immortal SV40-transformed cell line with a normal cell line that was at the end of its *in vitro* life-span.[17] We observed that the hybrid colonies proliferated for various numbers of PDs and then stopped dividing. About 70% of the colonies were able to go through fewer than 8 PDs, while the other 30% were able to go through a range of PDs varying from 30 to 60, but they all stopped dividing. We also showed that all of the clones expressed the SV40 large T-antigen which is thought to be the immortalizing agent for normal human diploid fibroblasts infected with SV40 virus. In order to investigate the generality of this phenomenon, we fused a number of different cell lines with normal human diploid fibroblasts and observed the same results in all cases.[18] The hybrids had finite proliferative potential. In all the fusion experiments, we found that immortal variants arose in the culture at a frequency of approximately 1 per 10^5 to 10^6 cells. This is a much greater frequency of immortalization than that observed in normal diploid fibroblasts. The tentative explanation for this is that in hybrids, chromosomal segregation takes place and the hybrids lose a chromosome which encodes a gene that causes the finite proliferative potential. Conclusions from these experiments are that the limited life-span of normal cells in culture is dominant over the phenotype of cellular immortality and that cells become immortal because they lose some of the program that is necessary to impose a limited proliferative potential on normal cells in culture.

C. FUSION OF IMMORTAL CELLS WITH OTHER IMMORTAL CELL LINES

If cellular immortality results from recessive changes in the cellular genome, then we might expect that different defects could occur to render a cell immortal. If that is the case, then fusion of cell lines having one defect with cell lines having another defect could result in complementation, giving a hybrid that has finite proliferative potential. On the other hand, fusion of cell lines having the same defect would not result in complementation and would give rise to hybrids that could divide indefinitely. Therefore, we would predict that hybrids resulting from fusions among different immortal cell lines would give two different kinds of results. In one case, some hybrids would have a finite life-span and the other hybrids would have an indefinite life-span. In a series of experiments, Pereira-Smith and Smith fused different cell lines with each other and observed the proliferative phenotype (either finite or indefinite),[19] and assigned more than 30 different cell lines to four different complementation groups. In order to begin the process of complementation group assignment, one SV40-transformed cell line was chosen at random to be representative of complementation group A. Other cell lines were fused with it. Those that had an indefinite proliferative potential also assigned to complementation group A; those that had a finite proliferative potential assigned to a different complementation group. Using HeLa as a prototypic cell line for complementation group B, we repeated the process and assigned cell lines to complementation group B. Cell lines were assigned to other complementation groups in a similar fashion. This process involved numerous cell fusions, and in no case did we find a cell line that assigned to more than one complementation group. This indicated that the processes resulting in cellular immortality were very rare, with no cell lines carrying two different defects. We looked at a large number of different cell lines resulting from different kinds of tumors, different cell types of origin, cell lines derived from different embryonic layers, and cell lines that contained activated oncogenes, and in no cases did these parameters affect complementation group assignment. The only parameter that did affect assignment was immortalization by the SV40 large T-antigen. Seven out of eight of the SV40-immortalized cell lines assigned to complementation group A. One of the SV40 cell lines failed to assign to complementation group A. The reason for this is not known. We can speculate that the SV40 T-antigen was not the actual immortalizing agent in this case, but was only coincidental in the transformation process. The assignment of cell lines to different complementation groups allows us to take a systematic approach to trying to understand what processes might have occurred to result in cellular immortalization. We speculate at the present time that those cell lines which assign to the same complementation group have become immortalized by the same genetic defect. The case for this interpretation is strengthened by the fact that almost all of the SV40-transformed immortalized cell lines fall into the same complementation group. However, it appears that not all DNA tumor viruses immortalize cells by the same mechanism, because we found that cell lines immortalized by adeno, papilloma, and herpes virus fell into different complementation groups. Efforts are currently underway to find the genetic defect that leads to immortalization in the case of SV40 T-antigen.

D. MICROCELL HYBRID EXPERIMENTS

The introduction of single normal human chromosomes into immortal cell lines represents a considerable refinement over the techniques of somatic cell hybridization involving whole cells discussed above. The use of microcell hybrid techniques has allowed us to assign genes coding for normal cellular aging processes to one particular human chromosome.

Ning et al.[20,21] introduced chromosome 11 from a normal human cell line into immortal cell lines representative of all four complementation groups. They observed no effect on growth rate or the immortal phenotype of these cells. There was some minor and variable effect on tumorigenicity when cells carrying the intact human chromosome 11 were injected into nude mice. There was, in some cases, suppression of tumorigenicity and, in other cases, a delay in the formation of tumors.

Ning et al.[22] further showed that introduction of a normal human chromosome 4 into cell lines assigned to complementation group B restored the phenotype of limited proliferative potential. However, when the human chromosome 4 was introduced into cell lines assigning to the other complementation groups (A, C, and D), there was no decrease in proliferation potential and the phenotype of immortality was retained. Thus, it seems clear that genes on chromosome 4 code for some part of the genetic program that limits the division potential of normal cells in culture. Disruption of these genes leads to cells with an immortal phenotype. Sugawara et al.[23] found a similar result in studies in which they introduced the normal human chromosome 1 into Chinese hamster cells. Human chromosome 1 was able to restore the cellular aging phenotype in these immortal hamster cells. It remains to be seen whether chromosome 1 plays a role in the immortalization of human cells.

III. CELLULAR AGING IS AN ACTIVE PROCESS

A. HETEROKARYON EXPERIMENTS

One of the first experiments that gave us an idea of the kinds of processes that might be responsible for cellular senescence was performed by Norwood et al.[24] and independently by Stein and Yanishevsky.[25] They fused senescent cells that had reached the end of their *in vitro* life-span with normal cells that were still able to proliferate and asked whether the nuclei contained in the heterokaryon were able to synthesize DNA. When senescent cells were fused with young cells, it appeared that the senescent cell was able to suppress the initiation of DNA synthesis in the young cell nucleus, i.e., neither the young cell nucleus nor the senescent cell nucleus synthesized DNA in the heterokaryons up to 72 h after fusion. However, if young cells were fused with each other, there was no decrease in the ability of the young cell nuclei to synthesize DNA in the homodikaryon. From these results, it was concluded that senescent cells produce an inhibitor of DNA synthesis which is able to act in *trans* to inhibit the initiation of DNA synthesis in the young nucleus. Furthermore, it has been shown that senescent cells, when fused with various immortal cell lines,[25] suppress DNA synthesis in the nucleus of the immortal cell. This indicates that the inhibitor produced by senescent cells is able to also inhibit the initiation of DNA synthesis in certain immortal cell lines. However, other immortal cell lines,[26] in particular those that have been immortalized by DNA tumor viruses, e.g., SV40-transformed cells of HeLa cells (HeLa is known now to have part of the herpes virus DNA integrated into its genome), are able to induce DNA synthesis, in the short term, in the senescent cell nucleus. This indicates that although senescent cells produce an inhibitor of DNA synthesis, it is possible, through the intervention of DNA tumor viruses, to temporarily override this inhibitor.[27]

Yanishevsky and Stein showed that the inhibitor of DNA synthesis present in senescent cells could not interrupt ongoing DNA synthesis, but acted to block the initiation of DNA synthesis. They found that if senescent cells were fused to young cells that were more than 3 or 4 h from the S-phase, then initiation of DNA synthesis was blocked. If they were closer to the S-phase, then initiation of DNA synthesis was not blocked.[28]

B. MEMBRANE-ASSOCIATED DNA SYNTHESIS INHIBITORS OF SENESCENT CELLS

Although the production of an inhibitor by senescent cells is a simple and attractive explanation for the above results, there may be other explanations. For example, if senescent nuclei were depleted of some factors that were needed for induction of DNA synthesis and competed with the young cell nucleus for those factors, then the concentration of positive regulatory factors could fall below a critical threshold in the heterokaryons. In order to rule out that possibility, we initiated experiments in which we prepared enucleated cytoplasms from senescent cells and fused them to whole young cells, and then asked whether the senescent cytoplasts could cause inhibition of the initiation of DNA synthesis in the resulting cybrids. We found that senescent cytoplasts were indeed capable of inhibiting the initiation of DNA synthesis in young-cell cybrids.[29-31] There was an approximately 50% decrease in the number of young-cell nuclei synthesizing DNA in the senescent-young cell cybrids compared to cybrids from young-cell cytoplasts fused with young whole cells. We next asked the location of the inhibitor of DNA synthesis in senescent cytoplasts. By treating the senescent cytoplasts with trypsin under conditions that would limit the penetration of the trypsin into the cell and limit intracellular damage by trypsin (4°C for 1 min), we were able to show that the inhibitory activity resided on the outside surface of the membrane.

Further evidence for this conclusion was obtained by preparing membrane-enriched fractions from senescent cells and adding them to young-cell cultures. These membrane-enriched fractions were very effective in inhibiting the initiation of DNA synthesis in young-cell cultures.[31,32] Furthermore, proteins extracted from the membrane preparations and added directly to cultures of young cells were also effective in inhibiting the initiation of DNA synthesis.[31] We next examined the role of protein synthesis in the production of this inhibitor by treating the cells with cyclohexamine or puromycin at concentrations which would inhibit protein synthesis by at least 90%. We found that a relatively short treatment, approximately 2 h, with cyclohexaminc or puromycin was sufficient to eliminate the inhibitory activity from senescent cells. Upon removal of cyclohexamine from the culture and incubation of the cytoplasts in the absence of cyclohexamine, inhibitory activity was regained in about 4 h,[31] indicating that cytoplasts were still active and able to synthesize DNA. Further, this indicated that the messenger RNA coding for the inhibitor was relatively long-lived.

C. INHIBITION OF DNA SYNTHESIS BY POLY(A)$^+$ RNA

In order to explore the feasibility of searching for cDNA clones coding for the inhibitor, we microinjected poly(A)$^+$ RNA from senescent cells into young proliferation-competent cells to determine whether inhibitory activity could be conferred by the messenger RNA. We found that it was strongly inhibitory.[33] By microinjecting different amounts of RNA into the cells, we were able to calculate that the inhibitor RNA was present in relatively large abundance (0.1 to 1% of the total messenger RNA consisted of inhibitor RNA according to our calculations).

We also studied the possibility that nongrowing tissues, e.g., rat liver, might produce an inhibitor. Rat liver was chosen because it can exist either in a state of nonproliferation or in a state of proliferation. Lumpkin et al.[34] found that nonregenerating liver RNA was able to block the initiation of DNA synthesis when microinjected into young human fibroblasts, whereas the RNA isolated from regenerating rat liver had no inhibitory activity; indeed, it had a stimulatory activity. This raised the possibility of using rat liver as a source of RNA to carry out syntheses and screening of a cDNA library to isolate genes that were expressed in nonregenerating liver but not expressed in regenerating liver. Inhibitory RNA has been hybrid selected by cDNA clones isolated by differential screening of a cDNA library made from nonregenerating rat liver poly(A)$^+$ RNA.[35] However, these clones do not code for a messenger RNA that is upregulated in senescent cells (unpublished data). Other investigators have confirmed these results using RNA from human liver[36] and have extended them to show that RNA isolated from resting human T-lymphocytes also will inhibit initiation of DNA synthesis when injected into proliferation-competent human fibroblast cells.[37] The results from these microinjection experiments indicate that it would be feasible to search for cDNAs coding for the inhibitory messenger RNA and the inhibitory protein. Construction of cDNA libraries and screening by differential screening using probes made from young cells and senescent cells is now underway in several laboratories. Using this approach, various cDNA clones have been isolated; however, as of yet, no clones have been isolated that have been shown to code for the inhibitory activity expressed in senescent cells. Although the cDNA cloning of inhibitor genes has not been fruitful to date, investigators are still optimistic that straightforward (+)/(−) screening of cDNA libraries produced from senescent cells or cells from premature aging syndrome patients will yield the inhibitor genes that are associated with cellular senescence.

Another approach to this problem is to produce monoclonal antibodies by immunizing mice with surface membrane preparations from senescent cells. We have isolated monoclonal antibodies in this way.[38] Of approximately 6000 hybridoma cultures screened, three yielded antibodies that reacted preferentially with senescent cells but not with young cells. All the antibodies react with an epitope on the fibronectin molecule. Although they react with fibronectin from various sources, when it is denatured, they only react with fibronectin distributed on the surface of senescent cells, not with fibronectin distributed on the surface of young cells. This indicates that senescent cells are producing either a fibronectin that is altered in its primary structure or posttranslationally modified in a way different from that of young cells. The senescent cell-produced fibronectin would then have a different conformation than that of young cells. Another possibility is that the fibronectin protein produced by senescent cells and young cells is the same, but the interaction of fibronectin with other molecules produced by senescent cells and young cells is different, thus leading to an altered confirmation of fibronectin associated with senescent cells, exposing an epitope which is sequestered in young cells.

IV. DISCUSSION

The experimental results reviewed here indicate that cellular aging is an active process, perhaps part of a genetic program, and that this genetic program can be disrupted in various ways, giving rise to cellular immortality. We have shown that there are at least four different ways that the normal cell processes can be disrupted to lead to cellular immortality. One of the processes that takes place in normal cells seems to be production of an inhibitor of the initiation of DNA synthesis. This inhibitor is produced in quiescent cells, but is reversible by the addition of growth factors or serum mitogens. The inhibitor produced by senescent cells is not reversible. We do not know at the present time whether the inhibitor produced by quiescent cells is the same as the inhibitor produced by senescent cells. It is possible that they are the same, but the control of expression of the inhibitors is different between young cells and senescent cells. In senescent cells, the inhibitor is being produced constitutively, whereas in young cells it is modulated in different parts of the cell cycle and by growth factors.

Recently, changes in a number of cell cycle genes as cells become senescent have been reported. These include failure to express c-*fos* or *cdc*2 when the cells are mitogenically stimulated,[39,40] and the failure to phosphorylate the *Rb* gene.[41] It may be that the defects in the regulation of these cell cycle-related genes are responsible for senescent cells being unable to enter into the S-phase. However, according to the data presented above, these changes would be secondary to the expression of a cell surface inhibitor of

DNA synthesis by senescent cells. The mechanism by which this inhibitor changes the expression of cell cycle genes remains to be elucidated. One possibility is that the cell cycle genes are controlled by other events which occur during the cell cycle, and since senescent cells are not cycling, these genes are not triggered by the proper series of events.

The full significance of cellular aging *in vivo* is not known. However, it may be that small decrements in various systems act synergistically. For example, a decline in lung capacity coupled with a decline in cardiac output, a decline in hemoglobin content, or the oxygen-carrying capacity of the blood could lead to a significant decline in the total oxygen available. It is clear that loss of cell proliferative capacity in some organ systems can have serious consequences for the organism as a whole. For example, if the cells lining the vascular system were not able to proliferate in response to injury, denudation of the vascular system could result. This could cause thrombosis or atherosclerosis.

On the other hand, it seems likely that the limited proliferative potential of normal cells *in vivo* is a powerful inhibitor of tumorigenesis. Even when some of the changes leading to tumor formation have occurred, the limited proliferative potential of nonimmortalized cells severely limits the damage caused by these potentially tumorigenic cells.

REFERENCES

1. **Liotta, L. A., Steeg, P. S., and Stetler-Stevenson, W. G.,** Cancer metastasis and angiogenesis: an imbalance of positive and negative regulation, *Cell,* 64, 327, 1991.
2. **Fearon, E. R. and Vogelstein, B.,** A genetic model for colorectal tumorigenesis, *Cell,* 61, 759, 1990.
3. **O'Brien, W., Stenman, G., and Sager, R.,** Suppression of tumor growth by senescence in virally transformed human fibroblasts, *Proc. Natl. Acad. Sci. U.S.A.,* 1986, 83, 8659.
4. **Hayflick, L. and Moorhead, P. S.,** The serial cultivation of human diploid cell strains, *Exp. Cell Res.,* 25, 585, 1961.
5. **Hayflick, L.,** The limited in vitro lifetime of human diploid cell strains, *Exp. Cell Res.,* 37, 614, 1965.
6. **Martin, G. M., Sprague, C. A., and Epstein, C. J.,** Replicative life-span of cultivated human cells: effects of donor age, tissue, and genotype, *Lab. Invest.,* 23, 86, 1970.
7. **Norwood, T. H. and Smith, J. R.,** The cultured fibroblast-like cell as a model for the study of aging, in *Handbook of Biological Aging,* Finch, C. E. and Schneider, E. L., Eds., Van Nostrand Reinhold, New York, 1985, 291.
8. **Macieira-Coehlo, A.,** Implications of the reorganization of the cell genome for aging or immortalization of dividing cells in vitro, *Gerontology,* 26, 276, 1980.
9. **Meek, R. L., Bowman, P. D., and Daniel, C. W.,** Establishment of mouse embryo cells in vitro. Relationship of DNA synthesis, senescence and malignant transformation, *Exp. Cell Res.,* 125, 453, 1977.
10. **Rothfels, F. H., Kupelweiser, E. B., and Parker, R. C.,** Effects of x-irradiated feeder layers on mitotic activity and development of aneuploidy in mouse embryo cells in vitro, *Can. Cancer Conf.,* 5, 191, 1963.

11. **Swim, H. E. and Parker, R. F.**, Culture characteristics of human fibroblasts propagated serially, *Am. J. Hyg.*, 66, 235, 1957.
12. **Rohme, D.**, Evidence for a relationship between longevity of mammalian species and life-spans of normal fibroblasts in vitro and erythrocytes in vivo, *Proc. Natl. Acad. Sci. U.S.A.*, 78, 5009, 1981.
13. **Duthu, G. S., Braunschweiger, K. I., Pereira-Smith, O. M., Norwood, T. H., and Smith, J. R.**, A long-lived human diploid fibroblast line for cellular aging studies: applications in cell hybridization, *Mech. Aging Dev.*, 20, 243, 1982.
14. **Goldstein, S.**, Aging in vitro: growth of cultured cells from the Galapagos tortoise, *Exp. Cell Res.*, 83, 297, 1974.
15. **Gorman, S. D., Hoffman, E., Nichols, W. W., and Cristofalo, V. J.**, Spontaneous transformation of a cloned cell line of normal diploid bovine vascular endothelial cells, *In Vitro*, 20, 339, 1984.
16. **Littlefield, J. W.**, Attempted hybridization with senescent human fibroblasts, *J. Cell Physiol.*, 82, 129, 1973.
17. **Pereira-Smith, O. M. and Smith, J. R.**, Expression of SV40 T antigen in finite life-span hybrids of normal and SV40-transformed fibroblasts, *Som. Cell Genet.*, 7, 411, 1981.
18. **Pereira-Smith, O. M. and Smith, J. R.**, Evidence for the recessive nature of cellular immortality, *Science*, 221, 964, 1983.
19. **Pereira-Smith, O. M. and Smith, J. R.**, Genetic analysis of indefinite division in human cells: identification of four complementation groups, *Proc. Natl. Acad. Sci. U.S.A.*, 85, 6042, 1988.
20. **Ning, Y., Shay, J. W., Lovell, M., Taylor, L., Ledbetter, D. H., and Pereira-Smith, O. M.**, Tumor suppression by chromosome 11 is not due to cellular senescence, *Exp. Cell Res.*, 192, 220, 1991.
21. **Ning, Y. and Pereira-Smith, O. M.**, Molecular genetic approaches to the study of cellular senescence, *Mutn. Res.*, in press.
22. **Ning, Y., Weber, J. L., Killary, A. M., Ledbetter, D. H., Smith, J. R., and Pereira-Smith, O. M.**, Genetic analysis of indefinite division in human cells: evidence for a cell senescence related gene(s) on human chromosome 4, *Proc. Natl. Acad. Sci. U.S.A.*, 88, 5635, 1991.
23. **Sugawara, O., Oshimura, M., Koi, M., Annab, L. A., and Barrett, J. C.**, Induction of cellular senescence in immortalized cells by human chromosome 1, *Science*, 247, 707, 1990.
24. **Norwood, T. H., Pendergrass, W. R., Sprague, C. A., and Martin, G. M.**, Dominance of the senescent phenotype in heterokaryons between replicative and post-replicative human fibroblast-like cells, *Proc. Natl. Acad. Sci. U.S.A.*, 71, 2231, 1974.
25. **Stein, G. H. and Yanishevsky, R. M.**, Entry into S phase is inhibited in two immortal cell lines fused to senescent human diploid cells, *Exp. Cell Res.*, 120, 155, 1979.
26. **Norwood, T. H., Pendergrass, W. R., and Martin, G. M.**, Reinitiation of DNA synthesis in senescent human fibroblasts upon fusion with cells of unlimited growth potential, *J. Cell Biol.*, 64, 551, 1975.
27. **Stein, G. H., Yanishevsky, R. M., Gordon, L., and Beeson, M.**, Carcinogen-transformed human cells are inhibited from entry into S phase by fusion to senescent cells but cells transformed by DNA tumor viruses overcome the inhibition, *Proc. Natl. Acad. Sci. U.S.A.*, 79, 5287, 1982.
28. **Yanishevsky, R. M. and Stein, G. H.**, Ongoing DNA synthesis continues in young human diploid cells (HDC) fused to senescent HDC, but entry into S phase is inhibited, *Exp. Cell Res.*, 126, 469, 1980.
29. **Drescher-Lincoln, C. K. and Smith, J. R.**, Inhibition of DNA synthesis in proliferating human diploid fibroblasts by fusion with senescent cytoplasts, *Exp. Cell Res.*, 144, 455, 1983.
30. **Drescher-Lincoln, C. K. and Smith, J. R.**, Inhibition of DNA synthesis in senescent-proliferating human cybrids is mediated by endogenous proteins, *Exp. Cell Res.*, 153, 208, 1984.

31. **Pereira-Smith, O. M., Fisher, S. F., and Smith, J. R.**, Senescent and quiescent cell inhibitors of DNA synthesis. Membrane-associated proteins, *Exp. Cell Res.*, 160, 297, 1985.

32. **Stein, G. H. and Atkins, L.**, Membrane-associated inhibitor of DNA synthesis in senescent human diploid fibroblasts: characterization and comparison to quiescent cell inhibitor, *Proc. Natl. Acad. Sci. U.S.A.*, 83, 9030, 1986.

33. **Lumpkin, C. K. J., McClung, J. K., Pereira-Smith, O. M., and Smith, J. R.**, Existence of high abundance antiproliferative mRNA's in senescent human diploid fibroblasts, *Science*, 232, 393, 1986.

34. **Lumpkin, C. K. J., McClung, J. K., and Smith, J. R.**, Entry into S phase is inhibited in human fibroblasts by rat liver poly(A) + RNA, *Exp. Cell Res.*, 160, 544, 1985.

35. **Nuell, M. J., Stewart, D. A., Walker, L., Friedman, V., Wood, C. M., Owens, G. A., Smith, J. R., Schneider, E. L., Dell'Orco, R., Lumpkin, C. K., Danner, D. B., and McClung, J. K.**, Prohibitin, an evolutionarily conserved intracellular protein that blocks DNA synthesis in normal fibroblasts and HeLa cells, *Mol. Cell. Biol.*, 11, 1372, 1991.

36. **Pepperkok, R., Schneider, C., Philipson, L., and Ansorge, W.**, Single cell assay with an automated capillary microinjection system, *Exp. Cell Res.*, 178, 369, 1988.

37. **Pepperkok, R., Zanetti, M., King, R., Delia, D., Ansorge, W., Philipson, L., and Schneider, C.**, Automatic microinjection system facilitates detection of growth inhibitory mRNA, *Proc. Natl. Acad. Sci. U.S.A.*, 85, 6748, 1988.

38. **Porter, M. B., Pereira-Smith, O. M., and Smith J. R.**, Novel monoclonal antibodies identify antigenic determinants unique to cellular senescence, *J. Cell Physiol.*, 142, 425, 1990.

39. **Seshadri, T. and Campisi, J.**, Repression of c-fos transcription and an altered genetic program in senescent human fibroblasts, *Science*, 247, 205, 1990.

40. **Stein, G. H., Drullinger, L. F., Robetorye, R. S., Pereira-Smith, O. M., and Smith, J. R.**, Senescent cells fail to express the cdc2 gene in response to mitogen stimulation, submitted.

41. **Stein, G. H., Beeson, M., and Gordon, L.**, Failure to phosphorylate the retinoblastoma gene product in senescent human fibroblasts, *Science*, 249, 666, 1990.

Chapter 2

DETECTION OF GROWTH FACTOR EFFECTS AND EXPRESSION IN NORMAL AND NEOPLASTIC HUMAN BRONCHIAL EPITHELIAL CELLS*

Jill M. Siegfried, Michael J. Birrer, and Frank C. Cuttitta

TABLE OF CONTENTS

*   The opinions and assertions contained herein are the private views of the authors and are not
     to be construed as official or reflecting the views of the Department of the Navy or the
     Department of Defense.

I. INTRODUCTION

Both normal and neoplastic human epithelial and stromal cells have been shown to produce peptides which are capable of stimulating the producing cells themselves (autocrine growth factors[1-4]) or the surrounding cell types which make up the tissue architecture *in vivo* (paracrine growth factors[5-7]). While these factors may play a critical role in the growth and development of normal tissues and organs, their local release in tumors may also be an important factor in the uncontrolled growth of neoplasia. Normal cells have been demonstrated to produce factors such as transforming growth factor α (TGF-α) and insulin-like growth factor I (IGF-I) under conditions of cell proliferation.[8-13] Thus, the basic mechanism of producing and responding to growth factors through signal transduction in many cases may not be fundamentally different in normal and neoplastic cells. In fact, most of the autocrine factors produced by tumors have some role in normal physiology, although structural aberrations or mutations in growth factors or their receptors are also known which are oncogenic.[14-16] Alternate forms of growth factors may also be produced by tumors,[17] for which the role in normal physiology is yet to be elucidated.

Neoplastic cells may therefore utilize existing mechanisms that bring about cell proliferation to produce a microenvironment which supports continuous growth, while lacking responses to inhibitory mechanisms which would control growth in normal tissues. Regardless of whether autocrine or paracrine secretion is a cause or an effect of cell transformation, growth factors or their receptors may be targets for new types of cancer therapy.[18-20] A knowledge of the biology of growth factors in normal and neoplastic human cells is important in designing such therapeutic strategies. This chapter will discuss some of the techniques which have been used by us and others to detect growth factor production and secretion in human epithelial cells. We will also discuss some of the specific growth factors which have been detected in normal and neoplastic human epithelial cells *in situ* and in culture.

II. PURIFICATION OF GROWTH FACTORS

The classical approach for detection of secreted growth factors from a tumor cell line was described by Marquardt and Todaro in 1982.[21] These investigators used biochemical methods to purify TGF-α from large volumes of medium conditioned by sarcoma virus-transformed rat embryo fibroblasts or a human melanoma cell line. TGF-α was shown to bind to the epidermal growth factor (EGF) receptor and to stimulate anchorage-independent growth of fibroblasts in consort with another secreted growth factor, transforming growth factor β, which was also present in conditioned medium.[22] These bioassays were used to monitor TGF-α activity during purification. Through laborious work, they were able to purify 1.5 μg from 9.2 l of medium conditioned by melanoma cells. A human cDNA coding for TGF-α was subsequently isolated by Derynck et al. in 1984.[23] TGF-α was shown to have only 30 to 40% homology with EGF, but to fold into an almost identical tertiary structure, allowing it to occupy the EGF receptor.[24] TGF-α is produced as a large prohormone which spans the cellular membrane; the mature 50-amino acid peptide is probably cleaved by an elastase-like enzyme.[25] This transmembrane structure is shared by a family of proteins related to EGF which have diverse functions, including growth factor activity, adhesion, and protease activity.[25]

TGF-α has many functions in addition to stimulation of cell growth and, for many, is more potent than EGF. For example, its effects on stimulating cell motility, angiogenesis, and morphogenesis occur at lower concentrations than EGF.[25] It is expressed by normal cells during specific times in embryonic development, and in adults can also be found in basal keratinocytes in skin[8] and in proliferating mammary epithelial cells.[11] It is found in many human tumors and tumor cell lines and can be detected in the urine of patients with disseminated cancer.[26] In tumors, different high-molecular-weight forms of TGF-α have been described, suggesting either alternate processing of the prohormone or different levels of glycosylation.[17,25]

Another autocrine growth factor produced by many cell types is IGF-I.[3,12,13] IGF-I was first isolated from serum,[27-29] but was later found to be produced by the liver[30,31] as well as other tissues,[32] and to be locally released *in situ*[33] and in culture.[34-36] It appears to be an important autocrine factor in cancer of the lung[3,37,38] and breast.[39-41] Insulin-like growth factor II (IGF-II), which has a structure similar to that of IGF-I, appears to only be released normally in embryonic tissues.[28,32,42] It also may be an important autocrine/paracrine growth factor in human tumors.[43] Basic fibroblast growth factor (bFGF) or the related *int*-2 gene product has also recently been shown to be expressed by some prostate tumors,[44] and to stimulate both mesenchymal[45] and some epithelial cells[46] in culture. Since bFGF is one of the major factors responsible for the growth-promoting effects of bovine pituitary extract (BPE) toward epithelial cells,[47] it is also a candidate autocrine/paracrine factor for carcinomas. The gastrin-releasing peptide (GRP)/bombesin family has been shown to act as an autocrine factor in small-cell carcinoma,[4] but since fibroblasts also respond to these neuropeptides, the paracrine effects of bombesin-like peptides may also be important in tumor growth. Another candidate paracrine factor which transformed epithelial cells may secrete is platelet-derived growth factor (PDGF). PDGF is highly growth-stimulatory for mesenchymal cells,[48] and may be produced by some carcinomas.[6]

III. APPROACHES TO THE DETECTION OF AUTOCRINE GROWTH FACTORS BY EPITHELIAL CELLS

A complication in applying large-scale biochemical purification to the identification of secreted growth factors from primary epithelial tissues or cultured epithelial cells is that many tumors, as well as normal tissues, produce a mixture of factors which may include inhibitory as well as stimulatory peptides. The question becomes how to separate and identify each of these, determine if any are unique, and demonstrate a cellular response to them, using very small amounts of material. How, then, to further establish an autocrine or paracrine role for these factors?

We have approached this problem in the study of normal and neoplastic human bronchial epithelial cells in a number of different ways. First, we have used primary human bronchial epithelial cells (HBE cells) and primary and secondary cultured non-small-cell lung carcinoma cells (NSCLC cells) in colony assays to screen known peptides for the ability to stimulate growth. We have examined peptides shown to be autocrine growth factors for small-cell lung cancer and for carcinomas derived from other tissues, as well as ectopic peptides known to be expressed clinically in non-small-cell lung tumors. Second, we have examined the ability of known peptides to activate signal transduction pathways in normal and neoplastic lung epithelial cells. These two approaches are intended to identify peptide hormones which are mitogens for lung cells and are therefore candidate autocrine factors which might also be produced by lung tissues. Third, we have adapted an NSCLC cell line to grow in completely serum-free conditions, without any exogenous growth factors, and have used medium conditioned by these cells as a source of growth factor activity. This approach creates conditions in which NSCLC cells are forced to produce growth factors which are needed for their own proliferation. Media conditioned by tumor cells under these conditions can be concentrated and used as a source of autocrine or paracrine factors. We have assayed both crude extracts and fractions separated by high-pressure liquid chromatography for growth-stimulating activity and for the presence of known mitogenic peptides. We have also used monoclonal antibodies raised against growth factor receptors to block effects of conditioned medium in order to demonstrate that the peptides detected are responsible for the observed growth stimulation.

We have also applied our *in vitro* system for detecting proliferation of HBE cells and lung tumor cells in response to novel growth factors. For example, we have examined the known sequences of peptide prohormones for motifs downstream of the mature peptides that are indicative of protein-processing enzymes. The presence of such motifs implies that the sequence found within the motif is a potential new growth factor, released during processing of the prohormone. We have synthesized putative peptide growth factors based on such sequences and have been able to demonstrate that they have biological effects in our system.[49,50] Below we will illustrate how each of these approaches has been useful in defining growth factors important in the proliferation of normal and neoplastic bronchial epithelial cells. As new growth factors are discovered, all of these approaches can be used to determine if they have a role in the growth of cells derived from the lung or other tissues.

TABLE 1
Effect of Known Peptides on Growth of Primary Normal Bronchial Epithelial Cells and Primary Non-Small-Cell Lung Carcinoma Cells

| Peptide | Degree of stimulation | |
	Bronchial epithelial cells	Carcinoma cells
TGFα	+ + +	+ + +
IGF-I	+ + +	+ + +
IGF-II	ND	+[a]
Bombesin	+[b]	+[c]
Nerve growth factor	−	ND
Human chorionic gonadotropin	−	ND
Prolactin	−	+[a]
ACTH	−	+[a]
PDGF	−	−
bFGF	ND	+[a]

Note: Growth stimulation was measured in colony-forming assays as described in text. ND, not done.

[a] One of three primary carcinoma specimens showed stimulation.
[b] Three of ten normal bronchial specimens showed stimulation.
[c] One of four primary carcinoma specimens showed stimulation.

IV. EFFECT OF KNOWN PEPTIDES ON GROWTH OF BRONCHIAL EPITHELIAL CELLS

For culture of primary and secondary normal and neoplastic bronchial epithelial cells, two different culture conditions are utilized which have been optimized for each cell type. HBE cells are cultured in a defined medium,[51] adapted from the formulation first described by Lechner et al.[52] NSCLC cells are grown as primary and secondary cultures using medium conditioned by the bronchiolo-alveolar carcinoma cell line A549, containing 1% fetal bovine serum.[53] Cells from solid tumors are plated onto 3T3 fibroblasts, which provide a superior surface for attachment and growth as three-dimensional colonies. This technique has allowed us to establish cell lines from a number of solid tumors and to maintain surgical tumor specimens for several months in culture, even if a cell line is not isolated.[54]

Colony assays on 3T3 feeder cells are used routinely to measure cell proliferation. The colony-forming efficiency of HBE and tumor cells is easily quantitated in multiwell dishes 7 to 10 d after plating at low density under different conditions.[55] Table 1 summarizes the response of normal and neo-plastic bronchial cells to known peptides. Either recombinant or synthetic peptides were used in these studies. TGF-α and IGF-I were found to stimulate proliferation to the greatest extent of all peptides examined in both normal and neoplastic cells (up to tenfold over control, depending on how the "basal" control condition is defined). These effects have been reproducible in cells from different individuals and in non-small-cell tumor cells of different histologic types. Other peptides examined showed lesser effects, which were not demonstrable in cells from every individual. For instance, bombesin showed up to a threefold stimulation of colony-forming units using HBE cells, but only in cells from a few individuals tested. Only one in four non-small-cell carcinomas showed stimulation by bombesin. Likewise, IGF-II, prolactin, ACTH, and bFGF were stimulatory in only one case. Nerve growth factor (NGF), human corionic gonadotropin (HCG), and PDGF did not increase human bronchial epithelial cell proliferation. We conclude from these findings that TGF-α and IGF-I are two of the most important candidate autocrine peptides for bronchial epithelial cells. Bombesin may also be important, perhaps in only certain stages of differentiation. More studies are needed to clarify the role of IGF-II, bFGF, and other peptides, including potentially important peptides such as bradykinin and substance P.

V. EFFECT OF KNOWN PEPTIDES ON SIGNAL TRANSDUCTION PATHWAYS

The early response genes *fos* and *jun* have been examined as a means of determining whether peptide hormones have elicited signal transduction in bronchial epithelial cells. These gene, which encode transcription factors, are important nuclear transducers of cell signals resulting from a hormone-receptor interaction. This technique allows the rapid determination of a cellular response to a putative growth factor and can be used to detect responses without detecting differences in cell growth. Although the activation of *fos* and *jun* may not be sufficient to cause cell proliferation, their increased expression upon addition of hormone or growth factor implies the presence of receptors for that peptide and propagation of a cell signal upon binding of peptide. Epithelial cells are made quiescent by removal of all growth factors and hormones for 18 h. Cells are then stimulated by addition of a complete medium or of individual growth factors. Cells are lysed at different times after addition of stimulatory medium, RNA is extracted, and expression of early-response genes is detected by Northern analysis.[56]

Figure 1 illustrates the induction of these two genes by growth factors in normal bronchial epithelial cells in culture. Cells were starved of all growth factors for 18 h and then were refed with complete medium. By 30 min, induction of both genes is seen; at 2 h, messenger RNA levels are already declining. This result demonstrates that both *fos* and *jun* are transcribed in NBE cells and that it may be feasible to use early-response gene induction in culture to determine if NBE cells or NSCLC cells respond to specific peptide hormones.

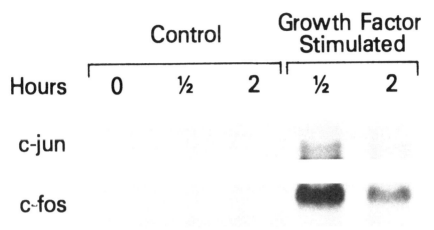

FIGURE 1.   Activation of *jun* and *fos* by addition of growth factors to quiescent NBE cells. RNA (5 μg) was added to each lane. Time zero denotes beginning of refeeding period after deprivation by growth factors. Complete medium elicited increased transcription of *jun* and *fos*.

VI. CULTURE OF A NON-SMALL-CELL LUNG CARCINOMA IN GROWTH FACTOR-FREE MEDIUM

We have previously demonstrated the presence of TGF-α and IGF-I in medium conditioned by A549 cells.[53,55] Our results also indicated that other unknown growth-stimulating peptides were present in A549 cell-conditioned medium. In an effort to simplify the concentration and purification of secreted growth factors from cultured non-small-cell lung tumors, and in an attempt to maximize the peptides produced from these cells, we have adapted A549 cells to grow in completely serum-free conditions, without exogenous peptides or hormones. The growth medium used was RPMI 1640, supplemented only with glutamine and selenium (designated R_0 medium). This medium is also phenol-red free in order to eliminate any possible estrogenic activity from phenol red and any co-concentration of the pH indicator with secreted peptides. This technique has previously been successful with small-cell lung carcinoma cells.[57] A549 cells were chosen because we had previously demonstrated that they secrete high levels of TGF-α and IGF-I, the two peptides we have already identified as potent mitogens in HBE cells and NSCLCs.

Although initially there was much cell death in the cultures, a small number of cells remained viable. After 2 months in culture, actively growing cells were detected that could proliferate in R_0 medium. These cells grow as a combination of attached and floating cells (Figure 2). The A549 cells growing in R_0 medium were shown to be the same as parent A549 cells by cytogenetics, indicating that we did not select out a subpopulation. Figure 3 illustrates the growth-stimulating effect of A549 cell-conditioned R_0 medium on cells from a secondary culture of a NSCLC. The assay was performed in the absence of serum or any other added growth factor. Final concentration of conditioned medium per well was 25%. We determined that 180 μg of

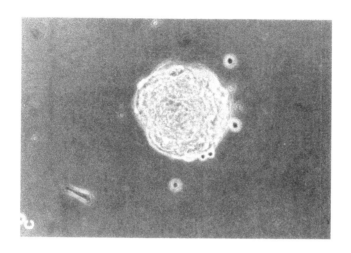

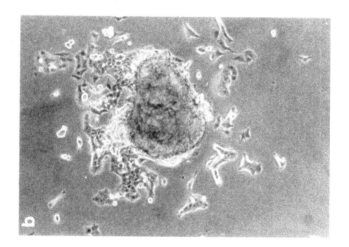

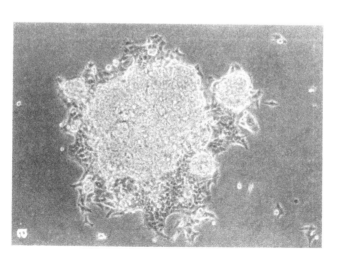

FIGURE 2.   Phase-contrast photomicrograph of A549 cells growing in R_0 medium in the absence of serum. Cells grow as a mixture of attached cells with three-dimensional outgrowths (a and b) and detached floating balls of cells (c). A549 cells have been propagated in this manner for over 1 year.

Stimulatory Effect of R_0 Medium Collected Over Time

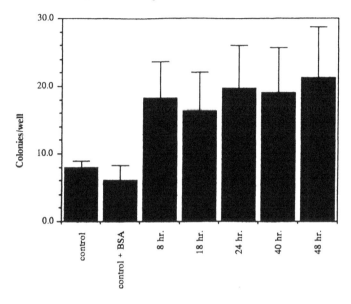

FIGURE 3. Results of colony assay using cells derived from a poorly differentiated squamous cell carcinoma of the lung. Assay was performed in the absence of serum with 25% conditioned R_0 added per well. Conditioned medium was collected at the times indicated.

protein was present in this amount of conditioned medium added per well. Addition of an equivalent amount of BSA did not cause growth stimulation. The medium caused an approximately threefold increase in colony formation, and although there was a trend toward an increased effect depending on the time medium was conditioned by A549 cells, it was not significant. This implies that by 8 h, the growth-stimulating peptides have already been released and there may be an equilibrium between secretion, processing to active forms, and degradation.

Figure 4 shows the effect of increasing amounts of R_0 medium in the colony assay. Again, no serum is present in this assay. Here, 50% conditioned medium is seen to be optimal. Previous experiments have shown that no additional effect is seen above 50%, and often there is a decline at higher concentrations. This is probably because fresh nutrients also are needed to support cell growth. This figure also illustrates that even though TGF-α and IGF-I have been measured in the medium, supplying them to the culture does not give the same growth stimulation as crude conditioned medium. We have recently determined that GRP immunoreactivity is present in R_0 medium; approximately 2.7 ng/ml was present in medium conditioned for 48 h. In order to assay R_0 medium for growth factors, and to separate the peptides present, we concentrated 2.5 l of conditioned medium and fractionated it by

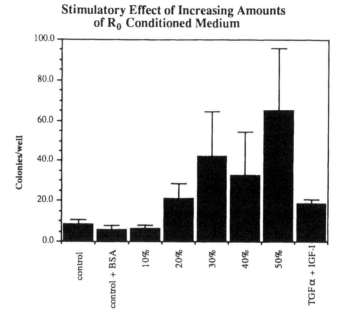

FIGURE 4.   Results of colony assay using cells as described in Figure 3 to detect effect of increasing amounts of R_0-conditioned medium. For last condition shown, 2 ng/ml TGF-α and 10 ng/ml IGF-I were added per well. This approximates the amount of these peptides detected in conditioned medium.

reverse-phase HPLC. Figure 5 shows the growth-promoting activity of 1-ml fractions of the medium concentrate. The results indicate that multiple growth-stimulating peaks are present in R_0-conditioned medium, some of which coincide with elution of the growth factors IGF-I and transferrin. Because secreted growth factors may differ in their processing from tumors, this does not prove the presence of these peptides. We have also not yet tested for the elution of other peptides such as TGFα in this system. More work is being planned in identifying these peaks and testing the effects of other peptides.

Antibodies to growth factor receptors have been tested for ability to block the activity of R_0-conditioned medium. Table 2 shows the effect of antibodies to the EGF receptor and the IGF-I receptor on growth stimulation of R_0-conditioned medium. Cells derived from a squamous cell carcinoma were used in this assay, at passage three from the original tissue. From Table 2, it can be seen that both antibodies blocked the stimulation of the medium, by 60.5% for the antibody to the IGF-I receptor, and by 26.3% for the antibody to the EGF receptor.

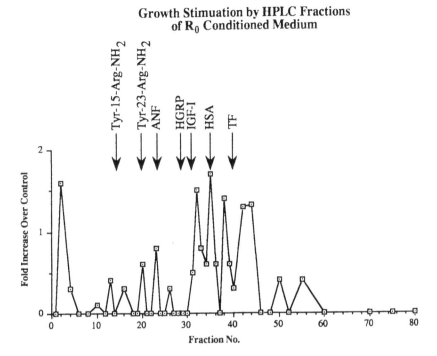

FIGURE 5.   Effect of HPLC fractions on growth of a squamous cell lung carcinoma cell line. A large volume of conditioned medium (2.5 l) was reduced to 2 ml by pumping onto a Waters Seppak reverse-phase precolumn and elution with acidified acetonitrile. The concentrate was loaded onto a Vydac C-18 reverse-phase column and eluted with an acetonitrile gradient, at a flow rate of 1 ml/min. One-ml fractions were collected. A series of standard peptides were eluted separately: Tyr-15-Arg-NH$_2$ and Tyr-23-Arg-HH$_2$ (synthetic analog of IBE$_2$ and IBE$_1$, respectively, as described in text); ANF, human atrial natriuretic factor; HGRP, human gastrin-releasing peptide; IGF-I; HSA, human serum albumin; TF, human transferrin. 200 μl of each fraction was freeze dried and stored frozen until reconstitution for colony assay. Each fraction was reconstituted in 200 μl of sterile water. 5 μl was used per colony well in duplicate to assess growth promotion of fractions. Medium used contained 0.1% fetal bovine serum in order to ensure a measurable level of colony formation. Data are expressed as fold increase over control; onefold increase indicates a 100% increase in colony formation.

VII. DETECTION OF NOVEL GROWTH FACTOR ACTIVITY

It is well established that many peptide hormones are produced as large precursor molecules which require proteolytic cleavage to release the active, mature growth factors or hormones. Several peptide hormones are known to produce multiple active peptides from one precursor. We examined the IGF-I prohormone, transcribed from the IGF-IB mRNA transcript, for the presence of possible proteolytic cleavage sites. In Figure 6, we illustrate the sequence within the E domain of the prohormone which encodes repeated basic amino acids. We hypothesize that these sequences are cleaved by proteases, releasing two small peptides which are terminally amidated, IBE$_1$ and

TABLE 2
Effect of Antibodies to Growth Factor Receptors on Stimulation of R_0-Conditioned Medium from A549 Lung Carcinoma Cells

Condition	Colonies per well
Control	0.0
Conditioned medium	119.5
Conditioned medium + IGF-I receptor antibody	47.0
Conditioned medium + EGF receptor antibody	67.7

Note: Effects were measured in a colony-forming assay using cells derived from a squamous cell carcinoma at passage 3. Control medium was 50% RPMI with selenium and glutamine supplementation (R_0) and 50% Basal Medium Eagles'. No serum was present in the assay. Conditioned medium was 50% RPMI as above, conditioned for 48 h by A549 cells, centrifuged, and filtered, and 50% Basal Medium Eagles'. Again, no serum was present under these conditions. Either 1 μg of IGF-I receptor antibody (αIR3, Oncogene Science) or 1 μg of EGF receptor antibody (Ab-1, Oncogene Science) was added to measure inhibition.

```
543: guc cgu gcc cag cgc cac acc gac aug ccc
     Val Arg Ala Gln Arg His Thr Asp Met Pro

573: aag acc cag aag uau cag ccc cca ucu acc
     Lys Thr Gln Lys Tyr Gln Pro Pro Ser Thr

603: aac aag aac acg aag ucu cag aga agg aaa
     Asn Lys Asn Thr Lys Ser Gln Arg Arg Lys

633: ggu ugg cca aag aca cau cca gga ggg gaa
     Gly Trp Pro Lys Thr His Pro Gly Gly Glu

663: cag aag gag ggg aca gaa gca agu cug cag
     Gln Lys Glu Gly Thr Glu Ala Ser Leu Gln

693: auc aga gga aag aag aaa gag cag agg agg
     Ile Arg Gly Lys Lys Lys Glu Gln Arg Arg

723: gag auu gga agu aga aau gcu gaa ugc aga
     Glu Ile Gly Ser Arg Asn Ala Glu Cys Arg

753: ggc aaa aaa gga aaa uga
     Gly Lys Lys Gly Lys •
```

FIGURE 6.   cDNA base sequence of the E domain of the IGF-IB transcript and the predicted amino acid composition of the transcribed E domain. Numerical locations of base and amino-acid sequences are indicated. Boxed areas indicate sites of potential proteolytic cleavage. The sequence of IBE$_1$, the proposed peptide amide, is highlighted by dashed lines (IGF-IB$_{103-124}$). A second potential peptide amide is indicated by the dotted line (IGF-IB$_{129-142}$, IBE$_2$).

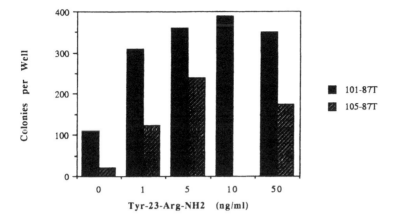

FIGURE 7. Effect of increasing amounts of Tyr-23-Arg-NH$_2$ (analog of IBE$_1$) on colony formation of secondary cultures derived from two NSCLCs. A threefold stimulation is observed with cell culture 101-87T, and an 11-fold stimulation is observed with culture 105-87T.

IBE$_2$.[49,50] We have found synthetic analogs of these sequences to have biological activity.[49,50] Figure 7 illustrates the growth-promoting effect of a synthetic analog to IBE$_1$, Tyr-23-Arg-NH$_2$, on primary cells isolated from two NSCLCs. The synthetic peptide is the sequence of IBE$_1$ as shown in Figure 6, with addition of Tyr at the zero position. It is terminally amidated, since we predict IBE$_1$ is amidated at the terminal Arg.

VIII. CONCLUSIONS

We have developed an *in vitro* model for detection of the effects and secretion of potential autocrine growth factors in normal and neoplastic human epithelial cells derived from the bronchus. Using this model, we have determined that TGF-α and IGF-I are two of the most important autocrine factors for these cells. GRP/bombesin may also be of importance. Assays for increased cell proliferation and expression of early-activation genes can be used to monitor effects in human normal and neoplastic bronchial epithelial cells. NSCLC cells adapted to grow in hormone and growth factor-free medium provide an excellent source of secreted growth factors. TGF-α-, IGF-I-, and GRP-related activities were detected in this conditioned medium. The culture system described here is also useful for detecting effects of novel growth factors.

ACKNOWLEDGMENTS

This work was supported in part by a grant from the National Institutes of Health (R01 CA50694), a Research Starter Grant from the Pharmaceutical Manufacturers Foundation Association, and by Co-operative Agreement

No. CR-816188 from the U.S. Environmental Protection Agency. Jill M. Siegfried is the recipient of an American Cancer Society Junior Faculty Research Award. The authors thank Sara Owens, Suzanne Hansen, Dr. Phyllis Andrews, and Autumn Gaither, who contributed to technical aspects of this research, and Dr. Jay Hunt, who prepared the figures.

REFERENCES

1. **DeLarco, J. E. and Todaro, G. J.,** Growth factors from murine-sarcoma virus transformed cells, *Proc. Natl. Acad. Sci. U.S.A.,* 75, 4001, 1978.
2. **Sporn, M. B. and Todaro, G. J.,** Autocrine secretion and malignant transformation of cells, *N. Engl. J. Med.,* 303, 878, 1980.
3. **Macaulay, V. M., Everard, M. J., Teale, J. D., Trott, P. A., VanWyk, J. J., Smith, I. E., and Millar, J. L.,** Autocrine function for insulin-like growth factor I in human small cell lung cancer cell lines and fresh tumor cells, *Cancer Res.,* 50, 2511, 1990.
4. **Cuttitta, F., Carney, D. N., Mulshine, J., Moody, T. W., Fedorko, J., Fishler, A., and Minna, J. D.,** Bombesin-like peptides can function as autocrine growth factors in human small cell-lung cancer, *Nature,* 316, 823, 1985.
5. **Heldin, C.-H., Betscholtz, C., Johnsson, A., and Westermark, B.,** Role of PDGF-like growth factors in malignant transformation, *Cancer Rev.,* 2, 34, 1986.
6. **Peres, R., Betsholtz, C., Westermark, B., and Heldin, C.-H.,** Frequent expression of growth factors for mesenchymal cells in human mammary carcinoma cell lines, *Cancer Res.,* 47, 3425, 1987.
7. **Finch, P. W., Rubin, J. S., Miki, T., Ron, D., and Aaronson, S. A.,** Human KGF is FGF-related with properties of a paracrine effector of epithelial cell growth, *Science,* 245, 752, 1989.
8. **Coffey, R. J., Jr., Derynck, R., Wilcox, J. N., Bringman, T. S., Goustin, A. S., Moses, H. L., and Pittelkow, M. R.,** Production and auto-induction of transforming growth factor-α in human keratinocytes, *Nature,* 328, 817, 1987.
9. **Madtes, D. K., Raines, E. W., Sakariassen, K. S., Assoian, R. K., Sporn, M. B., Bell, G. I., and Ross, R.,** Induction of transforming growth factor-α in activated human alveolar macrophages, *Cell,* 53, 285, 1988.
10. **Malden, L. T., Novak, U., and Burgess, A. W.,** Expression of transforming growth factor alpha messenger RNA in the normal and neoplastic gastrointestinal tract, *Int. J. Cancer,* 43, 380, 1989.
11. **Bates, S. E., Valverius, E. M., Ennis, B. W., Bronzert, D. A., Sheridan, J. P., Stampfer, M. R., Mendelsohn, J., Lippman, M. E., and Dickson, R. B.,** Expression of the transforming growth factor-α/epidermal growth factor receptor pathway in normal human breast epithelial cells, *Endocrinology,* 126, 596, 1990.
12. **Stiles, A. D. and D'Ercole, A. J.,** The insulin-like growth factors and the lung, *Am. J. Respir. Cell Mol. Biol.,* 3, 93, 1990.
13. **Daughaday, W. H.,** Fetal lung fibroblasts secrete and respond to insulin-like growth factors, *Am. J. Respir. Cell Mol. Biol.,* 1, 11, 1989.
14. **Downward, J., Yarden, Y., Mayes, E., Scrace, G., Totty, N., Stockwell, P., Ullrich, A., Schlessinger, J., and Waterfield, M. D.,** Close similarity of epidermal growth factor receptor and v-*erbB* oncogene protein sequences, *Nature,* 307, 521, 1984.
15. **Sherr, C. J., Rettenmier, C. W., Sacca, R., Roussel, M. F., Look, A. T., and Stanley, E. R.,** The c-*fms* proto-oncogene product is related to the receptor for the mononuclear phagocyte growth factor, CSF-1, *Cell,* 41, 665, 1985.

16. Doolittle, R. F., Hunkapiller, M. W., Hood, L. E., Devare, S. G., Robbins, K. C., Aaronson, S. A., and Antoniades, H. N., Simian sarcoma virus *onc*-gene, v-*sis*, is derived from a gene (or genes) encoding a platelet-derived growth factor, *Science*, 221, 275, 1983.

17. Lupu, R., Dickson, R. B., and Lippman, M., Biologically active glycosylated TGF-α released by an estrogen receptor negative human breast cancer cell line, *J. Cell. Biochem.*, 13B, 132, 1989.

18. Ennis, B. W., Valverius, E. M., Bates, S. E., Lippman, M. E., Bellot, F., Kris, R., Schlessinger, J., Masui, H., Goldenberg, A., Mendelsohn, J., and Dickson, R. B., Anti epidermal growth factor receptor antibodies inhibit the autocrine-stimulated growth of MDA-468 human breast cancer cells, *Mol. Endocrinol.*, 3, 1830, 1989.

19. Masui, H., Kawamoto, T., Sato, J. D., Wolk, B., Sato, G., and Mendelsohn, J., Growth inhibition of human tumor cells in athymic mice by anti-epidermal growth factor receptor monoclonal antibodies, *Cancer Res.*, 44, 1002, 1984.

20. Harris, A. L., The epidermal growth factor receptor as a target for therapy, *Cancer Cells*, 2, 321, 1990.

21. Marquardt, H. and Todaro, G. J., Human transforming growth factor. Production by a melanoma cell line, purification and initial characterization, *J. Biol. Chem.*, 257, 5220, 1982.

22. Anzano, M. A., Roberts, A. B., Smith, J. M., Sporn, M. B., and DeLarco, J. E., Sarcoma growth factors from conditioned medium of virally transformed cells is composed of both type α and type β transforming growth factors, *Proc. Natl. Acad. Sci. U.S.A.*, 80, 6264, 1983.

23. Derynck, R., Roberts, A. B., Winkler, M. E., Chen, E. Y., and Goeddel, D. V., Human transforming growth factor-α: precursor structure and expression in E. coli, *Cell*, 38, 287, 1984.

24. Nestor, J. J., Newman, S. R., DeLustro, B., Todaro, G. J., and Schreiber, A. B., A synthetic fragment of rat transforming growth factor α with receptor binding and antigenic properties, *Biochem. Biophys. Res. Commun.*, 129, 226, 1985.

25. Salomon, D. S., Kim, N., Saeki, T., and Ciardiello, F., Transforming Growth Factor-α: an oncodevelopmental growth factor, *Cancer Cells*, 2, 389, 1990.

26. Stromberg, K., Hudgins, W. R., and Orth, D. N., Urinary TGFs in neoplasia: immunoreactive TGF-alpha in the urine of patients with disseminated breast carcinoma, *Biochem. Biophys. Res. Commun.*, 144, 1059, 1987.

27. Salmon, W. D. and Daughaday, W. H., A hormonally controlled serum factor which stimulates sulfate incorporation by cartilage *in vitro*, *J. Lab. Clin. Med.*, 49, 825, 1957.

28. Van Wyk, J. J., The somatomedins: biological actions and physiologic control mechanisms, in *Hormonal Proteins and Peptides*, Vol. 12, Li, C. H., Ed., Academic Press, New York, 1984, 81.

29. Svoboda, M. E., Van Wyk, J. J., Klapper, D. G., Fellows, R. E., Grissom, F. E., and Schlueter, R. J., Purification of somatomedin-C from human plasma: chemical and biological properties, partial sequence analysis, and relationship to other somatomedins, *Biochemistry*, 19, 790, 1980.

30. Vassillopoulou-Sellin, R. and Phillips, L. S., Extraction of somatomedin activity from rat liver, *Endocrinology*, 110, 582, 1982.

31. Schlach, D. S., Heinrich, U. E., Draznin, B., Johnson, C. J., and Hiller, L. L., Role of the liver in regulating somatomedin activity: hormonal effects on the synthesis and release of insulin-like growth factor and its carrier protein by the isolated perfused liver, *Endocrinology*, 104, 1143, 1979.

32. Daughaday, W. H. and Rotwein, P., Insulin-like growth factors I and II. Peptide, messenger ribonucleic acid and gene structures, serum, and tissue concentrations, *Endocr. Rev.*, 10, 68, 1989.

33. Han, V. K., Lund, P. K., Lee, D. C., and D'Ercole, A. J., Expression of somatomedin/insulin-like growth factor messenger ribonucleic acids in the human fetus: identification, characterization, and tissue distribution, *Science*, 236, 193, 1987.

34. **Atkinson, P. R., Weidman, E. R., Bhaumick, B., and Bala, R. M.**, Release of somatomedin-like activity by cultured WI-38 human fibroblasts, *Endocrinology,* 106, 2006, 1980.

35. **Clemmons, D. R. and Shaw, D. S.**, Variables controlling somatomedin production by cultured human fibroblasts, *J. Cell. Physiol.,* 115, 137, 1983.

36. **Snyder, J. M. and D'Ercole, A. J.**, Somatomedin-C/insulin-like growth factor I production by human fetal lung tissue maintained *in vitro, Exp. Lung Res.,* 13, 449, 1985.

37. **Minuto, F., Del Monte, P., Barreca, A., et al.**, Evidence for increased somatomedin-C/insulin-like growth factor I content in primary human lung tumors, *Cancer Res.,* 46, 985, 1986.

38. **Siegfried, J. M., Hansen, S. K., Lawrence, V. L., and Owens, S. E.**, Secretion of autocrine growth factors by cultured human lung tumors: effects on neoplastic lung epithelial cells, *Lung Cancer,* 4, 205, 1988.

39. **Huff, K. K., Kaufman, D., Gabbay, K. J., Spencer, E. M., Lippman, M. E., and Dickson, R. B.**, Human breast cancer cells secrete an insulin-like growth factor-I related polypeptide, *Cancer Res.,* 46, 4613, 1986.

40. **Yee, D., Paik, S., Lebovic, G., Marcus, R., Favoni, R., Cullen, K., Lippman, M. E., and Rosen, N.**, Analysis of IGF-I gene expression in malignancy — evidence for a paracrine role in human breast cancer, *Mol. Endocrinol.,* 3, 509, 1989.

41. **Cullen, K. J., Yee, D., Sly, W. S., Perdue, J., Hampton, B., Lippman, M. E., and Rosen, N.**, Insulin-like growth factor receptor expression and function in human breast cancer, *Cancer Res.,* 50, 48, 1990.

42. **Brice, A. L., Cheetham, J. E., Bolton, V. N., Hill, N. C., and Schofield, P. N.**, Temporal changes in the expression of the insulin-like growth factor II gene associated with tissue maturation in the human fetus, *Development,* 106, 543, 1989.

43. **Yee, D., Paik, S., Lebovic, G., Marcus, R., Favoni, R., Cullen, K., Lippman, M. E., and Rosen, N.**, Analysis of IGF-I gene expression in malignancy — evidence for a paracrine role in human breast cancer, *Mol. Endocrinol.,* 3, 509, 1989.

44. **Thompson, T. C.**, Growth factors and oncogenes in prostate cancer, *Cancer Cells,* 2, 345, 1990.

45. **Gospodarowicz, D., Neufeld, G., and Schweigerer, L.**, Molecular and biological characterization of fibroblast growth factor, an angiogenic factor which also controls the proliferation and differentiation of mesoderm and neuroectoderm derived cells, *Cell Differ.,* 19, 1, 1986.

46. **Takahashi, K., Suzuki, K., Kawahara, S., and Ono, T.**, Growth stimulation of human breast epithelial cells by basic fibroblast growth factor in serum-free medium, *Int. J. Cancer,* 43, 870, 1989.

47. **Gospodarowicz, D., Cheng, J., Lui, G. M., Baird, A., and Böhlent, P.**, Isolation of brain fibroblast growth factor by heparin-Sepharose affinity chromatography: identity with pituitary fibroblast growth factor, *Proc. Natl. Acad. Sci. U.S.A.,* 81, 6963, 1984.

48. **Heldin, C.-H., Wasteson, Å, and Westermark, B.**, Platelet-derived growth factor, *Mol. Cell. Endocrinol.,* 39, 169, 1985.

49. **Cuttitta, F., Kasprzyk, P. G., Treston, A. M., Avis, I., Jensen, S., Levitt, M., Siegfried, J., Mobley, C., and Mulshine, J. L.**, Autocrine growth factors that regulate the proliferation of pulmonary malignancies in man, in *Biology, Toxicology, and Carcinogenesis of Respiratory Epithelium,* Thomassen, D. G. and Netteshein, P., Eds., Hemisphere, New York, 1990, 228.

50. **Siegfried, J. M., Kasprzyk, P. G., Treston, A. M., Mulshine, J. L., and Cuttitta, F.**, A novel mitogenic peptide amide contained within the E peptide of insulin-like growth factor-IB prohormone, *Proc. Natl. Acad. Sci. U.S.A.,* submitted.

51. **Siegfried, J. M. and Nesnow, S.**, Cytotoxicity of chemical carcinogens towards human bronchial epithelial cells evaluated in a clonal assay, *Carcinogenesis,* 5, 1317, 1984.

52. **Lechner, J. F., Haugh, A., McClendon, I. A., and Pettis, E. W.**, Clonal growth of normal adult human bronchial epithelial cells in a serum-free medium, *In Vitro,* 18, 633, 1982.

53. **Siegfried, J. M.**, Detection of human lung epithelial cell growth factors produced by a lung carcinoma cell line: use in culture of primary solid human lung tumors, *Cancer Res.*, 37, 2903, 1987.

54. **Siegfried, J. M.**, Culture of primary lung tumors using medium conditioned by a lung carcinoma cell line, *J. Cell. Biochem.*, 41, 91, 1989.

55. **Siegfried, J. M. and Owens, S. E.**, Response of primary human lung carcinomas to autocrine growth factors produced by a lung carcinoma cell line, *Cancer Res.*, 48, 4976, 1988.

56. **Birrer, M. J., Alani, R., Brown, P. H., Cuttitta, F., Preis, L. H., Sanders, B. A., Siegfried, J. M., and Szabo, E.**, Early events in the neoplastic transformation of respiratory epithelium, *J. Natl. Cancer Inst.*, in press.

57. **Reeve, J. R., Jr., Cuttitta, F., Zigna, S. R., Huebner, V., Lee, T. D., Shively, J. E., Ho, F. J., Fedorko, J., Minna, J. D., and Walsh, J. H.**, Multiple gastrin-releasing peptide gene-associated peptides are produced by a human small cell lung cancer line, *J. Biol. Chem.*, 263, 1928, 1988.

Chapter 3

HUMAN CELL METABOLISM AND DNA ADDUCTION OF POLYCYCLIC AROMATIC HYDROCARBONS

Hudson H. S. Lau and William M. Baird

TABLE OF CONTENTS

I. INTRODUCTION

The Millers initially reported that many classes of carcinogenic chemicals react with the nucleophiles present in cellular macromolecules and that it is these interactions which initiate the cancer induction process.[1] Although some known chemical carcinogens such as β-propiolactone are reactive electrophiles capable of directly reacting with cellular nucleophiles,[2] many major classes of carcinogens to which humans are exposed require metabolism to form reactive electrophile "ultimate" carcinogens. Some of the major classes of carcinogens that require metabolic activation in human tissues include (1) the polycyclic aromatic hydrocarbons, a group of widespread environmental contaminants formed by incomplete combustion,[3,4] (2) the mycotoxins such as aflatoxin B_1 that are found in mold-contaminated foods in many temperate climates,[2] (3) the aromatic amines such as benzidine which are dye intermediates and many of which are well established as human carcinogens[5] and (4) the nitrosamines which are found in various foods and can be formed from the reaction of nitrite and secondary amines in the body.[2,6] Autrup[7] has recently reviewed the metabolism of each of these classes of carcinogens in human cells and cultured human tissues. This chapter will focus on one class of carcinogenic chemicals, the polycyclic aromatic hydrocarbons (PAHs), and the methods used for analysis of their metabolism in human cells in culture. It will also examine the common method for detection and identification of the reactive "ultimate" carcinogenic hydrocarbon metabolites, analysis of their covalent interaction products with DNA. Since this chapter will focus on studies in human cells, many of the original observations carried out in microsomal reaction mixtures and cells from other species will of necessity be omitted. For further details about the metabolic activation of hydrocarbons, the reader is referred to recent reviews by Baird and Pruess-Schwartz,[8] Harvey,[9] and Yang and Silverman.[10] Similarly, it is not the intention of this chapter to provide a detailed listing of all studies of PAH metabolism carried out in human cell and organ cultures. For a more detailed listing of these studies, the reader is referred to reviews by Autrup[7] and Harris.[11,12] This chapter will use selected examples of studies of PAH metabolism and DNA binding in human cells, mainly from our laboratory, to illustrate the techniques presently available for such studies and their application to understanding mechanisms by which carcinogens transform human epithelial cells.

PAHs are metabolized by numerous cellular enzymes. The PAH whose metabolism is best characterized is benzo(a)pyrene (BaP), and its pathways of metabolism are shown in Figure 1. One of the most important groups of hydrocarbon-metabolizing enzymes is the cytochrome P450 monooxygenase family.[13,14] These enzymes oxidize PAHs to form epoxides on a number of bonds[9,10] and in certain cases may generate radical cations capable of reacting with cellular macromolecules.[15] Although the epoxides themselves are highly reactive electrophiles, most do not reach the nucleophilic sites in cellular macromolecules such as DNA, and very few if any DNA interaction products

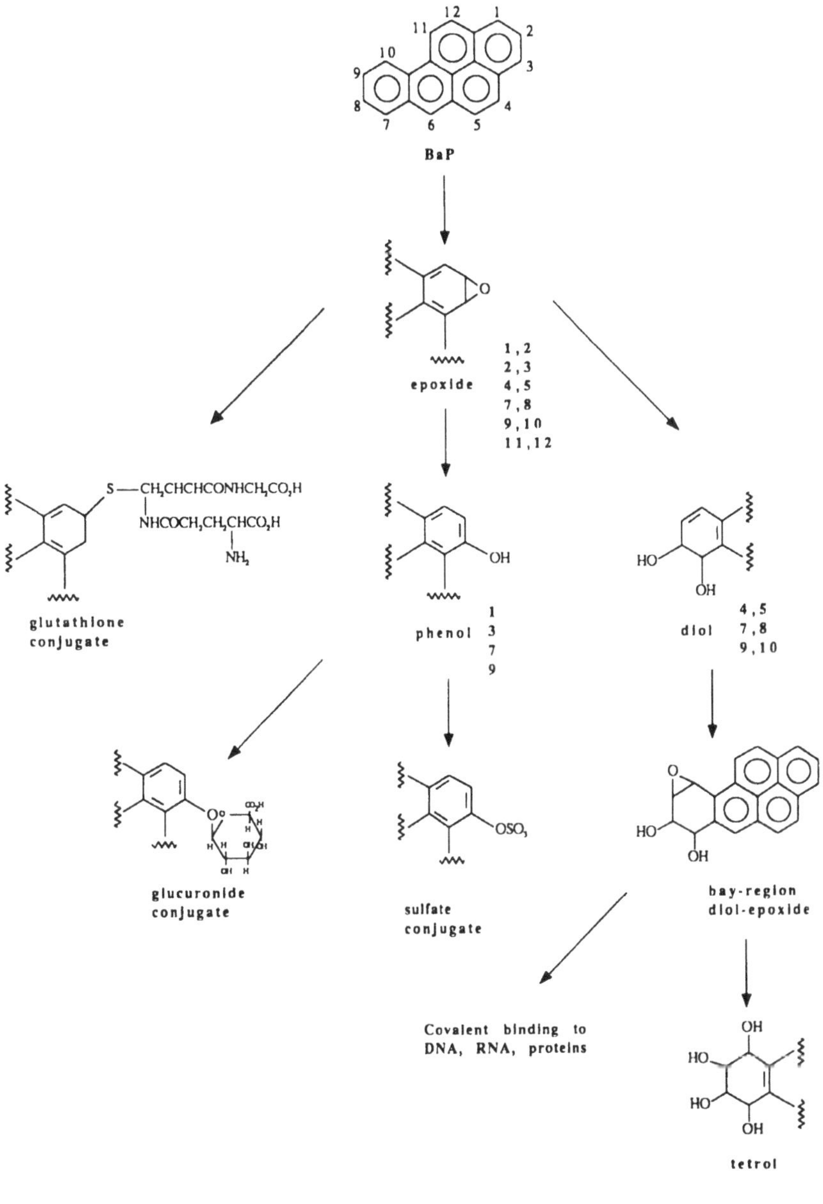

FIGURE 1. Metabolic pathways of BaP.

in cells are formed from these arene oxides. One possible exception is the epoxide formed on the "K-region"[16] of certain PAHs, which is relatively stable and can in certain cases reach the DNA.[17] The majority of epoxides either undergo chemical rearrangement to form phenols or are metabolized by epoxide hydrase to form *trans*-dihydrodiols or by glutathione-*S*-transferase

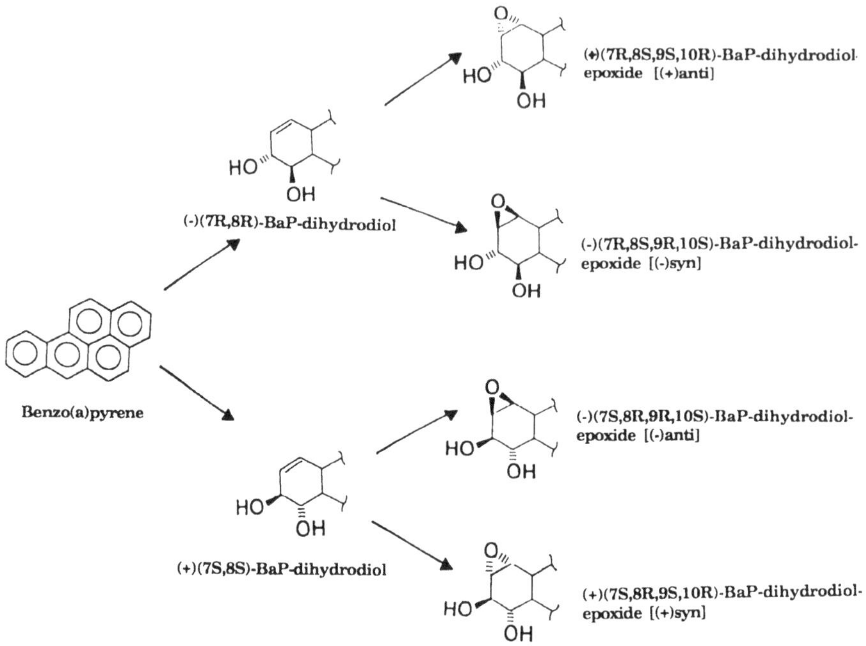

FIGURE 2.   Metabolic activation of BaP to the "bay region": BaP-7,8-diol-9,10-epoxide.

to form glutathione conjugates.[2,9,10] The phenols and diols are substrates for conjugation by UDP-glucuronosyl transferase to form glucuronide conjugates and by sulfottransferase to form sulfate conjugates (Figure 1). These metabolites are also subject to further oxidation by cytochrome P450 to form numerous multioxygen derivatives.[4] One of the most important groups of these multiple oxidation products results from a second oxidation of particular diols to form "bay-region" diol epoxides. On the hydrocarbon shown in Figure 1, the *trans*-7,8-diols can be oxidized to give benzo(a)pyrene-7,8-diol-9,10-epoxide (BaPDE).[18] This has been shown to be one of the major DNA-binding metabolites of BaP in a number of cells and tissues.[8] Since epoxide hydrase forms a *trans*-7,8-diol, four optical isomers of BaPDE can be formed in cells (Figure 2). These have been synthesized and tested for their relative mutagenic and tumorigenic potencies. In mammalian cell mutation assays and tumorigenicity assays in mouse skin and lung, one optical isomer, (+)*anti*-BaPDE with the 7R, 8S, 9S, 10R configuration, has much greater activity than the other three.[19-25] Bay-region diol epoxides of other hydrocarbons have also been shown to be potent tumor initiators, and the activity has been found to differ between optical isomers of the diol epoxides tested.[26] Therefore, characterization of hydrocarbon metabolism in human epithelial cells must take into account the multiple enzymatic pathways that are involved and ultimately be able to define not only the position of modification of the hydrocarbon, but also the stereospecificity of the metabolic pathway. Examples of the types of analyses necessary and their application to human cells will be described.

II. METABOLISM OF CARCINOGENIC POLYCYCLIC AROMATIC HYDROCARBONS

A. ORGANIC SOLVENT EXTRACTION OF CULTURE MEDIUM

Initial studies of hydrocarbon metabolism depended upon the fluorescence of the hydrocarbon molecule for detection of the metabolites in studies of the metabolites present in hydrocarbon-treated rodents.[27] Unfortunately, such studies detected only a small portion of the total metabolites formed. The introduction by the late Charles Heidelberger of the use of radioisotopes to studies of hydrocarbon metabolism allowed accurate quantitation of specific types of metabolites.[28] Using tritium-labeled hydrocarbons, Diamond and co-workers examined the metabolism of hydrocarbons in cell cultures derived from various species, including humans, by an organic solvent extraction technique.[29,30] She found that cell lines derived from human tissues were capable of metabolizing BaP and 7,12-dimethylbenz(a)anthracene (DMBA).[30] The basic technique remains one of the most rapid and widely used for initial assessment of the hydrocarbon-metabolizing capacity of cell cultures. It is based upon the concept that unmetabolized hydrocarbon as well as diols, phenols, and quinones are extracted from the cell culture medium into an organic solvent.[30] The extraction procedure used was based upon a lipid extraction method of Bligh and Dyer.[31] By mixing appropriate proportions of water (0.8 ml, including the cell medium), methanol (2 ml), and chloroform (1 ml), it is possible to create a single-phase system ideal for complete extraction of the hydrocarbon and its phase I metabolites. Addition of 1 ml of water and 1 ml of chloroform results in a two-phase system in which the organic phase is essentially chloroform. After centrifugation at low speed to separate the phases, the amount of radioactivity in aliquots of the chloroform and the methanol-water phase is determined by liquid scintillation counting. This allows calculation of the percentage of water-soluble hydrocarbon metabolites formed (these include glutathione conjugates, glucuronide conjugates, sulfate conjugates, and multiple oxidation products). This rapid procedure provides a quantitative estimate of the relative hydrocarbon-metabolizing capacity of a particular type of cell culture.[30] Since many cells in culture, especially cell lines,[29] have low or varying cytochrome P450 levels, this type of assay provides a good procedure for rapidly assessing the hydrocarbon-metabolizing capacity of a culture without requiring the more extensive analytical procedures necessary to measure individual PAH metabolites or PAH-DNA adducts.

A number of other organic solvent extraction techniques have been described for analysis of the metabolism of PAH to water-soluble metabolites. One of the most commonly used procedures is extraction of the medium with ethyl acetate or ethyl acetate and acetone.[32] This procedure has been applied to analysis of BaP metabolites formed in cell and explant cultures from a number of human tissues, including trachea, lung, liver, colon, mammary, and endometrium.[11,32-40] The advantages of this extraction technique are the elimination of the need to use chloroform and the ability to carry out the extraction with only one vortex step. The disadvantages are the two-phase extraction, which may be less efficient in extracting certain metabolites that complex to culture medium components and the potential for oxidation of hydrocarbon phenols to quinones during the extraction procedure. Antioxidants such as butylated hydroxyltoluene[35] or ascorbic acid are frequently added to the extraction to minimize the latter problem. One difference between this ethyl acetate procedure and the chloroform-methanol-water procedure is in the extraction of hydrocarbon-phenol glucuronides. These metabolites are retained in the aqueous phase at both pH 7 and 4.5 in the chloroform-methanol-water procedure.[41] In contrast, with the ethyl acetate procedure, they are retained in the aqueous phase at pH 7, but extract into the ethyl acetate phase if the sample pH is adjusted to 4.5 prior to extraction.[41] This pH-determined differential extractability can be used to advantage in certain studies of glucuronide formation. Sulfate conjugates of BaP phenols also behave differently with these two extraction procedures. BaP phenol sulfate conjugates have been found to extract into ethyl acetate,[37] but not into chloroform.[42]

Other types of organic solvent extractions have been used less frequently to measure hydrocarbon metabolism. Duncan and Brookes[43] extracted 1 ml of medium with 5 ml of cyclohexane. Analysis of the cyclohexane phase by TLC revealed only unmetabolized BaP; thus, this method may provide a good measure of unmetabolized hydrocarbon. The presence of only the parent hydrocarbon in the cyclohexane phase should be verified by HPLC for each hydrocarbon and cell type to insure that all the metabolites remain in the medium phase.

B. CHROMATOGRAPHIC ANALYSIS OF ORGANIC SOLVENT-EXTRACTABLE METABOLITES

The majority of the chromatographic analyses of hydrocarbon metabolism in cells in culture are carried out with the organic solvent-extractable material. The solvent extraction removes many of the proteins, salts, and nutrients present in cell culture medium and greatly facilitates chromatographic analyses. In the first study of the metabolites formed from [3H]DMBA in cell cultures by thin-layer chromatography (TLC), Diamond et al.[30] found that the organic solvent-extractable metabolites were similar in the human cervical carcinoma cell line, HeLa, and in rodent embryo cell cultures. Interestingly, in this study they also analyzed the water-soluble metabolites by TLC and found that the major water-soluble DMBA metabolite(s) formed in the human HeLa cell line differed from those formed in the primary hamster embryo cell cultures.[30] Although they were unable to identify these metabolites, this study provided evidence for differences in the pathways of hydrocarbon metabolism between rodent and human cells. This TLC technique was used for studies of the metabolism of other hydrocarbons such as [3H]BaP in rodent[40] cells, but was satisfactory mainly for analyzing classes of primary metabolites such as diols, phenols, and quinones rather than determining the amount of specific isomers of each metabolite.

The major technique presently used for analysis of the organic solvent-extractable metabolites is reverse-phase high-pressure liquid chromatography (HPLC) using methanol-water gradients to elute the individual metabolites as described by Selkirk et al.[44] This procedure allows complete separation of the three major BaP diols formed in cells as well as several quinones and the two major phenols.[45] This technique proved useful for analysis of the BaP metabolites formed in human lymphocytes in suspension.[46] Similar HPLC techniques have subsequently been used in a large number of studies of the metabolism of PAHs, especially BaP, in cell and explant cultures from a number of human tissues.[40,47-50] Reverse-phase HPLC analysis is the standard analytical technique for separation of the organic solvent-extractable metabolites formed from a number of hydrocarbons (reviewed in Reference 10). This technique will be useful for characterizing the major pathways of PAH metabolism in human epithelial cell cultures. The use of radioisotope-labeled hydrocarbons provides high sensitivity and accurate quantitation if the PAH of interest is available with, or can be synthesized with, a radioisotope label. Tritium is usually used because of the ability to prepare [³H]PAH of high specific radioactivity relatively easily. Some hydrocarbons have been synthesized with ¹⁴C labels, which prevents loss of radioisotope during metabolism (as can happen with [³H]), but the lower specific radioactivity reduces the sensitivity of detection of metabolites formed in small amounts. Hydrocarbon metabolites may also be detected by fluorescence or with a lower sensitivity by UV absorption. These methods are generally less sensitive than radioisotope studies, and quantitation of metabolites can be difficult due to differences in extinction coefficients for different metabolites. However, these methods offer additional information about the identity of the metabolites based upon spectral characterization. With radioisotope studies, the identification of most metabolites is based upon cochromatography with synthetic standards. The lack of availability of such standards may limit identification of metabolites for many PAHs.

C. CHROMATOGRAPHIC ANALYSIS OF WATER-SOLUBLE METABOLITES

Most analyses of the water-soluble conjugates of BaP metabolites have been based upon enzymatic cleavage of the conjugates, followed by organic solvent extraction and TLC or HPLC analysis of the primary BaP oxidation products that are released. β-Glucuronidase is available free of sulfatase, and treatment of media samples with β-glucuronidase prior to organic solvent extraction can be used to selectively cleave the hydrocarbon glucuronide conjugates formed in cells.[51,52] For identification of the glucuronides, the medium may be extracted first with the organic solvent to remove all unmetabolized BaP and BaP diols, phenols, and quinones. After removal of any organic solvent remaining in the aqueous phase and adjustment of the pH to 4.5, β-glucuronidase treatment and subsequent organic solvent extraction gives an organic phase containing the BaP metabolites that were formerly conjugated to glucuronic acid.[51] For more rapid quantitation of hydrocarbon glucuronides, equal aliquots of medium can be treated directly with β-glucuronidase or buffer alone, extracted, and then analyzed by HPLC. Subtraction of the PAH metabolites in the control incubation sample from those in the β-glucuronidase-treated samples allows quantitation of the glucuronide conjugates present in the original medium.[52] It is also advisable to include a medium sample containing β-glucuronidase plus saccharo-1,4-lactone, an inhibitor of β-glucuronidase.[51] This should prevent cleavage of the water-soluble metabolites to organic solvent-extractable metabolites and verify that the conjugates being cleaved are glucuronides.

Sulfate conjugates of PAH metabolites can also be analyzed by cleavage with aryl sulfatase, followed by organic solvent extraction.[37,53,54] Sulfatase incubations are normally carried out in the presence of saccharo-1,4-lactone to inhibit any β-glucuronidase activity, a common contaminate of sulfatase preparations.

Enzymatic cleavage of PAH glucuronides and PAH sulfates provides a rapid method to detect these conjugates and identify the PAH metabolite(s) that is conjugated. One disadvantage is that any PAH metabolite-glucuronide or sulfate that is resistant to enzymatic cleavage will not be detected. A more common problem is that many PAHs form glutathione conjugates and presently there are no enzymatic methods available for cleavage to release the PAH metabolite.

Autrup[36] described a column chromatographic procedure for the separation of the three major classes of BaP conjugates formed in human colon explant cultures — glutathiones, glucuronides, and sulfates — as the intact conjugates. This was carried out on an alumina column from which the unmetabolized BaP and primary metabolites were eluted with ethanol, the BaP sulfates were eluted with water, the BaP glucuronides were eluted with ammonium phosphate buffer (pH 3), and the BaP-glutathione conjugates were eluted with 25% formic acid.[36] This technique has been widely used for studies of BaP conjugate formation in explant cultures from human tissues:[34,35] in general, the human tissue explant cultures formed mainly glutathione and sulfate conjugates,[38,42,55] whereas many rodent cells formed high proportions of glucuronides. Subsequently, a number of HPLC techniques were reported for the isolation of specific classes of BaP conjugates, but these did not separate all three classes in a single HPLC run.[56-59]

Our laboratory developed an ion-pair HPLC procedure for the separation of BaP-glucuronide, -sulfate, and -glutathione conjugates in a single HPLC run.[60] This procedure uses a C_{18} reverse-phase HPLC column. The glutathione conjugates were eluted with a 1:1 mixture of methanol and 0.04 M tetrabutyl-ammonium bromide (TBAB) in aqueous solution; then, the glucuronide and sulfate conjugates were eluted with a 7:3 mixture of methanol and 0.04 M TBAB solution.[60] The separation of these conjugates is shown in Figure 3. Application of this technique to analysis of the BaP metabolites present in the human hepatoma cell line HepG2 demonstrated the presence of both

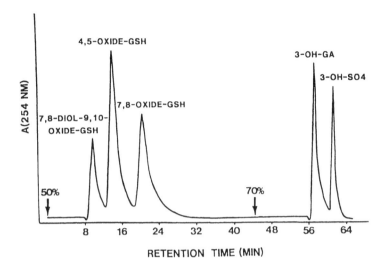

FIGURE 3. Ion-pair HPLC analysis of BaP conjugate standards. BaP conjugates were analyzed on a C_8 column at room temperature at a flow rate of 1.0 ml/min by step gradients of 50 and 70% methanol-TBAB in ammonium formate buffer (pH 6.4) and monitored by UV absorbance at 254 nm. The conjugate standards eluting in order of decreasing ion-pair polarity were GSH conjugates of BaP 7,8-diol-9,10-oxide, and BaP 4,5-oxide, BaP 7,8-oxide, followed by 3-benzo(a)pyrenyl-β-D-glucopyranosiduronic acid and BaP-3-sulfate. (From Plankunov, I., Smolarek, T. A., Fischer, D. L., Wiley, J. C., Jr., and Baird, W. M., *Carcinogenesis*, 8, 59, 1987. By permission of Oxford University Press.)

glutathione conjugates and a large peak of sulfate conjugates (Figure 4). Treatment of the major sulfate peak with aryl sulfatase released mainly 3-hydroxy-BaP and some 9-hydroxy-BaP, and treatment of the earlier-eluting sulfate peak released BaP-4,5-diol and BaP-7,8-diol. This technique was shown to be applicable to analysis of the BaP conjugates formed in media from cell cultures derived from many species[60] and should be useful for characterization of the BaP conjugates formed in human epithelial cell cultures. One limitation to this procedure, a problem common to all of the above analytical techniques involving radioisotopes, is the lack of standard reference compounds for hydrocarbons other than BaP. Although these can be synthesized, a direct analysis method that provides information about the structure of the PAH conjugate would be desirable.

One approach to the development of such a method is the use of very sensitive mass spectrometric techniques. Attempts to analyze the BaP-glucuronide conjugates collected from the ion-pair HPLC method described above were unsuccessful and it was found that the TBAB, even at the very low concentrations present after rechromatography in the absence of this reagent, reacted with the conjugates under desorption chemical ionization conditions and prevented analysis by mass spectrometry.[61] This problem occurs for glucuronides of other materials as well as PAHs.[61] Bieri and Greaves[62] were

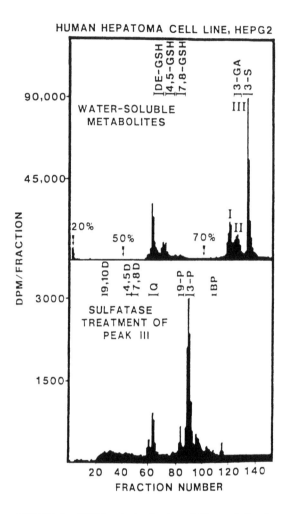

FIGURE 4.   HPLC analysis of water-soluble metabolites in medium from HepG2 cells after 24 h exposure to 0.5 μg[³H]BaP/ml medium. The medium samples were extracted with chloroform-methanol-water and the aqueous phase analyzed by ion-pair HPLC (top). The fractions corresponding to peak III were collected, treated with arylsulfatase, and the ethyl acetate-extractable metabolites eluted with linear gradients of 55 to 95% methanol-water (bottom). The elution positions of BaP metabolite markers are shown at the top of each panel. (From Plankunov, I., Smolarek, T. A., Fischer, D. L., Wiley, J. C., Jr., and Baird, W. M., *Carcinogenesis*, 8, 59, 1987. By permission of Oxford University Press.)

able to obtain mass spectra of BaP-glucuronide and -sulfate standards by HPLC-MS using negative chemical ionization, but failed to obtain molecular ions from these conjugates.

Recently, our lab developed a continuous-flow, high-resolution, fast atom bombardment mass spectrometric procedure for the analysis of sulfate conjugates in cell culture medium.[42] The BaP-sulfate peak from medium from BaP-treated human hepatoma HepG2 cells was isolated by reverse-phase HPLC on a C_{18} column eluted with a gradient of 5% acetonitrile in water with 0.1% trifluoroacetic acid- 10% acetonitrile in water with 0.1% trifluoroacetic acid.[42] This material was then dissolved in 50% acetonitrite- 50% water containing 3% glycerol and pumped at 3 μl/min into the ion source of a Kratos MS50 RF double-focusing mass spectrometer. Figure 5 shows the detection of BaP-SO$_4$ in medium from HepG2 cells using single-ion monitoring at a 200-ppm-wide window centered at 347.0378 with a resolution of 5000. The upper portion of the figure shows the UV absorbance of this material from medium from cells treated with BaP and control cells. Both have a UV-absorbing peak eluting at the same retention time as a BaP-3-sulfate standard. The medium from BaP-treated cells (part a) shows a strong signal at this retention time — it can be quantitated by comparison with a standard curve prepared by injection of known amounts of BaP-sulfate standard (part b). In contrast, the medium sample from cells not treated with BaP shows no mass spectrometric signal (part d). A solvent blank also shows no signal (part c). Thus, this technique is capable of analyzing BaP-sulfate conjugates directly in cell culture medium. As little as 1.5 pg of BaP-SO$_4$ can be detected with a signal-to-noise ratio of 10. This technique should be generally applicable to the analysis of sulfate conjugates formed from any PAH in human epithelial cells. One limitation is that although this will allow detection of any PAH sulfate at the picogram level, quantitation of the amount of this PAH sulfate requires the availability of a synthetic standard. In addition, this will not identify the PAH metabolite isomer present, so that cleavage experiments will be necessary to identify individual PAH phenols that are conjugated to sulfate. However, the wide applicability and sensitivity of this technique will allow one to determine if specific classes of conjugates are formed from a PAH in human cells and to direct the synthesis of reference standards to a limited number of necessary compounds.

This continuous-flow, fast atom bombardment mass spectrometric technique has now been extended to the analysis of BaP-phenol glucuronides formed in cells.[63] Using a resolution of 10,000, it is possible to detect 5 pmol of BaP-phenol glucuronide. This technique demonstrated that BaP-phenol glucuronides were formed in two rodent cell cultures.[63] In addition, it showed that no BaP glucuronides could be detected in the medium of human hepatoma HepG2 cells treated with 1 μg of BaP per milliliter of medium; therefore, these human cells contained less than 0.5 ng of BaP glucuronides per milliliter of medium.[63] Thus, this technique provides convincing evidence of how little, if any, BaP-phenol glucuronides are formed in this human cell line. Presently, we are working on a direct HPLC-continuous flow, fast atom bombardment mass spectrometry procedure for the analysis of the major classes of BaP conjugates formed in cells in culture.

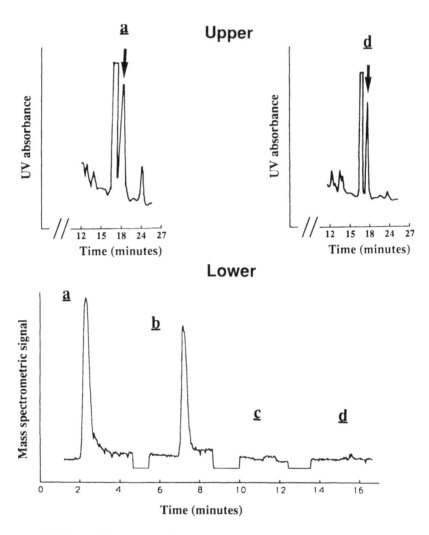

FIGURE 5. UV and CF FABMS detection of BaP-SO$_4$ in media of HepG2 cells treated with BaP for 24 h. Upper: UV recording of HPLC fractionation of medium from cells that were treated with BaP (a), and untreated cells (d). Arrows indicate the time at which fractions were collected and analyzed by CF FABMS. Lower: Single ion recordings for (a) medium from cells treated with BaP, (b) 15 mg of BaP-3-SO$_4$ standard, (c) solvent blank, and (d) medium from untreated cells. (Reprinted with permission from Teffera, Y., Baird, W. M., and Smith, D. L., *Anal. Chem.*, 63, 5, 1990. Copyright 1990, American Chemical Society.)

III. INTERACTIONS OF POLYCYCLIC AROMATIC HYDROCARBONS WITH DNA

PAHs are chemically inert and are not capable of reacting covalently with cellular macromolecules. As indicated in the preceding section, most of the metabolites are conjugated and/or detoxified. However, small proportions of these are highly reactive and capable of reacting with nucleophilic sites in proteins and nucleic acids. Since these electrophilic metabolites are highly reactive, isolation of these metabolites from cells is virtually impossible. The formation of PAH-DNA adducts represents a unique opportunity for the identification of these reactive electrophilic metabolites formed in cells. DNA serves as a trapping agent for these reactive metabolites formed within cells, and the electrophilic metabolites are covalently bound to DNA as PAH-DNA adducts. Analysis of PAH-DNA adducts in cells is difficult due to the very small amounts of PAH-DNA adducts formed in cells treated with the parent PAH. For example, BaP-treated human mammary epithelial cell cultures typically contain less than one adduct per 50,000 deoxyribonucleotides. Analysis of these low levels of PAH-DNA adducts by standard chemical techniques is not feasible. This resulted in the development of a number of highly specific and sensitive methods of PAH-DNA adduct analysis. These methods include (1) the use of radioisotope-labeled hydrocarbon, (2) fluorescence spectroscopy, (3) immunological assays, and (4) postlabeling of PAH-DNA adducts with radioisotopes such as ^{32}P or ^{35}S.

A. ANALYSIS OF PAH-DNA ADDUCTS FORMED FROM RADIOISOTOPE-LABELED HYDROCARBONS

Most studies of PAH-DNA adducts have been performed with hydrocarbons labeled with tritium at high specific radioactivity (up to 60 Ci/mmol).[64] In order to detect and analyze the PAH-DNA adducts formed, a variety of chromatographic techniques have been developed for separation of radiolabeled PAH-DNA adducts from the vast amounts of unmodified deoxyribonucleotides present in the sample. The general separation strategy involves degradation of the PAH-adducted DNA to deoxyribonucleosides through sequential enzymatic treatments prior to chromatographic separations.[65] Typically, PAH-DNA samples are degraded to deoxyribonucleosides by using bovine pancreatic deoxyribonuclease I, snake venom phosphodiesterase, and bacterial alkaline phosphatase.[65] For accurate quantitation of the PAH-DNA adducts, it is necessary to separate them from the unmodified deoxyribonucleosides, which can contain some tritium from tritium exchange.[65] The initial successful analyses of radiolabeled PAH-DNA adducts were carried out by the use of Sephadex LH-20 columns.[66] In this chromatographic system, the PAH-DNA adducts were separated from the unmodified deoxyribonucleosides by using a methanol-water gradient with increasing methanol content. By using this approach, Baird and Brookes[66] were able to completely resolve the hydrocarbon-DNA adducts from the unmodified deoxyribonucleotides in DNA samples from cells treated with 7-methylbenz(a)anthracene. Grover et al.[67] examined the formation of BaP-DNA adducts in human bronchial tissues and in mouse skin after treatment with tritiated BaP and BaP-7,8-dihydrodiol. By using the Sephadex LH-20 chromatography system, they found that the DNA adducts obtained from human bronchial mucosa and mouse skin after BaP treatments were indistinguishable from those obtained from the reaction of the 7,8-dihydro-7,8-dihydroxy 9,10-epoxide of BaP with DNA in solution.[67] This demonstrated that human bronchial mucosa was capable of metabolically activating BaP to a carcinogenic metabolite, BaP diol epoxide. Although this chromatographic approach has been used for a number of studies,[65-71] it requires a long chromatographic separation and has very limited ability to resolve individual hydrocarbon-deoxyribonucleoside adducts. This led to the development of more efficient separation techniques.

The chromatographic technique most commonly used for analysis of PAH-DNA adducts is HPLC. Jeffrey et al.[72] first reported the use of reverse-phase HPLC for analysis of individual hydrocarbon-nucleoside adducts formed from the reaction of DMBA with polyguanylic acid [poly(G)] and polyadenylic acid [poly(A)] by a microsomal activating system. The system allowed separation of a number of DMBA-deoxyribonucleosides within a 60-min chromatogram and is substantially more sensitive than the Sephadex LH-20 chromatographic system.[72] The same HPLC chromatographic separation technique was used to analyze the [³H]BaP-DNA adducts formed in [³H]BaP-treated human bronchial explant cultures.[73] By comparing the elution time of adducts formed with those of synthetic markers, Jeffrey et al.[73] determined that the majority of the BaP-DNA adducts formed in the BaP-treated explant cultures coelute with the (+)*anti*-BaPDE-deoxyguanosine adduct. Although it is possible to identify certain adducts formed in cells treated with the hydrocarbon by comparing elution time with that of standards, this identification strategy is sometimes frustrated by the complex PAH-DNA adduct patterns that are often encountered in cells.[74,75] There is often overlap between different adducts,[76] which makes accurate quantitation of adduct levels difficult. In order to improve the resolution of individual PAH-DNA adducts, other separation techniques have been developed and applied to the analysis of PAH-DNA adducts formed in human mammary epithelial cell cultures treated with PAHs. These techniques include (1) immobilized boronate chromatography,[77,78] (2) acid hydrolysis of PAH-DNA adducts to PAH-purine adducts,[79] and (3) acid hydrolysis of PAH-DNA adducts to PAH tetraols.[79]

During the metabolic activation of BaP, four different stereoisomers of BaPDE can be formed.[18,20,80] These BaPDEs are the (+)*anti*-, (−)*anti*-, (+)*syn*-, and (−)*syn*-7,8-diol 9,10-epoxide of BaP (Figure 2). In order to separate the DNA adducts formed from *anti*- and *syn*-BaPDE, a chromatographic separation technique based upon the use of immobilized dihydroxyboryl groups was developed.[77,78] The use of immobilized boronate groups in the separation of *cis*-vicinal hydroxyl-containing compounds was described by Weith and Gilham[81,82] and applied to the analysis of 7,12-DMBA-DNA adducts by Sawicki et al.[77] In this separation strategy, the molecules that contain *cis*-vicinal hydroxyls were retained by the boronate groups due to the formation of a complex, while the molecules that did not contain *cis*-vicinal hydroxyl groups were washed from the column by using a high pH (pH 8 to 9) binding buffer. The molecules that contained the *cis*-vicinal hydroxyl group and were retained by the column were subsequently eluted with a buffer that

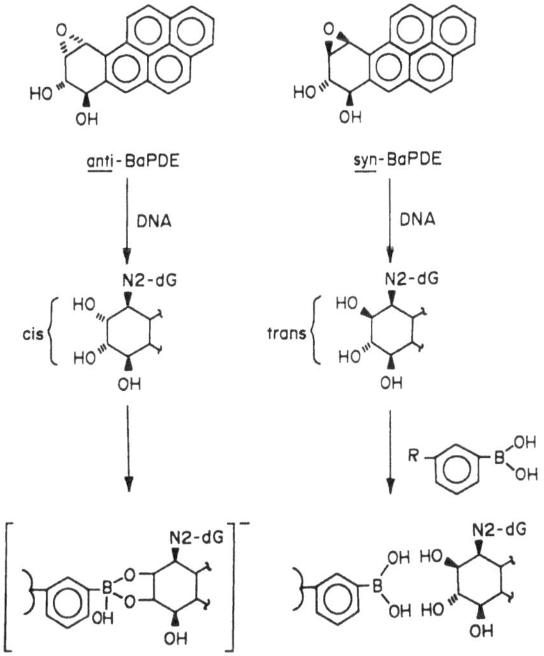

FIGURE 6. Analysis of BaP-DNA adducts by immobilized boronate chromatography.

contained sorbitol.[77,78,81] This separation technique has been proven successful in the analysis of ribonucleic acids,[81,82] as well as PAH-DNA adducts.[8,77,78,83] As shown in Figure 6, adducts derived from the reaction of *anti*-BaPDE contained *cis*-vicinal hydroxyl groups, and thus were retained by the boronate column. The adducts formed from *syn*-BaPDE did not contain *cis*-vicinal hydroxyls and were not retained. The boronate complex formed between the *anti*-BaPDE-DNA adducts and the boronate group was reversed by the application of an eluting buffer that contained 10% sorbitol. This technique makes it possible to completely resolve the *anti*-diol-epoxide-DNA adducts from the *syn*-diol-epoxide-DNA adducts and was used in the analysis of BaP- and DMBA-DNA adducts formed from human mammary epithelial cultures treated with these hydrocarbons.[83]

When normal human mammary epithelial cell cultures were treated with [³H]BaP for 24 h, extensive binding of [³H]BaP to DNA was detected. In order to quantitate and resolve the individual BaP-DNA adducts formed, the combination of immobilized boronate chromatography followed by reverse-phase HPLC was used.[79] By using immobilized boronate chromatography, 24% of the radioactivity was eluted in the morpholine washing buffer, while 76% of the radioactivity was retained by the boronate column and subsequently eluted in the morpholine-sorbitol eluting buffer. Reverse-phase HPLC analysis of BaP-DNA adducts from the morpholine fraction revealed the presence of a single major adduct peak which eluted at the same retention time as a (+)*syn*-BaPDE-deoxyguanosine adduct. Two major BaP-DNA adducts peaks were present in the reverse-phase HPLC profile of the morpholine:sorbitol fraction. The one that coeluted with a (+)*anti*-BaPDE-deoxyguanosine marker contained 93% of the radioactivity, while the other minor peak eluted at the same relative position as a (−)*anti*-BaPDE-deoxyguanosine adduct.[79] Similar experiments[79] have been performed with the human mammary carcinoma T47D cell line.[79] When the T47D cell cultures were exposed to [³H]BaP for 24 and 48 h, the level of BaP covalently bound to DNA was 182 and 327 pmol/mg DNA, respectively.[79] The BaP-DNA adducts were analyzed by immobilized boronate chromatography and reverse-phase HPLC. After 24 h of BaP exposure, 81% of the radioactivity was eluted in the morpholine-sorbitol buffer, whereas after 48 h of exposure, the amount of radioactivity eluted in the sorbitol fraction was decreased to 54%. Individual BaP-DNA adducts from the morpholine and the morpholine-sorbitol fractions were then analyzed by reverse-phase HPLC. After 24 h of BaP exposure, the morpholine buffer fraction contained three adduct peaks (Figure 7A: M0, M1, and M2). Peak M2 eluted at the same retention time as a (+)*syn*-BaPDE-deoxyguanosine adduct, while peak M1 eluted in the same position as both *syn*-BaPDE-deoxyguanosine and *syn*-BaPDE-deoxycytidine markers. Peak M0 eluted prior to any of the *syn*-BaPDE-DNA markers. The morpholine-sorbitol fraction contained two adduct peaks that were resolved by reverse-phase HPLC (Figure 7B: MS1 and MS2). Peak MS2, the major adduct, coeluted with the (+)*anti*-BaPDE-dGuo marker. MS1 eluted in the same relative position as the (−)*anti*-BaPDE-dGuo marker. After 48 h of exposure of T47D cells to BaP, similar results were obtained.[79]

Similar techniques were used to identify the DMBA-DNA adducts formed in human mammary epithelial cell cultures.[83] When the cell cultures were treated with 2 µ*M* [³H]DMBA for 24 h, the level of [³H]DMBA covalently bound to DNA was 12 pmol/mg DNA. This level of DMBA binding was significantly lower than that obtained from rat mammary epithelial cell culture after a similar DMBA treatment (86 pmol/mg DNA).[83] Analysis of individual DNA adducts by immobilized boronate chromatography and reverse-phase HPLC indicated that the majority of the DMBA-DNA adducts formed were derived from the *syn*-DMBA-diol epoxide; *syn*-DMBADE-deoxyadenosine was the major adduct present. Despite differences in the overall level of binding, DNA-adduct profiles obtained from DMBA-treated rat mammary epithelial cultures contained essentially the same proportion of specific adducts as those obtained from DMBA-treated human mammary epithelial cell cultures. The formation of PAH-deoxyadenosine adducts has been correlated with the carcinogenic potency of the hydrocarbon.[8,84,85] Even though the HPLC profiles of DMBA-DNA adducts found in rat and human epithelial cells are virtually identical, the overall level of DMBA-deoxyadenosine adducts was much higher in the rat mammary cells.[83] This higher level of binding correlates well with DMBA-induced mutagenesis in a cell-mediated assay,

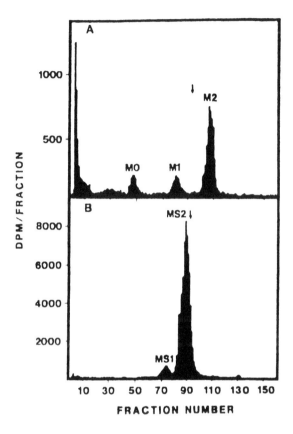

FIGURE 7. Reverse-phase HPLC elution profiles of BaP-deoxyribonucleoside adducts present in human mammary carcinoma T47D cells after exposure to 4 μM [³H]BaP (6 Ci/mmol) for 24 h. The BaP-deoxyribonucleoside adducts were isolated by chromatography on Sep-Pak C_{18} cartridges and analyzed by immobilized boronate chromatography. The adducts in the 1 M morpholine buffer fractions and the 1 M morpholine-10% sorbitol buffer fractions were concentrated and analyzed by reverse-phase HPLC on an Ultrasphere octyl column. (A) 1 M morpholine buffer fractions, 24 h; (B) 1 M morpholine-10% sorbitol buffer fractions, 24 h. Arrow elution position of a [¹⁴C]-(+)-*anti*-BaPDE:dGuo marker. (From Pruess-Schwartz, D., Baird, W. M., Nikbakht, A., Merrick, B. A., and Selkirk, J. K., *Cancer Res.*, 46, 2697, 1986. With permission.)

for the rat mammary epithelial cells activated DMBA to mutagenic metabolites more efficiently than human mammary epithelial cell cultures.[86]

Cochromatography with synthetic markers in a single chromatographic system provides only limited support for the identification of [³H]PAH-DNA adducts. To provide further information about the structure of the PAH-DNA adducts, two acid hydrolysis techniques have been developed[75,79,87] and applied to the analysis of BaP-DNA adducts present in human epithelial cell cultures.[79] The first involved a mild acid hydrolysis of BaP-DNA adducts to their respective BaP-purine adducts.[75,79] This was performed by treatment of the BaP-DNA adducts with 0.1 N HCl at 37°C for 24 h.[75,79] The BaP-DNA adducts formed from the (+) and (−) enantiomers of *anti*-BaPDE with deoxyguanosine (or deoxyadenosine) exist as diastereomers with the opposite relative stereochemistry at every position except the C-1 position of the deoxyribose moiety. Under normal conditions of reverse-phase HPLC, diastereomers can easily be separated: the (−)*anti*-BaPDE-deoxyguanosine adduct eluted over 20 fractions (6 min) prior to the (+)*anti*-BaPDE-deoxyguanosine adduct.[79] However, upon mild acid hydrolysis, the glycosidic bond of these adducts was cleaved,[75,79] converting these diastereomeric adducts into enantiomers (i.e., with the relative stereochemistry opposite at every position). Under normal reverse-phase HPLC conditions, enantiomers are extremely difficult to resolve, thus the (+)- and (−)*anti*-BaPDE-guanine adducts cochromatograph.[88] By HPLC analysis of the adducts after acid hydrolysis, it is possible to determine if two adducts are formed from enantiomeric diol epoxides. In BaP-DNA from T47D human mammary carcinoma cell cultures, two adduct peaks were obtained in the morpholine-sorbitol fraction after the immobilized boronate chromatography.[79] The one which contained the majority of the radioactivity coeluted with the marker (+)*anti*-BaPDE-deoxyguanosine adduct.[79] Upon acid hydrolysis to the purine adducts, this adduct peak again coeluted with the (+)*anti*-BaPDE-guanine marker, providing further evidence that the cellular adduct peak was the (+)*anti*-BaPDE-deoxyguanosine adduct.[79] The other minor adduct peak obtained from the morpholine-sorbitol fractions from the immobilized boronate chromatography eluted prior to the (+)*anti*-BaPDE-deoxyguanosine marker on HPLC. After acid hydrolysis to the BaP-purine adducts, this peak cochromatographed with the hydrolysis product obtained from the (+)*anti*-BaPDE-deoxyguanosine.[79] This study provided the first direct evidence, other than cochromatography as deoxyribonucleoside adducts, that (−)*anti*-BaPDE is formed in human cells in culture and reacts with DNA to form the (−)*anti*-BaPDE-deoxyguanosine adduct.[79]

Another acid hydrolysis technique for characterization of adducts involves a more vigorous treatment to hydrolyze the BaP-DNA adducts to their respective BaP tetraols.[75,79,87,88] In this technique, BaP-DNA adducts were treated with 0.1 N HCl for 6 h at 80°C. This treatment results in the cleavage of the hydrocarbon-deoxyribonucleoside bond to form the hydrocarbon tetraols, which are analyzed by reverse-phase HPLC.[75,79,87,88] The BaP tetraols formed by the hydrolysis of *anti*- and *syn*-BaPDE-DNA adducts can be easily separated by reverse-phase HPLC.[75,79,87] This allows identification of hydrocarbon metabolite responsible for forming each hydrocarbon-DNA adduct. Characterization of the major BaP-DNA adducts formed in the human mammary carcinoma cell line T47D by this technique provided additional evidence for the identity of the BaP-DNA adducts formed. These results, along with those obtained from hydrolysis of BaP-DNA adducts to BaP-purine adducts and immobilized boronate chromatography, demonstrated that the major DNA-binding metabolites of BaP in the T47D human mammary cell cultures were (+)*anti*-BaPDE (75%), (−)*anti*-BaPDE (5%), and (+)*syn*-BaPDE (5%).[79]

Despite the popularity and widespread use of radioisotope-labeled PAH for PAH-DNA adduct detection, this technique has several limitations.[89] One limitation is the cost and the availability of the radiolabeled PAH needed. Another is the necessity of using a radiolabeled PAH, which eliminates studies of DNA adducts in humans after exposure to environmentally occurring PAHs. A third is the limited sensitivity for the detection of adducts in a small number of cells. Therefore, in recent years, a number of highly sensitive techniques have been developed for the analysis of PAH-DNA adducts formed in human tissues.

B. FLUORESCENCE SPECTROSCOPIC METHODS OF PAH-DNA ADDUCT ANALYSIS

PAHs exhibit strong fluorescence[80] when irradiated by UV light, and this fluorescence response has been used in the detection and identification of the parent PAH. Initial attempts to detect PAH-DNA adducts by fluorescence spectroscopy were hampered by the relatively low sensitivity of most detectors. The development of photon-counting detectors and improved techniques allowed the measurement of low levels of fluorescence response and increased the use of this method for the detection of PAH-DNA adducts.[90-94] To overcome the problem of fluorescence quenching, PAH-DNA was hydrolyzed into its respective hydrocarbon tetraols by acid treatments, and the free tetraols were then isolated and analyzed by HPLC.[95] By using this method, a two- to tenfold enhancement in sensitivity was observed.[95] Another fluorescent spectroscopic method of PAH-DNA detection, synchronous fluorescence spectroscopy,[96-98] scans the excitation and emission wavelengths with a fixed wavelength difference. Vahakangas et al.[99] have determined that BaPDE-DNA adducts have an optimum wavelength difference of 34 nm and an emission maximum of 382 nm, while the maximum for BaP tetraols is 379 nm. Since different PAH-DNA adducts have different optimum wavelength difference and emission maxima, it is possible to distinguish individual PAH adduct components present in complex mixtures.[99] Manchester et al.[100] applied this technique to detect BaPDE-DNA adducts in human placenta and demonstrated the presence of BaPDE-DNA adducts; in a similar study,[101] they have also detected the presence of other PAH-DNA adducts in human placenta. Another approach to fluorescence spectroscopy for the detection of PAH-DNA adducts involves the use of low temperatures.[102] At normal room temperatures, the fluorescence spectra of PAH-DNA adducts are broad and highly quenched. However, at lower temperatures (77 K), the fluorescence response increases significantly.[95,98] With a further lowering of the temperature to 4.2 K, a phenomenon known as fluorescence line narrowing occurs.[103] After fluorescence line narrowing, the spectrum of each individual adduct is highly characteristic,[104-107] and this technique allows identification of individual adducts among mixtures of different adducts. Recently, Lu et al.[108] have combined fluorescence line narrowing spectroscopy (FLNS) with fluorescence spectroscopy at 77 K and fluorescence quench to characterize reaction products of (+)*anti*-BaPDE with polymers of deoxyribonucleotides. In general, fluorescence spectroscopy is capable of detecting femtomole levels of PAH adducts, and should be applicable for the detection of PAH-DNA adducts formed in human cells.

C. IMMUNOCHEMICAL METHODS OF PAH-DNA ADDUCT ANALYSIS

The use of antibodies for the detection of PAH-DNA adducts was pioneered by Poirier and co-workers.[109] The basic principle of this technique involves the preparation of antibodies specific for PAH-DNA adducts. This is usually performed by reacting DNA or polymers of deoxyribonucleotides with the respective reactive electrophilic hydrocarbon metabolites, followed by multiple injections of this PAH-adducted DNA into New Zealand White rabbits[109-111] or mice[111,112] for the production of polyclonal or monoclonal antibodies, respectively. The specificity and affinity of the PAH-DNA adduct-specific antisera is constantly monitored using noncompetitive ELISA assays. This technique combines the advantages of high sensitivity of adduct detection and elimination of the need for using radioisotope-labeled materials. A number of immunological assays with different limits of PAH-DNA adduct detection have been developed, including radioimmunoassay, competitive ELISA, noncompetitive ELISA, USERIA, and electron microscopic and immunoaffinity chromatography techniques.[109-124] These assays have proven valuable for the detection of PAH-DNA adducts found in human cells from a number of tissues.

D. POSTLABELING ANALYSIS OF PAH-DNA ADDUCTS BY ^{32}P

In recent years, the ^{32}P-postlabeling analysis technique developed by Randerath and co-workers has been used extensively in the analysis of carcinogen-DNA adducts.[125] This technique combines both extraordinarily high sensitivity for the detection of carcinogen-DNA adducts and general applicability for assaying DNA adducts formed by many different classes of carcinogens, including a number of PAHs. This technique involves the enzymic degradation of carcinogen-adducted DNA to deoxyribonucleoside-3'-phosphates with micrococcal nuclease and spleen exonuclease. The adducted nucleoside-3'-phosphates are then 5'-phosphorylated with ^{32}P-phosphate through the phosphorylation reaction mediated by T4 polynucleotide kinase and γ-^{32}P-labeled ATP.[125,126] The ^{32}P-postlabeled adducted nucleoside-3',5'-*bis*-phosphates were then resolved from the unused ATP and the ^{32}P-labeled normal nucleotides by using a four-dimensional TLC system.[126] A number of procedures have been developed to enhance the sensitivity of detection of PAH-DNA adducts. These enhancement procedures separate the vast amounts of unmodified deoxyribonucleotides from the PAH-adducted deoxyribonucleotides by either enzymic degradation (nuclease P1 treatment),[127] liquid extraction (1-butanol),[128] or prior separation of normal deoxyribonucleotides from the adducted nucleotides through chromatographic techniques (HPLC,[129] sep-pak c-18 cartridges).[130] These procedures, along with the high specific activity of ^{32}P ATP (up to 6000 Ci/mmol), allow detection of subfemtomole quantities of PAH-DNA adducts. PAH-DNA adduct frequencies as low as one adduct per 10^{10} nucleotides can be measured using only microgram quantities of DNA.[127,128]

This ultrasensitive technique has been applied in the analysis of PAH-DNA adducts from both BaP-treated human epithelial cells in culture and cells from human donors. Seidman et al.[131] examined the formation of BaP-DNA adducts from BaP-treated human mammary epithelial cell cultures and obtained results consistent with those obtained from previous studies with [^{3}H]BaP; (+)*anti*-BaPDE-deoxyguanosine adduct was the major adduct present.[83] Furthermore, a minor labeled adduct spot was detected and tentatively identified as the (−)*anti*-BaPDE-deoxyguanosine adduct which had also been found in HPLC experiments with [^{3}H]BaP.[79] They then used this technique to examine the endogenous adducts present in mammary cells isolated from breast tissue from ten donors. Seven out of the ten donors had no detectable adducts. Three donors had detectable levels of DNA adducts; however, the adduct TLC spots were not identical to those obtained from *anti*-BaPDE-dNp standards. The adduct levels in these positive donors ranged from three adducts per 10^8 nucleotides to two adducts per 10^9 nucleotides.[131] The difference in the adduct patterns obtained from cell culture studies and those from humans may be due to systemic activation of BaP, which may affect mammary adduct formation or to the exposure of the cell cultures to relatively large doses of BaP compared to the much lower circulating doses of PAH that may reach the mammary cells *in vivo*.[131] Schoket et al.[132] have used this postlabeling technique to examine the formation of hydrocarbon-DNA adducts in human skin after application of coal tar, creosote, or bitumen. After solutions of hydrocarbon mixtures were applied to human skin samples maintained in short-term organ culture, the DNA was isolated and the presence of aromatic hydrocarbon-DNA adducts was examined by using the ^{32}P-postlabeling technique. The results revealed the presence of multiple hydrocarbon-DNA adducts at levels of up to one adduct per 10^7 nucleotides after a single application.[132] This provides strong evidence that these agents, which are suspected of being carcinogenic to humans, can cause DNA damage in human skin. Ribovich et al.[133] have also examined the extent of BaPDE binding in replicating DNA compared to the levels in parental DNA. By using the postlabeling procedure to quantitate the levels of BaPDE-DNA binding, they found that replicating DNA contained 1.4 to 2.5 times more adducts than parental DNA.[133] This is of particular interest because transformation of human cells is known to be increased when the cells are treated with carcinogens during the early S-phase.[134] Therefore, it is possible that critical DNA sites responsible for initiation of transformation may be unmasked from the nucleosome core and thus be preferentially modified by the carcinogen.

This ^{32}P-postlabeling technique has been used to monitor PAH-DNA adducts in humans exposed to environmental pollutants.[135,136] These studies have recently been reviewed by Phillips.[89] Although the ^{32}P-phosphate labeling assay provides extremely high sensitivity for adduct detection, it does have several potential limitations.[89] One is that quantitation of adduct levels is subject to error if the carcinogen-adducted DNA is resistant to degradation,[137] or if the adducted nucleotide is unstable or a poor substrate for the phosphate transfer reaction mediated by T4 polynucleotide kinase.[138,139] In analyses of complex mixtures of adducts, it is possible that some of the adducts will become preferentially labeled and alter the quantitative results. Therefore, it is essential to define the labeling efficiency of different adducts and to determine their resistance to the DNA degradation procedures.[140] This aspect is particularly important in the analysis of PAH-DNA adducts obtained from human tissues, as adduct levels are often lower than one per 10^8 nucleotides and these samples often contained large numbers of unidentified adducts. The TLC separation procedure used to separate adducts as nucleoside *bis*phosphates can also vary in quality of separation and resolution of adducts, depending upon the care taken in the preparation and running of the TLC plates.

E. POSTLABELING ANALYSIS OF PAH-DNA ADDUCTS BY ^{35}S-PHOSPHOROTHIOATE

In order to combine both the high resolution of PAH-DNA adducts by reverse-phase HPLC and the advantages provided by the postlabeling technique, two procedures for the analysis of PAH-DNA adducts postlabeled with ^{35}S-phosphorothioate were developed.[141] In one procedure, as shown in Figure 8A, BaP-adducted DNA was enzymatically digested to deoxyribonucleoside-3'-phosphates with micrococcal nuclease and spleen phosphodiesterase. These adducted nucleotides were then enriched in the organic phase through the 1-butanol extraction procedure developed by Gupta.[128] The enriched adducted nucleotides were 5'-thiophosphorylated with ^{35}S by using γ-labeled [35]SATP and T4 polynucleotide kinase.[141] This thiophosphate transfer yields the ^{35}S-labeled BaP-deoxyribonucleoside-5'-phosphorothioate-3'-phosphate adduct. In order to facilitate separation of the labeled adducts by HPLC, it is necessary to selectively remove the unlabeled 3'-phosphate.[141] This is achieved by brief alkaline phosphatase treatment, which takes advantage of the resistance of thiophosphates to dethiophosphorylation by alkaline phosphatase.[142,143] The ^{35}S-monophosphorothioate-postlabeled PAH-DNA adducts were then resolved and quantitated by reverse-phase HPLC. In the other ^{35}S-labeling procedure,[141] as shown in Figure 8B, the nuclease P1/prostatic acid phosphatase DNA degradation-^{32}P-postlabeling method described by Randerath et al.[144] was adapted to use ^{35}S-labeled ATP instead of ^{32}P-ATP. With this nuclease P1 procedure, the final labeled adducts were the ^{35}S-labeled adducted nucleoside-5'-phosphorothioate adducts, the same adducts obtained from the alkaline phosphatase procedure.[141] The ability to obtain ^{35}S-labeled PAH-adducted nucleoside-5'-phosphorothioate adducts by two independent enzymic

A

1-Butanol/AKP procedure

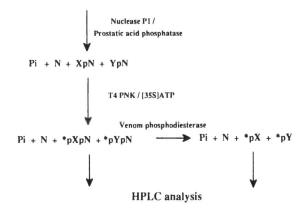

BaP-DNA

..NpNpNpXpNpNpYpNpNpNp..

Micrococcal Nuclease /
Spleen exonuclease

Np + Xp + Yp   1-Butanol extraction → Np (aqueous phase)

(1-butanol phase)
T4 PNK / [35S]ATP

*pXp + *pYp   Alkaline phosphatase → *pX + *pY

HPLC analysis

B

Nuclease P1/PAP procedure

BaP-DNA
..NpNpNpXpNpNpYpNpNpNp..

Nuclease P1 /
Prostatic acid phosphatase

Pi + N + XpN + YpN

T4 PNK / [35S]ATP

Pi + N + *pXpN + *pYpN   Venom phosphodiesterase → Pi + N + *pX + *pY

HPLC analysis

FIGURE 8.   Schemes of [^{35}S]phosphorothioate labeling of BaP-DNA adducts. (A) 1-butanol/alkaline phosphatase method; (B) nuclease P$_1$/prostatic acid phosphatase method.

degradation procedures allows one to determine the optimal conditions in the postlabeling analysis of PAH-DNA adducts and to reduce the possibility of failing to label a particular adduct. For quantitation of adduct levels, a control sample that contained known amounts of the carcinogen-DNA adducts of interest was labeled under identical conditions, and the levels of adduct recovery were determined and used for quantitation of cellular samples.[141] Studies of comparative recoveries of postlabeled carcinogen-DNA adducts have shown that the susceptibility of certain carcinogen adducts to 3'-dephosphorylation by nuclease P1 is highly dependent on the type and the structure of the adducts.[145,146] It is also possible that some carcinogen-DNA adducts may resist the enzymic degradation mediated by micrococcal nuclease and spleen phosphodiesterase.[125,126] For instance, Cheh et al.[140] have found selective release of benzo(c)phenanthrene-deoxyadenosine adducts by these two nucleases. Analysis of BaP-DNA adducts by the two[35]S-labeling procedures gave similar results, although the nuclease P1 procedure did give a higher efficiency of adduct recovery than the alkaline phosphatase procedure.[141] The adduct levels determined by both ^{35}S-postlabeling procedures closely matched those obtained from the [^{3}H]BaP nucleoside/HPLC analysis of the same sample.[141] The initial development of these ^{35}S-labeling procedures was based upon the detection of BaP-DNA adducts formed in hamster embryo cell cultures;[141] however, it should be possible to analyze PAH-DNA adducts obtained from PAH-treated human epithelial cells in culture. The ^{35}S-method reduces the risk of radiation exposure for personnel[147] and also facilitates the separation of labeled adducts as nucleoside monophosphorothioates by HPLC.[141]

The immobilized boronate chromatography procedure for the separation of *cis*-vicinal hydroxyl-containing diol epoxide DNA adducts prior to adduct analysis by HPLC greatly improved the analysis of specific PAH diol epoxide-DNA adducts in studies with [^{3}H]PAH-deoxyribonucleoside adducts. This technique has now been adapted for use in combination with the ^{35}S-postlabeling procedures for the analysis of PAH-DNA adducts formed in cells. In this procedure, PAH-adducted DNA is enzymatically degraded and labeled with ^{35}S as in the standard ^{35}S-labeling protocol. The ^{35}S-labeled adducts are then partially purified by chromatography on a sep-pak C-18 cartridge and applied to an immobilized boronate column. Under conditions of high pH and high salt concentration, the labeled adducts that contain *cis*-vicinal hydroxyl groups are retained by the column, while the adducts which do not have *cis*-vicinal hydroxyls are washed off. The retained labeled adducts are then eluted from the column with a sorbitol-containing buffer. The fractions from the column washings and the sorbitol-eluted fractions are then desalted and analyzed by reverse-phase HPLC. Initial optimization of separation conditions was performed on a sample of DNA from hamster embryo cell cultures treated with DMBA at 2 μM for 24 h. Previous characterization of the identity of the individual DMBA-DNA adduct allowed accurate assignment of the tentative structure of the postlabeled adduct peaks obtained. As shown in Figure 9A, three major labeled adduct peaks were obtained (P3, P4, and P5).

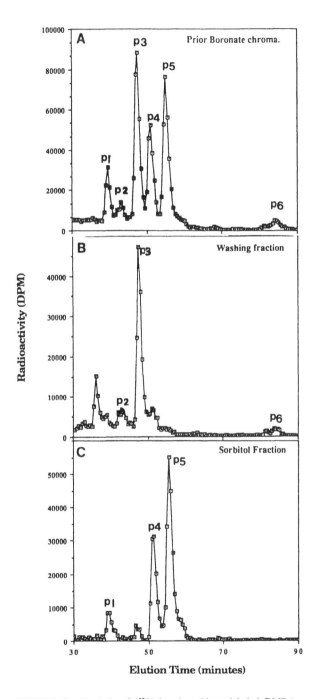

FIGURE 9. Analysis of [³⁵S]phosphorothioate-labeled DMBA-DNA adducts by an immobilized boronate chromatograph-HPLC procedure. See text for further details.

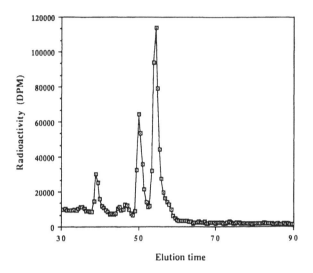

FIGURE 10.   Analysis of DMBA-DNA adducts formed in MCF-7 cells treated with racemic DMBA-3,4-diol by [^{35}S]phosphorothioate labeling.

Cochromatography with DMBADE-reacted deoxyribonucleotide standards allowed the tentative assignment of P3 as *syn*-DMBADE-dAdo, P4 as *anti*-DMBADE-dGuo, and P5 as *anti*-DMBADE-dAdo.[148] This assignment was further confirmed by the use of immobilized boronate chromatography. As shown in Figure 9B, which was obtained from the washing fraction of the column, only P3 and other minor peaks were present. Both P4 and P5 were present in the HPLC profile obtained from the sorbitol-containing eluting buffer (Figure 9C). These results were in good agreement with the tentative assignments obtained from DMBADE-deoxyribonucleotide standards.[148] This technique was then applied in the analysis of DMBA-DNA adducts formed in the human mammary carcinoma (MCF-7) cell cultures treated with the proximate carcinogenic 3,4 diol of DMBA. MCF-7 cells in culture were treated with 0.15 µ*M* of the racemic DMBA 3,4 diol for 24 h and the DNA was isolated. The adduct levels were then determined by ^{35}S-labeling 20 µg of the adducted DNA followed by analysis with reverse-phase ion-pair HPLC. As shown in Figure 10, three major adduct peaks were detected. All three major peaks resulted from the reaction of *anti*-DMBADE with DNA in cells, and the level of DNA adducts obtained from *syn*-DMBADE was much smaller. These results indicated that in MCF-7 cells, racemic DMBA-3,4-diol can be converted to both *anti*- and *syn*-DMBADE; the formation of *anti*-DMBADE-DNA adducts predominates. This finding is in contrast to the results obtained from the DMBA-treated normal human epithelial cell cultures; in these cultures, the major adduct was the *syn*-DMBADE-DNA adducts.[83] It is not clear if this difference is due to differences in the two human cell types being studied or because the MCF-7 cells are being treated with the racemic

DMBA-3,4 diol and not DMBA itself. Vericat et al.[149] found that the formation of *anti*-DMBADE-DNA adducts predominated when mouse embryo cells were treated with optically pure DMBA-(4R,3R)-dihydrodiol; thus, the high levels of *syn*-DMBADE-DNA adducts in DMBA-treated human mammary epithelial cells may result from limited metabolism of DMBA to the DMBA-(4R,3R)-dihydrodiol. The combination of immobilized boronate chromatography with ^{35}S-postlabeling has allowed us to detect and identify PAH-DNA adduct formed in human cells treated with PAH or PAH derivatives. This technique not only provides the high sensitivity of adduct detection and great adduct resolution obtained by HPLC, but is also generally applicable for detection of PAH-DNA adducts without the need of radioisotope-labeled PAH or PAH metabolites.

IV. PERSPECTIVES

The past decade has witnessed an incredible growth in our understanding of the molecular mechanisms of cell transformation. Other chapters describe our increased understanding of growth factors, oncogenes, and tumor suppressor genes. Great advances have been made in the development of culture procedures for growing and maintaining the differentiated state of human epithelial cell cultures from a number of tissues. Thus, the basic cell systems for studying mechanisms of chemical carcinogenesis in human epithelial cells have improved dramatically.

During this same period, major advances have been made in our understanding of the molecular mechanisms of carcinogen activation. Many of the enzymes involved in hydrocarbon metabolism have now been cloned, allowing detailed studies of their regulation during carcinogen exposure. Inhibitory antibodies against specific P450 isozymes allow assessment of the role of individual P450s in the activation and detoxification of PAHs. The development of more sensitive analytical techniques such as recent advances in mass spectrometry allow analysis of PAH metabolites from cell culture medium. The advances in chromatographic techniques for metabolite separation such as microbore HPLC will allow improved analysis with even smaller quantities of PAH metabolites.

During the past decade, there has been tremendous progress in the development of techniques for analysis of PAH-DNA adducts. In the 1970s, research in this area focused on identifying an ultimate carcinogenic metabolite of a hydrocarbon and led to the present knowledge of the role of diol epoxides in this process. This continues to be an important area, and other mechanisms of hydrocarbon activation were recognized in the past decade. Some of the greatest advances in the 1980s were in the detection of PAH-DNA adducts in humans at the levels produced by environmental exposure. Both antibody technology and ^{32}P-postlabeling technology have proven exceptionally valuable for such studies. Studies using combinations of the techniques such as that of Manchester et al.[101] provide even better opportunities for future

PAH-DNA adduct analysis. Improved resolution of postlabeled adducts prepared by both the ^{32}P and the more recent ^{35}S methods will be important for future studies of PAH adducts. These new techniques for adduct analysis will allow detection of PAH-DNA adducts at very low levels such as environmental exposure samples, but they also offer the ability to analyze PAH-DNA adducts from much smaller samples of DNA than in previous [^{3}H]PAH studies. This will greatly facilitate characterization of the role of PAH-DNA adducts in human epithelial cell transformation.

The development of PAH-DNA adduct analysis techniques with improved sensitivity will allow more detailed studies of how PAH interact with DNA. These studies have already defined the binding of PAH to replicating DNA, and will help to determine the role of chromatin structure and nucleotide sequences in PAH-DNA interactions and the repair of PAH-DNA adducts in specific regions of the genome. The techniques for studies of PAH metabolism and PAH-DNA adduct formation described in this chapter, as well as the advances which will be made in these analytical techniques over the next few years, provide the tools necessary to characterize the pathways of metabolic activation of PAH in human epithelial cells. This knowledge will increase our understanding of both the tissue and cell specificity of these carcinogens and the molecular mechanisms by which they transform human cells.

ACKNOWLEDGMENTS

The authors thank Marilyn Hines and Betty Leak for typing the manuscript and Sherry Brozich for preparing the chemical structures. This work was supported by Public Health Service Grants CA28825 and CA40228 from the National Cancer Institute, Department of Health and Human Services.

REFERENCES

1. **Miller, E. C.**, Some current perspectives on chemical carcinogenesis in human and experimental animals, *Cancer Res.*, 38, 1479, 1978.
2. U.S. Department of Health and Human Services, Fifth Annual Report on Carcinogens, NTP 89-239, NIEHS, Research Triangle Park, North Carolina.
3. **Grimmer, G.**, *Environmental Carcinogens: Polycyclic Aromatic Hydrocarbons*, CRC Press, Boca Raton, 1983.
4. **Gelboin, H. V. and Ts'o, P. O. P., Eds.**, *Polycyclic Hydrocarbons and Cancer*, Vol. 1-3, Academic Press, New York, 1978, 1981.
5. **Kadlubar, F. F. and Beland, F. A.**, Chemical properties of ultimate carcinogenic metabolites of arylamines and arylamides, in *Polycyclic Hydrocarbon and Carcinogenesis*, Harvey, R. G., Ed., American Chemical Society, Washington, D.C., 1985.
6. **Anselme, J.-P., Ed.**, *N-Nitrosamines*, American Chemical Society, Washington, D.C., 1979.
7. **Autrup, H.**, Carcinogen metabolism in cultured human tissues and cells, *Carcinogenesis*, 11, 707, 1990.

8. **Baird, W. M. and Pruess-Schwartz, D.**, Polycyclic aromatic hydrocarbon-DNA adducts and their analysis: a powerful technique for characterization of pathways of metabolic activation of hydrocarbons to ultimate carcinogenic metabolites, in *Polycyclic Aromatic Hydrocarbon Carcinogenesis: Structure-Activity Relationships,* Vol. 11, Yang, S. K. and Silverman, B. D., Eds., CRC Press, Boca Raton, 1988, 141.

9. **Harvey, R. G., Ed.,** *Polycyclic Hydrocarbons and Carcinogenesis,* ACS Symp. Ser. 283, American Chemical Society, Washington, D.C., 1985.

10. **Yang, S. K. and Silverman, B. D.,** *Polycyclic Aromatic Hydrocarbon Carcinogenesis: Structure-Activity Relationships,* Vol. 1 and 2, CRC Press, Boca Raton, 1988.

11. **Harris, C. C.**, Human tissues and cells in carcinogenesis research, *Cancer Res.,* 47, 1, 1987.

12. **Harris, C. C.**, Interindividual variation among humans in carcinogen metabolism, DNA adduct formation and DNA repair, *Carcinogenesis,* 10, 1563, 1989.

13. **Guengerich, F. P.**, Roles of cytochrome P-450 enzymes in chemical carcinogenesis and chemotherapy, *Cancer Res.,* 48, 2946, 1988.

14. **Ioannides, C. and Parke, D. V.**, The cytochrome P450 I gene family of microsomal hemoproteins and their role in the metabolic activation of chemicals, *Drug Metab. Rev.,* 22, 1, 1990.

15. **Rogan, E. G., RamaKrishna, N. V. S., Higginbotham, S., Cavalieri, E. L., Jeong, H., Jankowiak, R. and Small, G. J.**, Identification and quantitation of N^7-(benzo[a]pyrene-6-yl)guanine in the urine and feces of rats treated with benzo[a]pyrene, *Chem. Res. Toxicol.,* 3, 441, 1990.

16. **Pullman, A. and Pullman, B.**, Electronic structure and carcinogenic activity of aromatic molecules, *Adv. Cancer Res.,* 3, 117, 1955.

17. **King, H. W. S., Thompson, M. H., and Brookes, P.**, The role of 9-hydroxy-benzo(a)pyrene in the microsome mediated binding of benzo(a)pyrene to DNA, *Int. J. Cancer,* 18, 339, 1976.

18. **Sims, P., Grover, P. L., Swaisland, A., Pal, K., and Hewer, A.**, Metabolic activation of benzo(a)pyrene proceeds by a diol expoxide, *Nature,* 252, 326, 1974.

19. **Marquardt, H., Baker, S., Grover, P. L., and Sims, P.**, Malignant transformation and mutagenesis in mammalian cells induced by vicinal diol epoxides derived from benzo[a]pyrene, *Cancer Lett.,* 3, 31, 1977.

20. **Wood, A. W., Wislocki, P. G., Chang, R. L., Levin, W., Lu, A. Y. H., Yagi, H., Hernandez, O., Jerina, D. M., and Conney, A. H.**, Mutagenicity and cytotoxicity of benzo(a)yrene benzo-ring epoxides, *Cancer Res.,* 36, 3358, 1976.

21. **Huberman, E., Sachs, L., Yang, S. K., and Gelboin, H.**, Identification of mutagenic metabolites of benzo(a)pyrene in mammalian cells, *Proc. Natl. Acad. Sci. U.S.A.,* 73, 607, 1976.

22. **Newbold, R. F. and Brookes, P.**, Exceptional mutagenicity of a benzo(a)pyrene diol eponide in cultured mammalian cells, *Nature,* 261, 52, 1976.

23. **Levin, W., Wood, A. W., Wislocki, P. G., Kapitulnik, J., Yagi, H., Jerina, D. M., and Conney, A. H.**, Carcinogenicity of benzo-ring derivatives of benzo(a)pyrene on mouse skin, *Cancer Res.,* 37, 3356, 1977.

24. **Slaga, T. J., Bracken, W. M., Viaje, A., Levin, W., Yagi, H., Jerina, D. M., and Conney, A. H.**, Comparison of the tumor-initiating activities of benzo(a)pyrene arene oxides and diol-epoxides, *Cancer Res.,* 37, 4130, 1977.

25. **Slaga, T. J., Bracken, W. M., Gleason, G., Levin, W., Yagi, H., Jerina, D. M., and Conney, A. H.**, Marked differences in the skin tumor-initiating activities of the optical enantiomers of the diastereomeric benzo(a)pyrene 7,8-diol-9,10-epoxides, *Cancer Res.,* 39, 67, 1979.

26. **Jerina, D. M., Sayer, J. M., Agarwal, S. K., Yagi, H., Levin, W., Wood, A. W., Conney, A. H., Pruess-Schwartz, D., Baird, W. M., Pigott, M. A., and Dipple, A.**, Reactivity and tumorigenicity of bay-region diol epoxides derived from polycyclic aromatic hydrocarbons, in *Biological Reactive Intermediates-III,* Nelson, J. O. and Snyder, R., Eds., Plenum Press, New York, 1986, 11.

27. **Falk, H. L., Kotin, P., Lee, S. S., and Nathan, A.,** Intermediary metabolism of benzo(a)pyrene in the rat, *J. Natl. Cancer Inst.,* 28, 699, 1963.

28. **Wiest, W. G. and Heidelberger, C.,** The interaction of carcinogenic hydrocarbons with tissue constituents. II. 1,2,5,6-dibenzanthracene-9,10-C^{14} in skin, *Cancer Res.,* 13, 250, 1953.

29. **Diamond, L.,** Metabolism of polycyclic hydrocarbons in mammalian cell cultures, *Int. J. Cancer,* 8, 451, 1971.

30. **Diamond, L., Sardet, C., and Rothblat, G.,** The metabolism of 7,12-dimethylbenz(a)-anthracene in cell cultures, *Int. J. Cancer,* 3, 838, 1968.

31. **Bligh, E. G. and Dyer, W. J.,** A rapid method of total lipid extraction and purification, *Can. J. Biochem. Physiol.,* 37, 911, 1959.

32. **Selkirk, J. K.,** Analysis of benzo(a)pyrene metabolism by high-pressure liquid chromatography, *Adv. Chromatogr.,* 16, 1, 1978.

33. **Merrick, B. A., Mansfield, B. K., Nikbakht, P. A., and Selkirk, J. K.,** Benzo(a)pyrene metabolism in human T47D mammary tumor cells: evidence for sulfate conjugation and translocation of reactive metabolites across cell membranes, *Cancer Lett.,* 29, 139, 1985.

34. **Autrup, H.,** Carcinogen metabolism in human tissues and cells, *Drug Metab. Rev.,* 13, 603, 1982.

35. **Mass, M. J., Rodgers, N. T., and Kaufman, D. G.,** Benzo(a)pyrene metabolism in organ cultures of human endometrium, *Chem. Biol. Interact.,* 33, 195, 1981.

36. **Autrup, H.,** Separation of water soluble metabolites of benzo(a)pyrene formed by cultured human colon, *Biochem. Pharmacol.,* 28, 1727, 1979.

37. **Cohen, G. M., Hawes, S. M., Moore, B. P., and Bridges, J. W.,** Benzo(a)pyrene-3-yl-hydrogen sulfate, a major ethyl acetate-extractable metabolite of benzo(a)pyrene in human, hamster, and rat lung cultures, *Biochem. Pharmacol.,* 25, 2561, 1976.

38. **Mehta, R. and Cohen, G. M.,** Major differences in the extent of conjugation with glucuronic acid and sulphate in human peripheral lung, *Biochem. Pharmacol.,* 28, 2479, 1979.

39. **Autrup, H., Wefald, F. C., Jeffrey, A. M., Tate, H., Schwartz, R. D., Trump, B. F., and Harris, C. C.,** Metabolism of benzo(a)pyrene by cultured tracheobronchial tissues from mice, rats, hamsters, bovines and humans, *Int. J. Cancer,* 25, 293, 1980.

40. **Diamond, I., Kruszewski, F., Aden, D. P., Knowles, B. B., and Baird, W. M.,** Metabolic activation of benzo(a)pyrene by a human hepatoma cell line, *Carcinogenesis,* 1, 871, 1980.

41. **Baird, W. M., Chemerys, R., Erickson, A. A., Chern, C. J., and Diamond, L.,** Differences in pathways of polycyclic aromatic hydrocarbon metabolism as detected by analysis of the conjugates formed, in *Polynuclear Aromatic Hydrocarbons,* Jones, P. W. and Leber, P., Eds., Science Publishers, Ann Arbor, 1979, 507.

42. **Teffera, Y., Baird, W. M., and Smith, D. R.,** Determination of benzo(a)pyrene sulfate conjugates from benzo(a)pyrene-treated cells by continuous-flow fast atom bombardment mass spectrometry, *Anal. Chem.,* 63, 5, 1990.

43. **Duncan, M. E. and Brookes, P.,** The relation of metabolism to macromolecular binding of the carcinogen benzo(a)pyrene, by mouse embryo cells in culture, *Int. J. Cancer,* 6, 496, 1970.

44. **Selkirk, J. K., Croy, R. G., and Gelboin, H. V.,** Benzo(a)pyrene metabolites: efficient and rapid separation by high-pressure liquid chromatography, *Science,* 184, 169, 1974.

45. **Selkirk, J. K.,** High-pressure liquid chromatography: a new technique for studying metabolism and activation of chemical carcinogens, in *Environmental Cancer,* Kraybill, H. F. and Mehlman, M. A., Eds., Hemisphere, New York, 1977, 1.

46. **Selkirk, J. K., Croy, R. G., Whitlock, J. P., Jr., and Gelboin, H. V.,** *In vitro* metabolism of benzo(a)pyrene by human liver microsomes and lymphocytes, *Cancer Res.,* 35, 3651, 1975.

47. **Selkirk, J. K., Nikbakht, A., and Stoner, G. D.,** Comparative metabolism and macromolecular binding of benzo(a)pyrene in explant cultures of human bladder, skin, bronchus and esophagus from eight individuals, *Cancer Lett.,* 18, 11, 1983.

48. **Stoner, G. D., Harris, C. C., Autrup, H., Trump, B. F., Kingsbury, E. W., and Myers, G. A.,** Explant culture of human peripheral lung, *Lab. Invest.,* 38, 685, 1978.

49. **Autrup, H., Harris, C. C., Stoner, G. D., Selkirk, J. K., Schafer, P. W., and Trump, B. F.,** Metabolism of [^{3}H]benzo(a)pyrene by cultured human bronchus and cultured human pulmonary alveolar macrophages, *Lab. Invest.,* 38, 217, 1978.

50. **Levin, W., Wood, A., Chang, R., Ryan, D., and Thomas, P.,** Oxidative metabolism of polycyclic aromatic hydrocarbons to ultimate carcinogens, *Drug Metab. Rev.,* 13, 555, 1982.

51. **Baird, W. M., Chern, C.-J., and Diamond, L.,** Formation of benzo(a)pyrene-glucuronic acid conjugates in hamster embryo cell cultures, *Cancer Res.,* 37, 3190, 1977.

52. **Baird, W. M., Chemerys, R., Chern, C.-J., and Diamond, L.,** Formation of glucuronic acid conjugates of 7,12-dimethylbenz(a)anthracene phenols in 7,12-dimethylbenz(a)anthracene-treated hamster embryo cell cultures, *Cancer Res.,* 38, 3432, 1978.

53. **Cohen, G. M. and Moore, B. P.,** Metabolism of [^{3}H]benzo(a)pyrene by different portions of the respiratory tract, *Biochem. Pharmacol.,* 25, 1623, 1976.

54. **Moore, B. P. and Cohen, G. M.,** Metabolism of benzo(a)pyrene and its major metabolites to ethyl acetate-soluble and water-soluble metabolites by cultured rat trachea, *Cancer Res.,* 38, 3066, 1978.

55. **Autrup, H.,** Metabolic activation of chemical carcinogens in animal and human tissues, *Proc. N.Y. Acad. Sci.,* 534, 89, 1988.

56. **Hesse, S., Jernström, B., Martinez, M., Moldeus, P., Christodoulides, L., and Ketterer, B.,** Inactivation of DNA-binding metabolites of B(a)P and benzo(a)pyrene-7,8-dihydrodiol by glutathione and glutathione S-transferases, *Carcinogenesis,* 3, 757, 1982.

57. **Jernström, B., Babson, J. R., Moldeus, P., Holgren, A., and Reed, D. J.,** Glutathione conjugation and DNA-binding of (±)-7β,8α, dihydroxy-9α,10α-epoxy-7,8,9,10-tetrahydrozbenzo(a)pyrene in isolated rat hepatocytes, *Carcinogenesis,* 3, 861, 1982.

58. **Merrick, B. A. and Selkirk, J. K.,** H.P.L.C. of benzo(a)pyrene glucuronide, sulfate and glutathione conjugates and water-soluble metabolites from hamster embryo cells, *Carcinogenesis,* 6, 1303, 1985.

59. **Merrick, B. A., Mansfield, B. K., Nikbakht, P. A., and Selkirk, J. K.,** Benzo(a)pyrene metabolism in human T47D mammary tumor cells: evidence for sulfate conjugation and translocation of reactive metabolites across cell membranes, *Cancer Lett.,* 2, 139, 1985.

60. **Plakunov, I., Smolarek, T. A., Fischer, D. L., Wiley, J. C., Jr., and Baird, W. M.,** Separation by ion-pair high-performance liquid chromatography of the glucuronide, sulfate and glutathione conjugates formed from benzo(a)pyrene in cell cultures from rodents, fish and humans, *Carcinogenesis,* 8, 59, 1987.

61. **Emary, W. B., Shen, Z., and Cooks, R. G.,** Nucleophilic reactions between biological conjugates and tetraalkylammonium ion pair reagents during desorption chemical ionization, *Biomed. Environ. Mass Spectrom.,* 15, 571, 1988.

62. **Bieri, R. H. and Greaves, J.,** Characterization of benzo(a)pyrene metabolites by high performance liquid chromatography-mass spectrometry with a direct liquid introduction interface and using negative chemical ionization, *Biomed. Environ. Mass Spectrom.,* 14, 555, 1987.

63. **Teffera, Y., Smith, D. L., and Baird, W. M.,** Quantitation of benzo(a)pyrene-gluconic acid conjugates formed in benzo(a)pyrene-treated cell cultures by continuous-flow fast atom bombardment mass spectrometry, *Polycyclic Aromatic Compounds,* in press.

64. **Baird, W. M.,** The use of radioactive carcinogens to detect DNA modifications, in *Chemical Carcinogens and DNA,* Grover, P. L., Ed., CRC Press, Boca Raton, 1979, 59.

65. **Baird, W. M. and Brookes, P.,** Isolation of the hydrocarbon-deoxyribonucleoside products from the DNA of mouse embryo cells treated in culture with 7-methyl-benz[a]anthracene-3H, *Cancer Res.,* 33, 2378, 1973.

66. **Rayman, M. P. and Dipple, A.,** Structure and activity in chemical carcinogenesis: comparison of the reactions of 7-bromomethylbenz[a]anthracene and 7-bromomethyl-12-ethylbenz[a]anthracene with deoxyribonucleic acid *in vitro, Biochemistry,* 12, 1202, 1973.

67. **Grover, P. L., Hewer, A., Pal, K., and Sims, P.,** The involvement of a diol-epoxide in the metabolic activation of benzo(a)pyrene in human bronchial mucosa and in mouse skin, *Int. J. Cancer,* 18, 1, 1976.

68. **Dipple, A. and Slade, T. A.,** Structure and activity in chemical carcinogenesis: reactivity and carcinogenicity of 7-bromomethylbenz[a]anthracene and 7-bromomethyl-12-ethylbenz[a]anthracene, *Eur. J. Cancer,* 6, 417, 1970.

69. **Baird, W. M. and Diamond, L.,** The nature of benzo(a)pyrene-DNA adducts formed in hamster embryo cells depends on the length of time of exposure to benzo(a)pyrene, *Biochem. Biophys. Res. Commun.,* 77, 162, 1977.

70. **Ivanoic, V., Geacintov, N. E., Yamasaki, H., and Weinstein, I. B.,** DNA and RNA adducts formed in hamster embryo cell cultures exposed to benzo(a)pyrene, *Biochemistry,* 17, 1597, 1978.

71. **Osborne, M. R., Brookes, P., Lee, H., and Harvey, R. G.,** The reaction of 3-methylcholanthrene diol epoxide with DNA in relation to the binding of 3-methylcholanthrene to DNA of mammalian cells, *Carcinogenesis,* 7, 1345, 1986.

72. **Jeffrey, A. M., Blobstein, S. H., Weinstein, I. B., and Harvey, R. G.,** High-pressure liquid chromatography of carcinogen-nucleoside conjugates; separation of 7,12-dimethylbenzanthracene derivatives, *Anal. Biochem.,* 73, 378, 1976.

73. **Jeffrey, A. M., Weinstein, I. B., Jennette, K. W., Grzeskowiak, K., Nakanishi, K., Harvey, R. G., Autrup, H., and Harris, C. C.,** Structures of benzo(a)pyrene-nucleic acid adducts from human and bovine bronchial explants, *Nature,* 269, 348, 1977.

74. **Phillips, D. H., Hewer, A., and Grover, P. L.,** Abberant activation of benzo(a)pyrene in cultured rat mammary cells *in vitro* and following direct application to rat mammary glands *in vivo, Cancer Res.,* 45, 4167, 1985.

75. **Pruess-Schwartz, D. and Biard, W. M.,** BaP DNA adduct formation in early passage Wistar rat embryo cell cultures: evidence for multiple pathways of activation of BaP, *Cancer Res.,* 46, 545, 1986.

76. **Jennette, K. W., Jeffrey, A. M., Blobstein, S. H., Beland, F. A., Harvey, R. G., and Weinstein, I. B.,** Nucleoside adducts from *in vitro* reaction of benzo(a)pyrene-7,8-dihydrodiol-9,10-oxide or benzo(a)pyrene 4,5-oxide with nucleic acids, *Biochemistry,* 16, 932, 1977.

77. **Sawicki, J. T., Moschel, R. C., and Dipple, A.,** Involvement of both syn and anti-dihydroldiol-epoxides in the binding of 7,12-dimethylbenz(a)anthracene to DNA in mouse embryo cell cultures, *Cancer Res.,* 43, 3212, 1983.

78. **Pruess-Schwartz, D., Sebti, S. M., Gilham, P. T., and Baird, W. M.,** Analysis of benzo(a)pyrene: DNA adducts formed in cells in culture by immobilized boronate chromatography, *Cancer Res.,* 44, 4104, 1984.

79. **Pruess-Schwartz, D., Baird, W. M., Nikbakht, A., Merrick, B. A., and Selkirk, J. K.,** Benzo(a)pyrene: DNA adduct formation in normal human epithelial cell cultures and the human mammary carcinoma T47D cell line, *Cancer Res.,* 46, 2697, 1986.

80. **Osborne, M. R. and Crosby, N. T.,** *Benzopyrenes,* Cambridge University Press, Cambridge, 1987, 107.

81. **Weith, H. L., Weibers, J. L., and Gilham, P. T.,** Synthesis of cellulose derivatives containing digydroxyboryl group and a study of their capacity to form specific complexes with sugars and nucleic acid components, *Biochemistry,* 9, 4396, 1970.

82. **Ho, N. W. Y., Duncan, R. E., and Gilham, P. T.,** Esterification of terminal phosphate groups in nucleic acids with sorbitol and its application to the isolation of terminal polynucleotide fragments, *Biochemistry,* 20, 64, 1981.

83. **Moore, C. J., Pruess-Schwartz, D., Mauthe, R. J., Gould, M. N., and Baird, W. M.,** Interspecies differences in the major DNA adducts formed from benzo(a)pyrene but not 7,12-dimethylbenz[a]anthracene in rat and human mammary cell cultures, *Cancer Res.,* 47, 4402, 1987.

84. **DiGiovanni, J., Romson, J. R., Linville, D., and Juchau, M. R.,** Covalent binding of polycyclic aromatic hydrocarbons to adenine correlates with tumorigenesis in mouse skin, *Cancer Lett.,* 7, 39, 1979.

85. **Dipple, A., Pigott, M., Moschel, R. C., and Constantino, N.,** Evidence that binding of 7,12-dimethylbenzo(a)anthracene to DNA in mouse embryo cell cultures results in extensive substitution of both adenosine and guanine residues, *Cancer Res.,* 43, 4132, 1983.

86. **Gould, M. N., Grau, D. R., Seidman, L. A., and Moore, C. J.,** An interspecies comparison of human and rat mammary epithelial cell-mediated mutagenesis of polycyclic aromatic hydrocarbons, *Cancer Res.* 46, 4942, 1986.

87. **Shugart, L., Rahn, R. O., and Holland, J. M.,** Quantifying benzo(a)pyrene binding to DNA by fluorescent analysis, in *Polynuclear Aromatic Hydrocarbons: Formation, Metabolism and Measurements,* Cooke, M. and Dennis, A. J., Eds., Battelle Press, Columbus, Ohio, 1983, 1087.

88. **Osborne, M. R., Beland, F. A., Harvey, R. G., and Brookes, P.,** The reaction of $(+)-7\alpha,8\beta$-dihydroxy-9β,10β-epoxy-7,8,9,10-tetrahydrobenzo(a)pyrene with DNA, *Int. J. Cancer,* 18, 362, 1976.

89. **Phillips, D. H.,** Modern methods of DNA adduct determination, in *Handbook of Experimental Pharmacology,* Vol. 94, *Chemical Carcinogenesis and Mutagenesis I,* Cooper, C. S. and Grover, P. L., Eds., Springer-Verlag, New York, 1990, 503.

90. **Kriek, E., Den Engelse, L., Scherer, E., and Westra, J. G.,** Formation of DNA modifications by chemical carcinogens: identification, localization and quantification, *Biochim. Biophys. Acta,* 738, 181, 1984.

91. **Vigny, P., Duquesne, M., Coulomb, H., Tierney, B., Grover, P. L., and Sims, P.,** Fluorescence spectral studies on the metabolic activation of 3-methylcholanthrene and 7,12-dimethylbenz[a]anthracene in mouse skin, *FEBS Lett.,* 82, 278, 1977.

92. **Moschel, R. C., Baird, W. M., and Dipple, A.,** Metabolic activation of the carcinogen 7,12-dimethylbenz[a]anthracene for DNA binding, *Biochem. Biophys. Res. Commun.,* 76, 1092, 1977.

93. **Moschel, R. C., Pigott, M. A., Constantino, N., and Dipple, A.,** Chromatographic and fluorescence spectroscopic studies of individual 7,12-dimethylbenz[a]anthracene-deoxyribonucleoside adducts, *Carcinogenesis,* 4, 1201, 1983.

94. **Daudel, P., Duquesne, M., Vigny, P., Grover, P. L., and Sims, P.,** Fluorescence spectral evidence that BaP-DNA products in mouse skin arise from diolepoxides, *FEBS Lett.,* 57, 250, 1975.

95. **Rahn, R. O., Chang, S. S., Holland, J. M., and Shugart, L. R.,** A fluorometric-HPLC assay for quantitating the binding of BaP metabolites to DNA, *Biochem. Biophys. Res. Commun.,* 109, 262, 1982.

96. **Vo-Dinh, T.,** Multicomponent analysis by synchronous luminescence spectrometry, *Anal. Chem.,* 50, 396, 1978.

97. **Vo-Dinh, T.,** Synchronous luminescence spectroscopy: methodology and applicability, *Appl. Spectrosc.,* 36, 576, 1982.

98. **Rahn, R. O., Chang, S. S., Holland, J. M., Stephens, T. J., and Smith, L. H.,** Binding of benzo[a]pyrene to epidermal DNA and RNA as detected by synchronous luminescence spectrometry at 77 K, *J. Biochem. Biophys. Methods,* 3, 285, 1980.

99. **Vahakangas, K., Haugen, A., and Harris, C. C.,** An applied synchronous fluorescence spectrophotometric assay to study BaPDE-DNA adducts, *Carcinogenesis,* 6, 1109, 1985.

100. **Manchester, D. K., Weston, A., Choi, J.-S., Trivers, G. E., Fennessey, P., Quintana, E., Farmer, P. B., Mann, D. L., and Harris, C. C.,** Detection and characterization of benzo(a)pyrene diol epoxide-DNA adducts in human placenta, *Proc. Natl. Acad. Sci. U.S.A.,* 85, 9243, 1988.

101. **Manchester, D. K., Wilson, V. L., Hsu, L.-C., Choi, J.-S., Parker, N. B., Mann, D. L., Weston, A., and Harris, C. C.,** Synchronous fluorescence spectroscopic, immunoaffinity chromatography and ^{32}P-postlabeling analysis of human placental DNA known to certain benzo(a)pyrene diol epoxide adducts, *Carcinogenesis,* 11, 553, 1990.

102. **Ivanovic, V., Geacintov, N. E., and Weinstein, I. B.,** Cellular binding of BaP to DNA characterized by low temperature fluorescence, *Biochem. Biophys. Res. Commun.,* 70, 1172, 1976.

103. **Heisig, V., Jeffrey, A. M., McGlade, M. J., and Small, G. J.,** Fluorescence-line-narrowed spectra of polycyclic aromatic carcinogen-DNA adducts, *Science,* 223, 289, 1984.

104. **Jankowiak, R., Cooper, R. S., Zamzow, D., Small, G. J., Doscocil, G., and Jeffery, A. M.,** Fluorescence line narrowing-non-photochemical hole burning spectrometry: femtomole detection and high selectivity for intact DNA-PAH adducts, *Chem. Res. Toxicol.,* 1, 60, 1988.

105. **Jankowiak, R., Lu, P., Small, G. J., and Geacintov, N. E.,** Laser spectroscopic studies of DNA adduct structure types from enantiomeric diol epoxides of benzo(a)pyrene, *Chem. Res. Toxicol.,* 3, 39, 1990.

106. **Zamzow, D., Jankowiak, R., Cooper, R. S., and Small, G. J.,** Fluorescence line narrowing spectrometric analysis of benzo(a)pyrene-DNA adducts formed by one-electron oxidation, *Chem. Res. Toxicol.,* 2, 29, 1989.

107. **Sanders, M. J., Cooper, R. S., Jankowiak, R., Small, G. J., and Jeffrey, A. M.,** Identification of polycyclic aromatic hydrocarbon metabolites and DNA adducts in mixtures using fluorescence line narrowing spectrometry, *Anal. Chem.,* 58, 816, 1986.

108. **Lu, P., Jeong, H., Jankowiak, R., and Small, G. J.,** Comparative laser spectroscopic study of DNA and polynucleotide adducts from (+)*anti*-diol epoxide of benzo(a)pyrene, *Chem. Res. Toxicol.,* 4, 58, 1991.

109. **Poirier, M. C.,** Antibodies to carcinogen-DNA adducts, *J. Natl. Cancer Inst.,* 67, 515, 1981.

110. **Poirier, M. C.,** The use of carcinogen-DNA adduct antisera for quantitation and localization of genomic damage in animal models and the human population, *Environ. Mutagen.,* 6, 879, 1984.

111. **Van Schooten, F. J., Kriek, E., Steenwinkel, M.-J. S. T., Noteborn, H. P. J. M., Hildebrand, M. J. X., and Van Leeuwen, F. E.,** The binding efficiency of polyclonal and monoclonal antibodies to DNA modified with benzo(a)pyrene diol epoxide is dependent on the level of modification. Implications for quantitation of benzo(a)pyrene-DNA adducts *in vivo, Carcinogenesis,* 8, 1263, 1987.

112. **Santella, R. M., Lin, C. D., Cleveland, W. L., and Weinstsein, I. B.,** Monoclonal antibodies to DNA modified by a benzo(a)pyrene diol epoxide, *Carcinogenesis,* 5, 373, 1984.

113. **Hsu, I. C., Poirier, M. C., Yuspa, S. H., Grunberger, D., Weinstein, I. B., Yolken, R. H., and Harris, C. C.,** Measurement of BaP-DNA adducts by enzyme immunoassays and radioassay, *Cancer Res.,* 41, 1091, 1981.

114. **Nakayama, J., Yuspa, S. H., and Poirier, M. C.,** BaP-DNA adduct formation and removal in mouse epidermis *in vivo* and *in vitro:* relationship of DNA binding to initiation of skin carcinogenesis, *Cancer Res.,* 44, 4087, 1984.

115. **Paules, R. S., Poirier, M. C., Mass, M. J., Yuspa, S. H., and Kaufman, D. G.,** Quantitation by electron microscopy of the binding of highly specific antibodies to BaP-DNA adducts, *Carcinogenesis,* 6, 193, 1985.

116. **Perera, F. P., Poirier, M. C., Yuspa, S. H., Nakayama, J., Jaretzki, A., Curnen, M. M., Knoweles, D. M., and Weinstein, I. B.,** A pilot project in molecular cancer epidemiology: determination of BaP-DNA adducts in animal and human tissues by immunoassays, *Carcinogenesis,* 3, 1405, 1982.

117. **Poirier, M. C., Santella, R., Weinstein, I. B., Grunberger, D., and Yuspa, S. M.,** Quantitation of BaP-deoxyguanosine adducts by immunoassay, *Cancer Res.,* 40, 412, 1980.

118. **Shamsuddin, A. K. M., Sinopolin, N. T., Hemminli, K., Boesch, R. R., and Harris, C. C.,** Detection of BaP: DNA adducts in human white blood cells, *Cancer Res.,* 45, 66, 1985.

119. **Wallin, H., Borrebaeck, C. A. K., Glad, C., Mattiasson, B., and Jergil, B.,** Enzyme immunoassay of BaP conjugated to DNA, RNA and microsomal proteins using a monoclonal antibody, *Cancer Lett.,* 22, 163, 1984.

120. **Santella, R. M., Hsieh, L.-L., Lin, C.-D., Viet, S., and Weinstein, I. B.,** Quantitation of exposure to benzo(a)pyrene with monoclonal antibodies, *Environ. Health Persspect.,* 62, 95, 1985.

121. **Slor, H., Mizusawa, H., Niehaart, N., Kakefuda, T., Day, R. S., and Bustin, M.,** Immunochemical visualization of binding of the chemical carcinogen benzo(a)pyrene diol-epoxide 1 to the genome, *Cancer Res.,* 41, 3111, 1981.

122. **Strickland, P. T. and Boyle, J. M.,** Immunoassay of carcinogen-modified DNA, *Prog. Nucleic Acid Res. Mol. Biol.,* 31, 1, 1984.

123. **Weston, A., Trivers, G., Vahakangas, K., Newman, M., Rowe, M., Man, D., and Harris, C. C.,** Detection of carcinogen-DNA adducts in human cells and antibodies to these adducts in human sera, in *Carcinogenesis and Adducts in Animals and Humans,* Vol. 31, Poirier, M. C. and Beland, F. A., Eds., S. Karger AG, Basel, 1987.

124. **Tierney, B., Benson, A., and Garner, R. C.,** Immunoaffinity chromatography of carcinogen DNA adducts with polyclonal antibodies directed against benzo(a)pyrene diol-epoxide-DNA, *J. Natl. Cancer Inst.,* 77, 261, 1986.

125. **Randerath, K., Reddy, M. W., and Gupta, R. C.,** ^{32}P-labeling test for DNA damage, *Proc. Natl. Acad. Sci. U.S.A.,* 78, 6126, 1981.

126. **Gupta, R. C., Reddy, M. V., and Randerath, K.,** ^{32}P-postlabeling analysis of non-radioactive aromatic carcinogen-DNA adducts, *Carcinogenesis,* 3, 1081, 1982.

127. **Reddy, M. V. and Randerath, K.,** Nuclease P1-mediated enhancement of sensitivity of ^{32}P-postlabeling test for structurally diverse DNA adducts, *Carcinogenesis,* 7, 1543, 1986.

128. **Gupta, R. C.,** Enhanced sensitivity of ^{32}P-postlabeling analysis of aromatic carcinogen: DNA adducts, *Cancer Res.,* 5, 343, 1985.

129. **Dunn, B. P. and San, R. H. C.,** HPLC enrichment of hydrophobic DNA adducts for enhanced sensitivity of ^{32}P-postlabeling analysis, *Carcinogenesis,* 9, 1055, 1988.

130. **Gorelick, N. J. and Wogan, G. N.,** Fluoranthene-DNA adducts: identification and quantification by an HPLC-^{32}P-postlabeling method, *Carcinogenesis,* 10, 1567, 1989.

131. **Seidman, L. A., Moore, C. J., and Gould, M. N.,** ^{32}P-Postlabeling analysis of DNA adducts in human and rat mammary epithelial cells, *Carcinogenesis,* 9, 1071, 1988.

132. **Schoket, B., Hewer, A., Grover, P. L., and Phillips, D. H.,,** Formation of DNA adducts in human skin maintained in short-term organ culture and treated with coal-tar, creosote or bitumen, *Int. J. Cancer,* 42, 622, 1988.

133. **Ribovich, M. L., Kurian, P., and Milo, G. E.,** Specific BPDE I modification of replicating and parental DNA in early S phase human foreskin fibroblasts, *Carcinogenesis,* 7, 737, 1986.

134. **Milo, G. E. and DiPaolo, J. A.,** Neoplastic transformation of human diploid cells *in vitro* after chemical carcinogen treatment, *Nature,* 275, 130, 1978.

135. **Hemminki, K., Grzybowska, E., Chorazy, M., Twardowska-Saucha, K., Sroczynski, J. W., Putman, K. L., Randerath, K., Phillips, D. H., Hewer, A., Santella, R. M., Young, T. L., and Perera, F. P.,** DNA adducts in humans environmentally exposed to aromatic compounds in an industrial area of Poland, *Carcinogenesis,* 11, 1229, 1990.

136. **Everson, R. B., Randerath, E., Santella, R. M., Cefalo, R. C., Avitts, T. A., and Randerath, K.,** Detection of smoking-related covalent DNA adducts in human placenta, *Science,* 231, 54, 1986.

137. **Reddy, M. V., Irvin, T. R., and Randerath, K.,** Formation and persistence of sterigmatocystin-DNA adducts in rat liver determined via ^{32}P-postlabeling analysis, *Mutat. Res.,* 152, 85, 1985.

138. **Koivisto, P. and Hemminki, K.,** ^{32}P-Postlabeling of 2-hydroxyethylated, ethylated and methylated adducts of 2'deoxyguanosine 3'monophosphate, *Carcinogenesis,* 11, 1389, 1990.

139. **Vodicka, P. and Hemminki, K.,** ^{32}P-postlabeling of N-7, N^2 and O^6 2′deoxyguanosine 3′-monophosphate adducts of styrene oxide, *Chem. Biol. Interact.,* 77, 39, 1991.

140. **Cheh, A. M., Yagi, H., and Jerina, D. M.,** Stereoselective release of polycyclic aromatic hydrocarbon-deoxyadenosine adducts from DNA by the ^{32}P postlabeling and deoxyribonuclease I/snake venom phosphodiesterase digestion methods, *Chem. Res. Toxicol.,* 3, 545, 1990.

141. **Lau, H. H. S. and Baird, W. M.,** Detection and identification of benzo(a)pyrene DNA adducts by ^{35}S phosphorothioate labeling and high performance liquid chromatography, *Carcinogenesis,* 12, 885, 1991.

142. **Chlebowski, J. F. and Coleman, J. E.,** Mechanisms of hydrolysis of O-phosphorothioates and inorganic thiophosphate by *Escherichia coli* alkaline phosphatase, *J. Biol. Chem.,* 249, 7192, 1974.

143. **Eckstein, F.,** Phosphorothioate analogues of nucleotides — tools for the investigation of biochemical processes, *Angew. Chem. Int. Ed. Engl.,* 12, 423, 1983.

144. **Randerath, K., Randerath, E., Danna, T. F., van Golen, K. L., and Putman, K.L.,** A new sensitive ^{33}P-postlabeling assay based on the specific enzymatic conversion of bulky DNA lesions to radiolabeled dinucleotides and nucleoside 5′monophosphates, *Carcinogenesis,* 10, 1231, 1989.

145. **Gupta, R. C. and Earley, K.,** ^{32}P-adduct assay: comparative recoveries of structurally diverse DNA adducts in the various enhancement procedures, *Carcinogenesis,* 9, 1687, 1988.

146. **Gallagher, J. E., Jackson, M. A., George, M. H., Lewtas, J., and Robertson, I. G. C.,** Differences in the detection of DNA adducts in ^{32}P-postlabeling assay after either 1-butanol extraction or nuclease P1 treatment, *Cancer Lett.,* 45, 7, 1989.

147. **Zoon, R. A.,** Safety with ^{32}P and ^{35}S labeled compounds, *Methods Enzymol.,* 152, 25, 1987.

148. **Lau, H. H. S. and Baird, W. M.,** unpublished data.

149. **Vericat, J. A., Cheng, S. C., and Dipple, A.,** Absolute stereochemistry of the major 7,12-dimethylbenz[a]anthracene-DNA adducts formed in mouse cells, *Carcinogenesis,* 10, 567, 1989.

Chapter 4

HUMAN ESOPHAGEAL EPITHELIAL CELLS: IMMORTALIZATION AND *IN VITRO* TRANSFORMATION

Gary D. Stoner, Zenya Naito, and George E. Milo

TABLE OF CONTENTS

I. INTRODUCTION

Esophageal cancer occurs worldwide; a recent estimate has placed it seventh in order of occurrence, in both sexes combined.[1] The geographic distribution of esophageal cancer varies considerably and is certainly more diverse than for any other cancer of the gastrointestinal tract.[2] The highest incidence rates have been reported from China,[1] the Caspian littoral in Iran,[3] parts of Central Asia,[4] and the Transkei in South Africa.[5] The disease occurs consistently among the poor in most areas of the world, where the diet is often restricted and nutritional imbalance is common.[6] Esophageal neoplasms account for only 10% of all cancers of the gastrointestinal tract, but they are responsible for 4% of all cancer deaths in the U.S. and are associated with a 5-year survival rate of 7% or less.[7]

Epidemiological studies indicate that the development of esophageal cancer in humans is associated with exposure to chemical carcinogens in the environment and in the diet.[8] An increased risk for development of the disease has been associated with the smoking and chewing of tobacco,[9] consumption of alcoholic beverages,[10] and of salt-cured, salt-pickled, and moldy foods, especially those contaminated with members of the *Fusarium* species, which produce several toxins,[11] and *Geotrichum candidum,* which promotes the formation of nitrosamines.[12] Other dietary factors implicated in the etiology of esophageal cancer are trace elements, vitamins, tannins, and hot beverages and foods.[13]

Histologically, about 50 to 70% of human esophageal cancers are either poorly or well-differentiated squamous cell carcinomas.[14] Another 5 to 10% are adenocarcinomas that originate either from the esophageal mucus glands or in regions of the esophagus where there is preexistent esophageal disease (e.g., esophagitis). The remainder are undifferentiated tumors. Like tumors that arise in other locations, those of the esophagus begin as inapparent *in situ* lesions in the mucosa. They extend with time along the long axis of the bowel and eventually encircle the lumen. From this point, three morphological patterns may evolve. The most common one (60%) is that of a polyploid fungating lesion that protrudes into the lumen. The second most common (25%) is a necrotic cancerous ulceration that invades deeply into the surrounding structures and may erode the respiratory tree and the aorta. The third morphologic variant is a diffuse, infiltrating lesion that spreads within the wall of the esophagus, causing thickening, rigidity, and narrowing of the lumen.

The conversion of normal human esophageal epithelial cells to cancer cells is associated with a variety of genotypic and phenotypic alterations. Cytogenetic studies using human esophageal carcinoma cell lines revealed frequent structural abnormalities (usually deletions) in chromosomes 1, 3, 9, and 11.[15] In addition, there was evidence of gene amplification in the form of homogeneously staining regions and double-minute chromosomes in primary and metastatic tumors.[16] Molecular studies revealed amplification of the epidermal growth factor receptor gene (c-*erb*B),[17] and coamplification of the *hst*-1 and *int*-2 genes in esophageal carcinomas.[18] Elevated levels of the EGF receptor appears to be associated with the malignant potential of these tumors.[19] There was no evidence for point mutations in codons 12, 13, or 61 in the H-, K- or N-*ras* genes in human esophageal carcinomas.[20]

Banks-Schlegel and Harris[21] and Grace et al.[22] have found that the patterns of keratins expressed in squamous cell carcinomas (SCCs) of the esophagus were consistently different from those of normal esophagus. SCCs typically expressed major keratins with molecular weights of 58, 56, 50, and 46 kDa, whereas normal tissues produced two major keratins with molecular weights of 58 and 52 kDa, and a minor keratin with a molecular weight of 56 kDa. The expression of another differentiated function of esophageal cells, cross-linked envelopes, in the carcinoma cells varies from unimpaired to severely restricted when compared to normal cells. Immunocytochemical studies revealed the following tumor-associated antigens in esophageal neoplasms: human chorionic gonadotrophin, human placental lactogen, α-fetoprotein, carcinoembryonic antigen, and nonspecific cross-reacting antigen.[7] Each antigen was detected in a variable percentage of esophageal tumors, and none were present in normal esophageal epithelium. These new keratins and tumor-associated antigens may prove to be early markers of esophageal carcinomas in humans.

Cell culture systems are widely used to elucidate the molecular and cellular events associated with the conversion of normal cells to cancer cells. The neoplastic transformation of cultured cells is thought to result from the accumulation of multiple genetic and cellular alterations.[23] Rodent cells undergo spontaneous transformation *in vitro,* and they are readily transformed by treatment with carcinogenic agents. In our laboratory, cultured rat esophageal epithelial cells were converted to the tumorigenic phenotype either spontaneously or following treatment with the esophageal carcinogen, *N*-nitroso-benzylmethylamine.[24,25] In contrast, due to their inherent genomic stability, normal human cells rarely undergo spontaneous transformation *in vitro,* and they are difficult to transform with carcinogenic agents.[26]

In recent years, an approach to the development of human cell systems for studies of *in vitro* transformation has involved (1) immortalizing the cells with viral genes introduced by one or more transfection procedures[27] and (2) treatment of the immortalized cells with chemical carcinogens or transfected oncogenes to achieve transformation to the tumorigenic endpoint.[28,29] In this chapter, we describe the development of an *in vitro* system for studies of the neoplastic transformation of normal human esophageal epithelial cells (NHE). NHE cells were immortalized by transfection with simian virus-40 (SV40) early-region genes, and the immortalized cells were transformed to the tumorigenic phenotype following treatment with *N*-methyl-*N*'-nitro-*N*-nitroso-guanidine (MNNG). This cell transformation system will be useful for investigating the molecular events involved in the conversion of normal human esophageal epithelial cells to cancer cells.

II. MATERIALS AND METHODS

A. CELL CULTURE

Normal human esophageal epithelial (NHE) cells were derived from outgrowths of autopsy tissue from noncancerous individuals.[30] The outgrowths were suspended with 1% polyvinylpyrrolidone-0.02% [ethylene*bis*-(oxyethylenenitrilo)]tetraacetic acid-0.2% crystalline trypsin in HEPES-buffered saline, pH 7.4, at room temperature and subcultured into coated T-flasks. NHE cells were cultured in esophageal growth medium (EGM) consisting of MCDB 153 basal medium[31] supplemented with 5 ng/ml epidermal growth factor, 1.4 μM hydrocortisone, 0.1 mM ethanolamine, 0.1 mM phosphoethanolamine, 5 μg/ml insulin, 40 μg/ml bovine pituitary extract, 250 μg/ml bovine serum albumin, and 0.5 μg/ml epinephrine. The Ca^{2+} concentration was 0.1 mM. Antibiotics were added as needed (100 U/ml penicillin G, 100 μg/ml kanamycin, 50 μg/ml gentamicin). Hormones and growth factors were from Sigma Chemical Co. (St. Louis, MO). Transforming growth factors were from R & D Systems, Inc. (Minneapolis, MN).

Cultures were monitored for *Mycoplasma* contamination by culture on anexic agar and by DNA fluorochrome staining of an indicator culture.[32] No contamination was detected.

B. TRANSFECTION

Subcultures of NHE cells were plated (3 to 5 $\times$ 10^5 per 100-mm coated dish) and transfected the next day with 10 μg of plasmid DNA coprecipitated with strontium phosphate. The plasmid, pRSV-T, obtained from Dr. Bruce Howard (National Cancer Institute), is an *ori* construct containing the SV40 early-region genes and the Rous sarcoma virus long-terminal repeat.[33] Four hours after transfection, the cells were shocked with 15% glycerol in HEPES-buffered saline, washed three times with LHC basal medium, and incubated in LHC-9 medium. After the appearance of transformed foci (3 to 4 weeks), the cells were subcultured (5 $\times$ 10^5 per 100-mm dish). The cultures were fed three times per week with fresh LHC-9 and transferred at weekly intervals at 5 $\times$ 10^5 cells per 100-mm dish or T-75 flask.

C. CLONAL GROWTH ASSAYS

Clonal response to growth stimulators or inhibitors was assessed by a clonal growth assay.[34] Subconfluent cultures were suspended with 1% polyvinyl-pyrrolidone-0.02% [ethylene*bis*(oxyethylenenitrilo)tetraacetic] acid-0.2% crystalline trypsin in HEPES-buffered saline, pH 7.4, and plated at 500 to 1000 cells per 60-mm coated dish containing 4 ml of EGM from which the factor under study has been omitted. After overnight incubation, the medium was removed and the experimental medium was added. Plates were fixed in Carnoy's fixative (ethanol:acetic acid, 3:1, containing 2.5% formaldehyde) and stained with iron hematoxylin after 6 to 8 d of incubation. Both colony-forming efficiency (CFE) and clonal growth rate (population doublings per day [PD/d]) were determined. Four replicate dishes per variable were used for the CFE assay, and at least 18 colony counts were averaged for determination of the PD/d.

D. CHROMOSOME AND ISOZYME ANALYSES

Chromosome studies were performed by Dr. Ward D. Peterson (Children's Hospital of Michigan, Detroit). Exact counts on 30 metaphases were made on banded chromosomes, and at least eight karyotypes per cell line were prepared. Analyses of eight isozymes were carried out using standard procedures.[35]

E. SLOT BLOT ANALYSIS

Cells were grown to 80 to 85% confluency, trypsinized, and collected by centrifugation. DNA was isolated by sodium dodecyl sulfate proteinase-K incubation followed by phenol/chloroform extraction and ethanol precipitation. DNAs (10 μg each) were loaded into individual wells of an S&S slot blot apparatus (Schleicher and Schuell, Keene, NH). DNAs were transferred onto a Hybond-N (Amersham) nylon membrane and probed with a nick-translated *Eco*Ri-*Hind*III fragment of the plasmid, pRSV-T, under stringent conditions. The membrane was washed and autoradiographed at $-75°C$.

F. DNA FINGERPRINT ANALYSIS

High molecular weight DNA was isolated from precrisis HE-457 and postcrisis HET-1A cell lines using Pronase-sodium dodecyl sulfate lysis. Following inorganic extraction and ethanol precipitation, the DNA samples were treated with RNase and additional inorganic extractions, and ethanol precipitations were performed. The quantity of DNA isolated was determined in a spectrophotometer at 260 nm, and its integrity by electrophoresis in a 0.8% agarose gel. DNA samples were digested with the restriction endonuclease *Hae*III (Promega, Madison, WI), electrophoresed in a horizontal 0.8% agarose gel (10 μg DNA per lane), and transferred to a nylon membrane. The blots were hybridized under stringent conditions (50% formamide-0.75 M NaCl-0.075 M sodium citrate, 42°C) to a ^{32}P-labeled DNA probe. The probe, designated pYNH24 (D2S44), was derived from a human genomic library.[36,37] pYNH24 recognizes a single locus, hybridizes to human-specific repeated DNA fragments, and has been used for determination of a genotype specific for a single individual.[36,37] The blots were washed to a final stringency in 0.1% sodium citrate-0.1% sodium dodecyl sulfate solution at 50°C. They were then exposed to film at $-70°C$ with intensifying screens.

G. IMMUNOFLUORESCENCE AND ELECTRON MICROSCOPY

Cells were fixed with 3% buffered glutaraldehyde for transmission electron microscopy or in absolute methanol for immunofluorescence. The cells were stained with anticytokeratin and antivimentin by the immunoperoxidase technique,[38] and with a monoclonal antibody to SV40 large T-antigen (Oncogene Science, Inc., Mineola, NY) by immunofluorescence.

H. TREATMENT WITH CHEMICAL CARCINOGENS

NHE cells were treated with methylmethane sulfonate (MMS) and MNNG in the laboratory of Dr. George Milo. Cells were grown routinely in Eagle's minimum essential medium (MEM) (Grand Island Biological Co., Grand Island, NY), buffered with 25 mM HEPES at pH 7.4, and supplemented with 1.0 mM sodium pyruvate, 2.0 mM glutamine, 0.1 mM nonessential amino acids, 50 μg/ml of gentamicin, and 10% fetal bovine serum (FBS) (lot 1111749, Hyclone, Logan, UT). Hereafter, the medium will be referred to as GM. For treatment with carcinogens, cells were seeded in 10 ml of GM at a density of 10,000 per square centimeter in 75 cm²-tissue culture flasks. After 24h, the cells were treated with either 50 μg/ml MMS or 0.01 μg/ml MNNG for 24 h. MMS was prepared in acetone, and the final concentration of acetone in the GM of MMS-treated and control cultures was ≤0.02%. MNNG was prepared in dimethyl sulfoxide (DMSO), and the final concentration of DMSO in the GM of MNNG-treated and control cultures was 0.5%. After treatment, the cultures were rinsed three times with GM minus FBS to remove the residual treatment medium, and the treated cells were allowed to grow in GM to 90% confluency (3 to 4 weeks). The cultures were then split 1 to 4 for three passages, after which they were treated for anchorage-independent growth and tumorigenicity as described below.

I. GROWTH IN SOFT AGAR

Immortalized and carcinogen-treated HET-1A cells were tested for anchorage-independent growth in agar as described previously.[39] Briefly, the cells were plated into 2 ml of 0.3% agar overlay at a seeding density of 1.0 × 10⁵ cells per 25-cm² well. The bottom agar layer was prepared by mixing 2 × GM prewarmed to 37°C with an equal volume of agar. Plates were incubated at 37°C in a humidified, CO_2-enriched air atmosphere and evaluated at 24 h for cell clumping and doublets. Colonies were counted after 14 d of incubation.

J. TUMORIGENICITY ASSAY

Male gnotobiotic nude NCr/sed mice (4 to 6 weeks old) were used to evaluate the tumorigenic potential of control and carcinogen-treated HET-1A cells. The mice were splenectomized, then received 0.1 ml of mouse anti-lymphocyte serum (Accurate Chemical and Scientific Co., Westbury, NY) twice weekly for 4 weeks. Within 1 week after splenectomy, 5 × 10⁶ cells in 0.1 ml of GM were inoculated s.c. into each flank of five to ten mice. All mice were observed weekly for a period of 1 year for the development of nodules at the injection site.

III. RESULTS

A. TRANSFECTION OF NHE CELLS

NHE cells were derived from outgrowths from autopsy tissue from a noncancerous 74-year-old male (case HE-457). Dispersed cells were plated

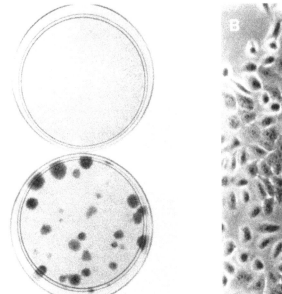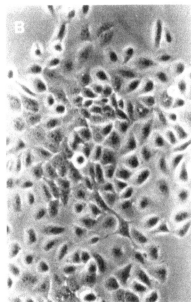

FIGURE 1. Transfection of normal human esophageal cells. (A) Control (top) and pRSV-T-transfected HE-457 cells (bottom) 3 weeks after treatment; (B) phase contrast photograph of HET-1A at passage 12.

at 3 to 5 × 10⁵ cells per dish and transfected with 10 μg of plasmid pRSV-T coprecipitated with strontium phosphate as described in Section II. After the appearance of transformed foci (Figure 1A), control and transfected cultures were subcultured at 2.5 × 10⁵ cells per 100-mm dish (Figure 1B). Control cells could be subcultured for no more than 20 PDs (data not shown), after which they senesced.

B. ESCAPE FROM CRISIS

pRSV-T-transfected cells (i.e., HE-457) grew exponentially for approximately 50 PDs, after which they went into "crisis" (Figure 2, inset). During this crisis period, which lasted for 6 to 8 months, the majority of cells terminally differentiated. The surviving cells formed two discreet colonies in a single flask. One of these colonies continued to grow after isolation and developed into a cell line designated HET-1A.[40] The growth of this line has accelerated, and it has doubled more than 320 times thus far.

C. CHARACTERIZATION

Precrisis HE-457 cells and the immortalized cell line HET-1A were characterized by immunohistochemistry. Keratin staining was intensely positive in HE-457 cells and in the HET-1A cell line especially in closely apposed foci of cells. Vimentin was also positive, although to a lesser extent than

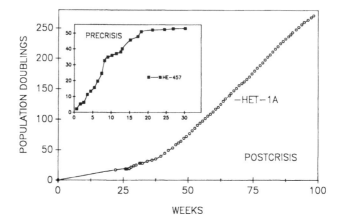

FIGURE 2. Cumulative growth of human esophageal cell lines transfected with plasmid, pRSV-T. Normal HE-457 cells at passage 2 were
transfected at time 0. Points, increases in cell number of each passage;
inset, growth of precrisis cells. No further increase in cell number was
seen after 50 population doublings. The postcrisis line, HET-1A, has
exceeded 250 population doublings.

keratin. The overall staining pattern of both pre- and postcrisis lines was
consistent with their epithelial origin. Transmission electron microscopy also
confirmed that all pre- and postcrisis cells are of epithelial origin, since they
contained tonofilaments and were joined by desmosomal junctions (data not
shown).

D. CELL IDENTIFICATION

To further establish that HET-1A cells are derived from HE-457 cells,
we characterized them by slot blot and DNA fingerprinting. Both lines had
integrated pRSV-T sequences, as expected (Figure 3). The DNA fingerprint
(Figure 4) shows that HET-1A (lane 2) is derived from HE-457 (lane 3).

E. CYTOGENETICS

HET-1A cells are hypodiploid, with only about 5% of the metaphases
examined in the hypotetraploid range. More than one half of the normal
autosomes are either absent or monosomic. Most of the missing copies of
normal chromosomes have been identified in the marker chromosomes (Figure
5). There are numerous other structural alterations which are seen with less
consistency, as well as random loss and/or gain of normal and marker chromosomes. There are no normal sex chromosomes in HET-1A cells; the Y
chromosome appears as a translocation in marker M5 [t(19pter>19q35::Yq12)].
Marker M7 contains the X chromosome with additional heterochromatin.

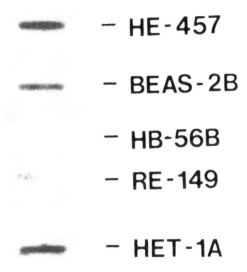

– HE-457

– BEAS-2B

– HB-56B

– RE-149

– HET-1A

FIGURE 3. Slot blot analysis of DNAs from cultured cell lines. HE-457 and HET-1A had been transfected with pRSV-T. BEAS-2B is a human bronchial epithelial cell line immortalized by infection with an Adeno12-SV40 hybrid virus. HB-56B, a human bronchial cell line, and RE-149, a rat esophageal cell line, served as untreated controls. Samples were probed with a nick-translated *Eco*R1-*Hind*III fragment of pRSV-T.

F. RESPONSE OF HET-1A TO CALCIUM, SERUM, AND TRANS-FORMING GROWTH FACTORS

Since both calcium and fetal bovine serum are known to have profound effects on the growth and differentiation of epithelial cells, the effects of these factors on the growth of HET-1A cells was assessed. Figure 6 illustrates the effect of calcium. Dose-dependent stimulation of the CFE was observed, with maximal growth occurring between 0.3 and 0.6 mM Ca^{2+}. On the other hand, the clonal growth rate (PD/d) became optimal at 0.1 mM Ca^{2+}. Figure 7 illustrates the effect of Ca^{2+} on the growth of HET-1A cells at high density. The growth rate was substantially the same at all three levels tested and was significantly greater than the calcium-free control. In contrast, FBS inhibited CFE and PD/d at all concentrations tested (Figure 8). Half-maximal inhibition of CFE was seen in medium containing 1% serum. Surviving colonies in serum grew at a somewhat slower rate than the serum-free controls.

The effect of TGF-β_1 and TGF-β_2 on the CFE of HET-1A cells is shown in Figure 9. With both factors, the CFE was inhibited in proportion to dose up to approximately 100 pg/ml, after which there was only a minimal inhibition.

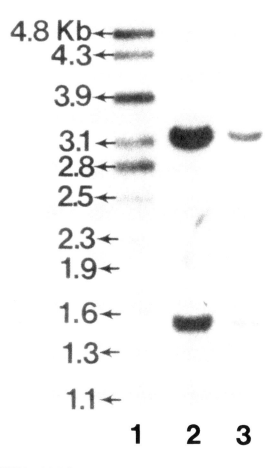

FIGURE 4.   DNA fingerprint analysis of transfected cell lines.
Lane 1, DNA markers; lane 2, HET-1A, p8; lane 3, HE-457
transfected at p 2 and analyzed at p 11. Samples were digested
with restriction endonuclease, *Hae*III, and probed with pYNH-
24 (D2S44). Kb, kilobase.

Surviving colonies grew at approximately the same rate in all concentrations
tested.

G. GROWTH IN SOFT AGAR

When plated in soft agar at passage levels 24 and 68, untreated HET-1A
cells exhibited CFEs as high as 25%. The CFEs of MMS- and MNNG-treated
cells were not appreciably higher than that of the controls.

H. TUMORIGENIC POTENTIAL IN MICE

At all passage levels tested, untreated HET-1A cells did not produce
tumors following injection into athymic nude mice (Table 1). Similarly, MMS-
treated HET-1A cells were nontumorigenic in nude mice. In contrast, three

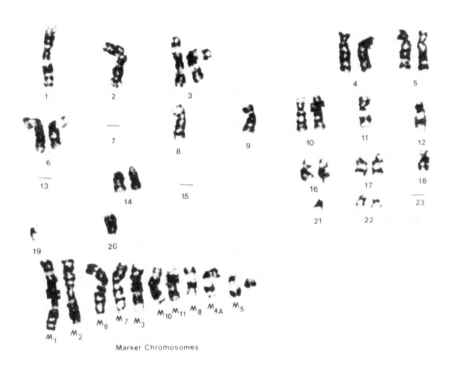

FIGURE 5. Karyotype of immortalized esophageal cell line HET-1A, p8, with 37 chromo-somes. The sex chromosomes appear in abnormal form in marker chromosomes M5 (Y) and M7 (X).

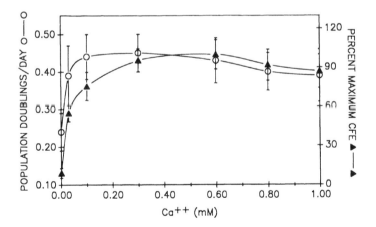

FIGURE 6. Effect of calcium on clonal growth of HET-1A. Cells (400/60-mm dish) were plated in EGM without calcium. The next day the medium was replaced with the experimental media. Cultures were fixed and stained after 6 d. Points, means; bars, SD.

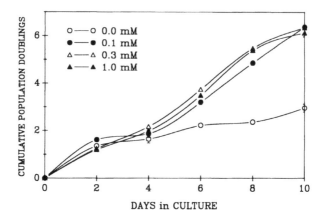

FIGURE 7.   Effect of calcium concentration on the growth of HET-
1A cells at high density. Replicate cultures were plated at 2×10^5
cells/60-mm dish. Every 2 d, quadruplicate cultures at each calcium
concentration were suspended and counted. Points, means; bars, SD.

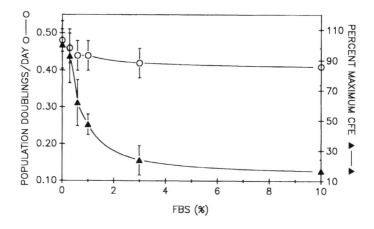

FIGURE 8.   Inhibition of clonal growth by fetal bovine serum (FBS). See
legend of Figure 6 for details. Points, means; bars, SD.

of nine mice injected with MNNG-treated HET-1A cells had tumors of 1 to
2 cm in diameter at the injection site at 4 to 6 weeks after treatment. These
tumors did not metastasize to other organ sites. Upon histopathological examination, the tumors were found to be undifferentiated carcinomas (data not
shown).

IV. DISCUSSION

The main objectives of this study were twofold: (1) develop an immortalized human esophageal epithelial cell line and (2) determine if the

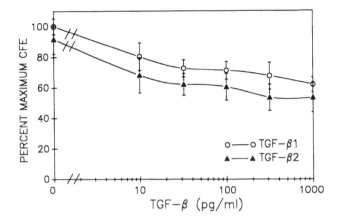

FIGURE 9. Effect of TGF-β_1 and TGF-β_2 on colony-forming efficiency of HET-1A cells. See legend of Figure 6 for details. Points, means; bars, SD.

TABLE 1
Tumorigenicity of Control and Carcinogen-Treated HET-1A Cells

Treatment	Passage number	Number of tumors/Number of mice injected
None	3	0/10
	27	0/10
	84	0/10
Acetone	29	0/4
DMSO	31	0/4
MMS	29—32	0/11
MNNG	31—33	3/9

Note: DMSO, dimethyl sulfoxide; MMS, methylmethane sulfonate; MNNG, *N*-methyl-*N'*-nitro-*N*-nitrosoguanidine.

immortalized cells could be transformed to the tumorigenic phenotype with chemical carcinogens. The first objective was achieved by strontium phosphate cotransfection with a plasmid carrying the SV40 early-region genes. Precrisis HE-457 cells transfected with pRSVT underwent 50 to 60 PDs before entering crisis and, after several months, one surviving colony eventually developed into an immortalized cell line (HET-1A) following subculture. These results indicate that the immortalization of NHE is, indeed, a rare event. Other human cell types either infected with SV40 or transfected with early-region genes have an extended life span and usually undergo a culture crisis.[41,42] This process is poorly understood and may be equivalent to senescence in nontransformed cells.

HE-457 and HET-1A cell lines express SV40 T-antigen by immunoperoxidase nuclear staining and have integrated SV40 early-region DNA, as shown by slot blot analysis. Both lines were confirmed to be epithelial by their positive reaction with anticytokeratin and the presence of desmosomal junctions and cytoplasmic tonofilaments in electron microscopic preparations.

The chromosomal alterations in the HET-1A cells are consistent with those of previously reported SV40-infected cells.[43] HET-1A is hypodiploid with a chromosome profile similar to that of pRSV-T-immortalized human bronchial epithelial cells[41] and prostatic epithelial cells.[44] Because of the known karyotypic instability of virally transformed cells,[43] an increase in chromosomal alterations with continued passage in culture can be expected. DNA-fingerprinting evidence clearly shows that HET-1A is derived from HE-457, and demonstrates the utility of this technique for confirming the origin of cell lines. HET-1A is nontumorigenic in athymic nude mice and did not induce transient carcinoma-like nodules at the injection site.[25,45] Others have also reported that SV40 T-antigen-immortalized human cells are nontumorigenic in nude mice.[41,45,46]

Epithelial cells, especially epidermal and epidermal-like cells, are usually induced to differentiate in serum-supplemented media.[31,47] The growth of HET-1A was also inhibited by serum. TGF-β is a major serum-derived inhibitor.[48] Both TGF-β_1 and TGF-β_2 inhibited the growth of HET-1A cells. This type of response to serum and to transforming growth factors is typical of both normal and SV40 T-antigen-immortalized human epithelial cells.

The calcium concentration in the growth medium also affects both growth and differentiation of epithelial cells.[31,47,49,50] Growth stimulation generally occurs at low concentrations, whereas high concentrations are growth-inhibitory. In the case of HET-1A cells, the Ca^{2+} response was population dependent. Both colony-forming efficiency and clonal growth rates were stimulated in a dose-dependent manner by Ca^{2+}. However, although Ca^{2+} stimulated growth at high cell density, there was no dose effect; equivalent stimulation was achieved at 0.1, 0.3, and 1.0 mM Ca^{2+}. No effect on terminal differentiation was apparent.

Cell lines immortalized by viral genes appear to represent an intermediate stage between normal and neoplastic in which altered growth control is offset by the capability of undergoing terminal differentiation when injected into nude mice.[45,51] This hypothesis is supported by the reported conversion of nontumorigenic immortalized human keratinocytes to tumorigenicity by the incorporation of additional viral genes[52] or treatment with chemical carcinogens.[53] A similar finding with SV40-immortalized human urothelial cells, in which tumorigenicity was induced by 3-methylcholanthrene, has been reported.[29]

In the present study, immortalization of HET-1A cells with pRSVT was associated with their ability for anchorage-independent growth, and treatment of HET-1A cells with MNNG, but not MMS, led to their conversion to the tumorigenic phenotype. The reason(s) for the ability of MNNG and not MMS to transform the cells is/are unknown, but may be related to the relative toxic effects of these two carcinogens. Studies are underway to characterize the immortalized (nontumorigenic) and tumorigenic HET-1A cell lines to determine the cellular and molecular events involved in the conversion of immortalized cells to tumorigenic cells.

ACKNOWLEDGMENTS

We thank Dr. Bruce Howard for providing the plasmid, pRSV-T, and Dr. Ward Peterson for the chromosome and isozyme analysis. We also thank the American Association for Cancer Research for permission to reproduce Figures 1A, 1B, 5, 6, 7, 8, and 9 from Stoner et al., *Cancer Res.*, 51, 365, 1991 (Reference 40). This research was supported by National Cancer Institute Grants CA28950 (Gary Stoner) and CA25907 (George Milo).

REFERENCES

1. **Parkin, D. M., Stjernsward, J., and Muir, C. S.,** Estimates of the worldwide frequency of twelve major cancers, *Bull WHO*, 62, 163, 1984.
2. **Silber, W.,** Carcinoma of the oesophagus: aspects of epidemiology and aetiology, *Proc. Nutr. Soc.*, 44, 101, 1985.
3. **Mahboubi, E., Kmet, J., Cook, P. J., Day, N. E., Ghadirian, P., and Salmasizadeh, S.,** Oesophageal cancer studies in the Caspian littoral of Iran: the Caspian registry, *Br. J. Cancer*, 28, 187, 1973.
4. **Wynder, E. L. and Bross, I. J.,** A study of etiological factors in cancer of the esophagus, *Cancer*, 14, 389, 1961.
5. **Warwick, G. P. and Harington, J. S.,** Some aspects of the epidemiology and etiology and esophageal cancer with particular emphasis on the Transkei, South Africa, *Adv. Cancer Res.*, 17, 81, 1973.
6. **Rothman, K. J.,** Alcohol, in *Persons at High Risk of Cancer: An Approach to Cancer Etiology and Control*, Fraumeni, J. F., Jr., Ed., Academic Press, New York, 1975, 139.
7. **Burg-Kurland, C. L., Purnell, D. M., Combs, J. W., Hillman, E. A., Harris, C. C., and Trump, B. F.,** Immunocytochemical evaluation of human esophageal neoplasms and preneoplastic lesions for β-chorionic gonadotropin, placental lactogen, α-fetoprotein, carcinoembryonic antigen, and nonspecific cross-reacting antigen, *Cancer Res.*, 46, 2936, 1986.
8. **Yang, C. S.,** Research on esophageal cancer in China: a review, *Cancer Res.*, 40, 2633, 1980.
9. **Tuyns, A. J.,** Epidemiology of esophageal cancer in France, in *Cancer of the Esophagus*, Vol. 1, Pfeiffer, C. J., Ed., CRC Press, Boca Raton, 1982, 3.
10. **Tuyns, A. J., Pe'quignot, G., and Jensen, O. M.,** Nutrition, alcohol et cancer de l'oesophage, *Bull. Cancer*, 65, 69, 1978.
11. **Hsia, C.-C., Tzian, B.-L., and Harris, C. C.,** Proliferative and cytotoxic effects of *Fusarium* T_2 toxin on cultured human fetal esophagus, *Carcinogenesis*, 4, 1101, 1983.
12. **Li, M. H., Ji, C., and Cheng, S.-J.,** Occurrence of nitroso compounds in fungi-contaminated foods: a review, *Nutr. Cancer*, 8, 63, 1986.
13. **Frank-Stromberg, M.,** The epidemiology and primary prevention of gastric and esophageal cancer, *Cancer Nurs.*, 12, 53, 1989.
14. **Robbins, S. L., Cotran, R. S., and Kumar, V.,** The gastrointestinal tract, in *Pathologic Basis of Disease*, W. B. Saunders, Philadelphia, 1984, 804.
15. **Whang-Peng, J., Banks-Schlegel, S. P., and Lee, E. C.,** Cytogenetic studies of esophageal carcinoma cell lines, *Cancer Genet. Cytogenet.*, 45, 101, 1990.

16. Rodriguez, E., Rao, P. H., Ladanyi, M., Altorki, N., Albino, A. P., Kelsen, D. P., Jhanwar, S. C., and Chaganti, R. S. K., 11p13-15 is a specific region of chromosomal rearrangement in gastric and esophageal adenocarcinomas, *Cancer Res.*, 40, 6410, 1990.

17. Hollstein, M. C., Smits, A. M., Galiana, C., Yamasaki, H., Bos, J. L., Mandard, A., Partensky, C., and Montesano, R., Amplification of epidermal growth factor receptor gene but not evidence of *ras* mutations in primary human esophageal cancer, *Cancer Res.*, 48, 5119, 1988.

18. Tsutsumi, M., Sakamoto, H., Yoshida, T., Kakizoe, T., Koiso, K., Sugimura, T., and Terada, M., Coamplification of the *hst*-1 and *int*-2 genes in human cancer, *Jpn. J. Cancer Res. (Gann,)* 79, 428, 1988.

19. Ozawa, S., Masakazu, U., Nobutoshi, A.,, Shimizu, N., and Abe, O., Prognostic significance of epidermal growth factor receptor in esophageal squamous cell carcinomas, *Cancer*, 63, 2169, 1989.

20. Victor, T., DuToit, R., Jordaan, A. M., Bester, A. J., and van Helden, P. D., No evidence for point mutations in codons 12, 13 and 61 of the *ras* gene in a high-incidence area for esophageal and gastric cancers, *Cancer Res.*, 50, 4911, 1990.

21. Banks-Schlegel, S. P. and Harris, C. C., Aberrant expression of keratin proteins and cross-linked envelopes in human esophageal carcinomas, *Cancer Res.*, 44, 1153, 1984.

22. Grace, M. P., Kim, K. H., True, L. D., and Fuchs, E., Keratin expression in normal esophageal epithelium and squamous cell carcinoma of the esophagus, *Cancer Res.*, 45, 841, 1985.

23. Barrett, J. C. and Fletcher, W. F., Cellular and molecular mechanisms of multistep carcinogenesis in cell culture models, in *Mechanisms of Environmental Carcinogenesis*, Vol. 2, Multistep Models of Carcinogenesis, Barrett, J. C., Ed., CRC Press, Boca Raton, 1987.

24. Stoner, G. D., Babcock, M. S., Cothern, G. A., Klaunig, J. E., Gunning, W. T., III, and Knipe, S. M., *In vitro* transformation of rat esophageal epithelial cells with N-nitrosobenzylmethylamine, *Carcinogenesis*, 3, 629, 1982.

25. Stoner, G. D., Babcock, M. S., McCorquodale, M. N., Gunning, W. T., III, Jamasbi, R., Budd, N., and Hukku, B., Comparative properties of untreated and N-nitrosobenzylmethylamine transformed rat esophageal epithelial cell lines, *In Vitro Cell. Dev. Biol.*, 25, 899, 1989.

26. McCormick, J. J. and Mayer, V. M., Towards an understanding of the malignant transformation of diploid human fibroblasts, *Mutat. Res.*, 199, 273, 1988.

27. Chang, S. E., *In vitro* transformation of human epithelial cells, *Biochem. Biophys. Acta*, 823, 161, 1986.

28. Newbold, R. F. and Overell, R. W., Fibroblast immortality is a prerequisite for transformation by EJ c-Ha-*ras* oncogene, *Nature*, 304, 651, 1983.

29. Reznikoff, C. A., Loretz, L. J., Christian, B. J., Wu, S.-Q., and Meisner, L. F., Neoplastic transformation of SV40-immortalized human urinary tract epithelial cells by *in vitro* exposure to 3-methylcholanthrene, *Carcinogenesis*, 9, 1427, 1988.

30. Stoner, G. D. and Klaunig, J., Selective methods for isolation of epithelial cells in primary explant cultures of human and animal tissues, in *Cell Separation: Methods and Selected Applications*, Pretlow, Vol. 2, T. G., III and Pretlow, T. P., Eds., Harcourt, Brace and Jovanovich, New York, 1983.

31. Boyce, S. T. and Ham, R. G.,, Normal human epidermal keratinocytes, in *In Vitro Models for Cancer Research*, Vol. 3, Webber, M. and Sekely, L., Eds., CRC Press, Boca Raton, 1985, 245.

32. DelGuidice, R. and Hopps, H. E., Microbiological methods and fluorescent microscopy for the direct demonstration of mycoplasma infection of cell cultures, in *Mycoplasma Infection of Cell Cultures*, McGarrity, G. J., Murphy, D. G., and Nichols, W. W., Eds., Plenum Press, New York, 1978, 57.

33. Brash, D. E., Reddel, R. R., Quanrad, M., Yang, K., Farrell, M. P., and Harris, C. C., Strontium phosphate transfection of human cells in primary culture: stable expression of the simian virus 40 large T-antigen in primary human bronchial epithelial cells, *Mol. Cell. Biol.*, 7, 2031, 1987.

34. Lechner, J. F. and Kaighn, M. E., Application of the principles of enzyme kinetics to clonal growth rate assays: an approach for delineating interactions among growth promoting agents, *J. Cell Physiol.*, 100, 519, 1979.

35. Peterson, W. D., Jr., Simpson, W. F., and Hukku, B., Cell culture characterization; monitoring for cell identification, *Methods Enzymol.*, 58, 164, 1979.

36. Nakamura, Y., Gillilan, S., O'Connell, P., Leppert, M., Lathrop, G. M., Lalouel, J.-M., and White, R., Isolation and mapping of polymorphic DNA sequence pYNH24 on chromosome 2 (D2S44), *Nucleic Acids Res.*, 15, 10073, 1987.

37. Nakamura, Y., Leppert, M., O'Connell, P., Wolff, R., Holm, T., Culver, M., Martin, C., Fujimoto, E., Hoff, N., Kumlin, E., and White, R., Variable number of tandem repeat (VNTR) markers for human genome mapping, *Science*, 235, 1616, 1987.

38. Katoh, Y., Stoner, G. D., Harris, C. C., McIntire, K. R., Hill, T., Anthony, R., McDowell, E., and Trump, B. F., Immunological markers of human bronchial epithelial cells *in vivo* and *in vitro*, *J. Natl. Cancer Inst.*, 62, 1177, 1979.

39. Milo, G., Yohn, J., Schuller, D., Noyes, I., and Lehman, T., Comparative stages of expression of human squamous carcinoma cells and carcinogen transformed keratinocytes, *J. Invest. Dermatol.*, 92, 848, 1989.

40. Stoner, G. D., Kaighn, M. E., Reddel, R. R., Resau, J. H., Bowman, D., Naito, Z., Matsukura, N., You, M., Galati, A. J., and Harris, C. C., Establishment and characterization of SV40 T-antigen immortalized human esophageal epithelial cells, *Cancer Res.*, 51, 365, 1991.

41. Reddel, R. R., Ke, Y., Gerwin, B. I., McMenamin, M. G., Lechner, J. F., Su, R. T., Brash, D. E., Park, J. B., Rhim, J. S., and Harris, C. C., Transformation of human bronchial epithelial cells by infection with SV40 or adenovirus-12 SV40 hybrid virus, or transfection via strontium phosphate coprecipitation with a plasmid containing SV40 early region genes, *Cancer Res.*, 48, 1904, 1988.

42. Ke, Y., Reddel, R. R., Gerwin, B. I., Reddel, H. K., Somers, A. N. A., McMenamin, M. G., LaVeck, M. A., Stahel, R. A., Lechner, J. F., and Harris, C. C., Establishment of human *in vitro* mesothelial cell model system for investigating mechanisms of asbestos-induced mesothelioma, *Am. J. Pathol.*, 134, 979, 1989.

43. Ohnuki, Y., Lechner, J. F., Bates, S. E., Jones, L. W., and Kaighn, M. E., Chromosomal instability of SV40-transformed human prostatic epithelial cell lines, *Cytogenet. Cell Genet.*, 33, 170, 1982.

44. Kaighn, M. E., Narayan, K. S., Ohnuki, Y., Jones, L. W., and Lechner, J. F., Differential properties among clones of simian virus 40-transformed human epithelial cells, *Carcinogenesis*, 1, 635, 1980.

45. Kaighn, M. E., Reddel, R. R., Lechner, J. F., Peehl, D. M., Camalier, R. F., Brash, D. E., Saffiotti, U., and Harris, C. C., Transformation of human neonatal prostate epithelial cells by strontium phosphate transfection with a plasmid containing SV40 early region genes, *Cancer. Res.*, 49, 3050, 1989.

46. Christian, B. J., Loretz, L. F., Oberley, T. D., and Reznikoff, C. A., Characterization of human uroepithelial cells immortalized *in vitro* by simian virus 40, *Cancer Res.*, 47, 6066, 1987.

47. Babcock, M. S., Marino, M. R., Gunning, W. T., III, and Stoner, G. D., Clonal growth and serial propagation of rat esophageal epithelial cells, *In Vitro*, 19, 403, 1983.

48. Masui, T., Wakefield, L. M., Lechner, J. F., LaVeck, M. A., Sporn, M. B., and Harris, C. C., Type B transforming growth factor is the primary differentiation-inducing serum factor for normal human bronchial epithelial cells, *Proc. Natl. Acad. Sci.*, 83, 2438, 1986.

49. Hennings, H., Michael, D., Cheng, C., Steinert, P., Holbrook, K., and Yuspa, S. H., Calcium regulation of growth and differentiation in mouse epidermal cells in culture, *Cell*, 19, 245, 1980.

50. **Trump, B. F. and Berezesky, I. K.**, Ion regulation, cell injury and carcinogenesis, *Carcinogenesis*, 8, 1027, 1987.
51. **Gaffney, E., Fogh, J., Ramos, L., Loveless, J. D., Fogh, H., and Dowling, A. M.**, Established lines of SV40-transformed human amnion cells, *Cancer Res.*, 30, 1668, 1970.
52. **Rhim, J. S., Jay, G., Arnstein, P., Price, F. M., Sanford, K. K., and Aaronson, S. A.**, Neoplastic transformation of human epidermal keratinocytes by AD12-SV40 and Kirsten sarcoma viruses, *Science*, 227, 1250, 1985.
53. **Rhim, J. S., Fujita, J., Arnstein, P., and Aaronson, S. A.**, Neoplastic conversion of human keratinocytes by adenovirus 12-SV40 virus and chemical carcinogens, *Science*, 232, 385, 1986.

Chapter 5

TRANSFORMATION OF HUMAN ENDOMETRIAL STROMAL CELLS *IN VITRO*

C. A. Rinehart, C. A. Carter, L. H. Xu, L. L. Barrett, T. D. Butler, C. H. Laundon, and D. G. Kaufman

TABLE OF CONTENTS

I. INTRODUCTION

In spite of its complexity, our understanding of mechanisms of cancer development has advanced enormously during the past several years.[1,2] A clearer picture has emerged of important roles for oncogenes, growth factors, transcriptional factors, and tumor suppressor genes, as well as the interaction of these factors in carcinogenesis.[3] Acknowledging the remarkable progress, we are still far from a complete understanding of the totality of carcinogenesis.

Steps of the process of neoplastic transformation have yet to be fully characterized for humans or rodents. While research progress is more rapid in rodent cells, fundamental differences between them and human cells, including vastly greater difficulty in transforming human cells with chemical carcinogens, make it imperative to place a high value on research involving human cells.[4] Neoplastic transformation is clearly a multistep process and it may involve qualitatively dissimilar events which together combine to cause cancer. The steps appear to include extended proliferative life-span beyond that of normal cells, increased production of positive growth factors, altered response to growth factors and tumor promoters, and a loss of contact inhibition which leads to a disruption of tissue organization as well as progressive growth of the tumor.[5] The steps may include loss of tumor suppressor gene functions and the development of aneuploidy. Other alterations which often characterize later stages in the process of malignant transformation include cellular heterogeneity, invasive growth, and finally, metastatic growth.

There may be many unique aspects of the process of malignant transformation as it applies to different cell types from different tissues. It is clear that there are unique features in the regulation of growth in different cells and tissues and in fetal or neonatal cells compared to adult cells. Therefore, it may be necessary to look at the unique aspects of transformation with regard to a wide spectrum of human cells to discover the major themes that characterize the transformation process. Our goal is to characterize the complete process as it occurs in the cells of one human tissue, endometrium.

The entire endometrium is derived from the mesodermal germ layer. It is composed primarily of two cell types, glandular epithelial cells and the cells that are specifically designated as "endometrial stromal cells". Stromal cells are the most numerous cells in the tissue and surround glands and blood vessels. The endometrial stromal cells differ from fibroblasts, which form the stroma of most tissues. They have steroid hormone receptors and respond to hormonal variations during the menstrual cycle with morphological and biochemical changes.[6-8] During pregnancy, stromal cells become the decidual cells at the placental implantation site, and in the process they change further to a more differentiated state. They produce large amounts of growth factors (the highest levels of TGF-α in the bodies of pregnant rodents[119] which may serve to facilitate fetal development.

Studies of transformation of endometrial tissue can also be seen as a paradigm for cancer development in tissues of mesodermal origin. These results may be applicable to mesothelium, kidney, and ovary, as well as endometrium, which together represent a notable fraction of human cancers. Endometrial stromal cells also represent a cell type different from the fibroblasts or epithelial cells used in most studies of neoplastic transformation. Although most human endometrial cancers are adenocarcinomas, as many as 3 to 10% of endometrial cancers are sarcomas of endometrial stromal cells or mixed Mullerian tumors (mixed carcinoma and sarcoma). A high incidence of mixed tumors is not unique to endometrium; it is common in other tissues in which the epithelium and mesenchymal cells are both of mesodermal origin (other parts of the female genital tract, mesothelium, etc.).

II. MATERIALS AND METHODS

Tissue culture — Endometrial tissue was obtained and processed as previously described.[9] Stromal cells were separated from the glandular epithelia and cultured in a 1:1 mixture of Opti-MEM and RPMI 1640 (GIBCO) supplemented with 1% fetal bovine serum (FBS), 3% bovine calf serum, 2 μg/ml insulin, 4 mM glutamine, and 2× nonessential amino acids. Most cell culture was antibiotic free. The cells were routinely subcultured at a 1-to-4 split ratio, and the medium was changed twice weekly.

Soft agar assays — The cells were plated at 50,000 per 60-mm dish in 0.33% Noble agar (DIFCO) over a 7-ml base layer of 1.0% agar. The medium was the same as for routine tissue culture, with the following alterations: 10% bovine calf serum, 5 μM putrescine, and 0.1 μg/ml DEAE-dextran (Sigma). The dishes were fed weekly with 1.0 ml of 1× medium. Colony-forming efficiency (CFE) was determined after 6 weeks by counting all colonies larger than 30 μm.

Transfection — The origin of the replication-defective[10] construct of the temperature-sensitive SV40 mutant A209[11-13] (tsSV40) cloned into plasmid pMK 16 was generously provided by S. P. Banks-Schlegel. Approximately 10^6 cells were transfected by electroporation with 100 μg/ml of plasmid DNA in Opti-MEM (GIBCO) at 4.0 kV/cm for 30 μs. Cells were replated, allowed to reach confluence, passaged once, and held at confluence until colonies of morphologically altered cells appeared. These were subcultured using cloning rings. Alternatively, the entire population was passaged until untransfected cells senesced and were overgrown.

DNA analysis — Integration of the transfected DNA was analyzed by the method of Southern.[14] High molecular weight DNA was isolated by CsCl gradient centrifugation and purified by phenol extraction. DNA (10 μg) was restricted with 20 U of *Eco*R1 (Promega) at 37°C for 18 h. It was electrophoresed in a 0.8% agarose gel at 50 mV for 20 h and transferred to nitrocellulose. The ^{32}P-labeled DNA probe was generated by random priming of the 2.6-kb *Stu*-1/*Bam*H1 fragment of transfected plasmid. This fragment includes the origin of replication and sequences coding for the small t- and large T-antigens.

RNA analysis — Total RNA was isolated by the guanidium-isothiocyanate method and then denatured. Twenty μg of RNA was electrophoresed through a 1% formaldehyde gel, and subsequently transferred to a nitrocellulose membrane.[15] Expression of c-*fos* was determined by hybridization with the 0.6-kb pst 1 fragment of v-*fos* generously provided by Inder Verma.[16] Hybridization was for 18 h at 42°C, and subsequently the filter was washed with $0.2 \times$ SSC at 56°C.

Temperature shift — The tsSV40-transfected cells were routinely cultured at 33°C. The nonpermissive temperature for growth was 39°C. The cells were incubated at the permissive or nonpermissive temperature for 3 d prior to initiating the experiments to evaluate c-*fos* and ODC expression.

ODS assays — ODC activity was determined by the release of $^{14}CO_2$ from 1-^{14}C-ornithine. Activity was determined in a 10,000 $\times$ g supernatant fraction (cytosol) of cell lysate. The CO_2 released was collected on an antibiotic sensitivity disc impregnated with 15 μl of Protosol (NEN) and quantitated by liquid scintillation spectroscopy as previously described.[17,18] Enzyme activity is expressed as nanomoles of CO_2 released per milligram of protein per hour.

DNA Content — Stromal cells (10^6) were harvested by trypsination, centrifuged, and resuspended in 0.1 ml of phosphate-buffered saline. Cells were fixed by adding 1.0 ml of cold (4°C) 70% ethanol. After pelleting and removal of the ethanol, ribonuclease A (2 mg/ml, Sigma) was added to remove RNA, and the DNA was stained with propidium iodide (50 μg/ml, Sigma) for 30 min at room temperature. The DNA content was then analyzed in an Ortho 50 H cytofluorometer using human lymphocytes as the standard. Cell cycle analysis was performed by the Ortho 2150 QuickEstimate™ program.

III. RESULTS

A. TREATMENTS WITH CHEMICAL CARCINOGENS AND TUMOR PROMOTERS

Initial efforts to understand the multistep nature of carcinogenesis in human endometrial stromal cells employed multiple treatments with relatively low doses of the chemical carcinogen N-methyl-N'-nitro-N-nitrosoguanidine (MNNG). This was done in an effort to simulate *in vitro* the features of the disease as it was assumed to occur *in vivo*. Human endometrial stromal cells treated repetitively with MNNG demonstrated the capacity to form colonies in soft agar and acquired increased gamma-glutamyltranspeptidase (GGT) activity. Compared to control cells, carcinogen-treated stromal cells displayed atypical morphology characterized by irregularities in cell and nuclear size and shape, increased nuclear-to-cytoplasmic ratios, and cellular crowding, and these abnormal features became more pronounced with increased numbers of treatments.[19,20]

Carcinogen-induced phenotypic changes were enhanced both by TPA[21] and diethylstilbestrol (DES).[22] Acute administration of DES did not provide a growth stimulus, nor did it alter the toxicity of MNNG. However, in stromal cells initiated by treatment with MNNG, chronic DES exposure altered the expression of GGT, increased the morphologic abnormality of cells, and enhanced the ability of cells to proliferate in restrictive medium. These results indicate that DES may act as a tumor promoter or cocarcinogen in these human cells. Chronic DES exposure furthers the process of cellular alteration such as that which occurs when cells are exposed to larger amounts of MNNG, and therefore it may be promoting the cells closer to full transformation.

Later experiments investigated the position of the cells in the cell cycle, and the interrelationship of MNNG and the DNA demethylating agent 5-azacytidine in chemical carcinogenesis.[23] Normal stromal cells in low passage were synchronized by density arrest and serum starvation. Cells were released back into a synchronized cycle by restoring serum and subculturing. Separate dishes of stromal cells were treated with MNNG during various phases of the cell cycle. One cell population was treated with 5-azacytidine at one passage prior to and one passage subsequent to treatment with MNNG. The cells treated with the combination of 5-azacytidine and MNNG exhibited morphological transformation evidenced by focus formation. The foci of transformed cells were characterized by cellular crowding and multilayering. Neither 5-azacytidine nor MNNG treatment alone produced morphologically transformed cells under these conditions. Of cells treated with MNNG and 5-azacytidine, the cell population that was treated with MNNG early in the S-phase produced morphologically altered cells, while the cell populations in either G_1 or G_2/M were not transformed. This result may be interpreted to suggest that altered gene expression due to demethylation can act in synergy with MNNG in the transformation of normal human cells, and that transformation occurred only when the MNNG was applied during the early S-phase of the cell cycle.

Despite great effort, the complete malignant transformation of diploid normal human adult endometrial stromal cells of finite life-span could not be achieved *in vitro* using chemical carcinogens alone or together with tumor promoters.[19-23] Our inability to achieve malignant transformation of these normal human cells with chemical carcinogens was confirmed for other types of human cells in other laboratories (reviewed in Reference 24). Perhaps the most important reason for this may be due to the limited cellular life-span of normal adult human cells *in vitro* and the low frequency with which they become immortalized spontaneously . For this reason, in subsequent studies we sought to extend the life-span of the endometrial stromal cells by transfecting them with a viral gene known to extend life-span.

Human cells derived from normal tissues have limited life-span in culture;[25] a proliferative phase is followed by a period of senescence during which the cells no longer proliferate, but remain viable. The senescent period may last for several months *in vitro*.[26] The large T-antigen of the DNA virus, simian virus-40 (SV40), has the ability to increase the life-span of human cells in culture. Transformation of cells with SV40 has become a standard approach in producing cell lines with extended life-spans.[27,28] SV40 infection fully transforms rodent cells, inducing immortality, anchorage independence, and tumorigenicity. Human cells infected with complete SV40 virus may integrate a high number of SV40 genomes. These cells have increased life-spans in culture, and they become genetically unstable[29] and aneuploid, but generally retain a finite life-span and usually are not tumorigenic. However, human mesenchymal cells transformed with SV40 may rarely become immortal as the result of a low-frequency, postcrisis event.[30,31] Cell fusion experiments indicate that this low-frequency event involves a second genetic alteration in addition to the presence of a functional SV40 T-antigen to achieve immortalization of human cells.[32]

Origin-defective SV40 constructs[10,34] produce a higher rate of transformation[35] and an increased frequency of production of immortalized populations.[36] In the following studies, an origin-defective construct of the SV40 temperature-sensitive mutant A209[11,12] (tsSV40) was used to transfect normal adult human endometrial stromal cells.

B. EXTENDED LIFE-SPAN RESULTING FROM TRANSFECTION WITH tsSV40

1. Morphological Alterations, Growth Kinetics

Endometrial stromal cells were transfected with tsSV40 either prior to primary culture or while in low passage (six or less population doublings [PDs]).[37] Cells from eight different donors were transfected, and transformants were isolated from all eight. The typical transfection resulted in 10 to 20 colonies per 100-mm dish containing approximately 2×10^6 cells, giving a stable transfection frequency of 0.5 to 1.0×10^{-5}.

Colonies of morphologically altered cells began to appear 4 to 6 weeks after transfection (Figure 1A). The transfected cells were smaller than their normal stromal cell parents, and continued to grow past confluence, eventually forming large multilayered colonies (Figure 1B). The growth pattern of the transfected cells retained a large degree of order, however, and exhibited little criss-cross growth pattern.

Southern blot analysis of tsSV40-transfected cells from the uncloned 87-062 line demonstrated the presence of the SV40 sequences (Figure 2). SV40 contains a single restriction site for *Eco*R1. Two bands are obtained from the transfected stromal cells (lane 1): one at approximately 4.9 kb and a weaker band at about 2.9 kb (not shown). The two hybridization bands in the DNA of transfected stromal cells indicate a single integration site and, based on the band intensity, suggests that a low number of copies, possibly a single copy, of the transferred virus has been integrated.

Proliferation of the tsSV40-transfected endometrial stromal cells is dependent upon a functional SV40 large T-antigen.[37] The untransfected stromal cells proliferate faster at the nonpermissive temperature (39°C) than at the permissive temperature (33°C). The tsSV40-transfected stromal cells cease proliferation upon shift to the nonpermissive temperature when propagated past their normal life-span of 20 PDs. Earlier studies with rodent cells have indicated that cell lines established with temperature-sensitive mutants of

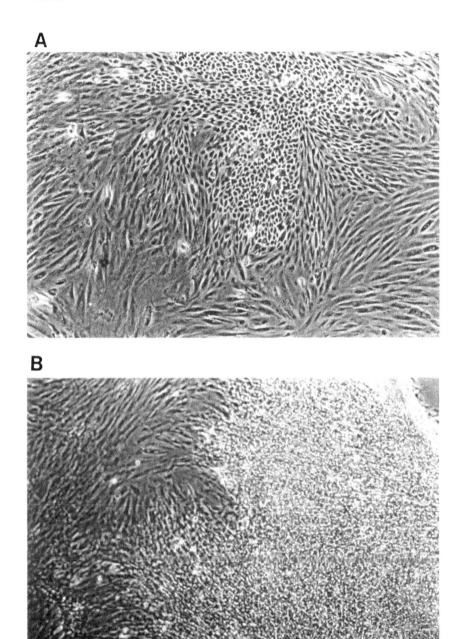

FIGURE 1.   Morphological alterations by tsSV40-transformed human endometrial stromal cells. (A) A small colony of tsSV40-transformed cells demonstrating the smaller size of the transformed cells; (B) a large focal area formed by the tsSV40-transformed cells. (Magnification × 100.)

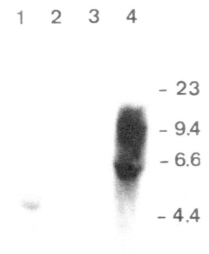

FIGURE 2. Analysis of integrated SV40 DNA. Lane 1, DNA from tsSV40-trans-
fected human endometrial stromal cells; lane 2, DNA from untransfected stromal
cells; lane 3, DNA from untransfected human monocytes; lane 4, DNA from immortal
SV40-infected human fibroblasts. (From Rinehart et al., *J. Virol.*, March, 1458,
1991. With permission.)

SV40 were growth restricted at the nonpermissive temperature.[38,39] The tsSV40-
transfected endometrial stromal cells continue to be viable for at least 2 weeks
at the nonpermissive temperature, and the cessation of growth is reversible.

At the permissive temperature, the tsSV40-transfected stromal cells continue to enter the S-phase of the cell cycle even under low-serum culture
conditions which prevent the proliferation of normal endometrial stromal cells
(Panel A, Figure 3). The percentage of cells entering the S-phase of the cell
cycle is unchanged during the course of the experiment at the permissive
temperature. The relief of serum dependence by SV40 transformation is a
well-known phenomenon.[12] At the nonpermissive temperature, the proportion
of the tsSV40-transfected stromal cells entering the S-phase declines from
26.6 to 16.3 and 10.5% after 24 and 48 h, respectively. However, this decline
appears to be insufficient to explain the complete cessation of growth at the
nonpermissive temperature (see Figure 8).[37] The cells in the S-phase may be
"trapped" there by failure of T-antigen function, but may not actually be
progressing through the cell cycle. Alternatively, it is also possible that cell
replication is evenly matched with cell death so that the total cell number
remains static. A third possibility is that a minority of the cells synthesize
DNA, but do not undergo mitosis. Other tsSV40-like mutants have been
shown to stimulate DNA synthesis without mitosis in senescent human fibroblasts at the nonpermissive temperature.[40] This result is not altered by the

addition of serum (data not shown). At the nonpermissive temperature, cells accumulate in both the G_0/G_1 and the G_2/M phases of the cell cycle, suggesting that there are at least two restriction points in the cell cycle which are overcome by the T-antigen. Similar results have been obtained with rat embryo fibroblasts transformed with tsSV40,[38] and the block in G_2/M was localized in the G_2 phase of the cell cycle.

2. Large T-Antigen Effects on c-*fos* and Ornithine Decarboxylase

Alterations in signal transduction may lead to a loss of the precise homeostatic control of cell growth, and this is a characteristic of neoplasms. To better understand alterations in the regulation of key gene products which may characterize the initial steps in the transformation of human endometrial stromal cells, we studied the effects of the large T-antigen (reviewed in References 41 and 42) on two important components of the signal transducing cascade, c-*fos* and ornithine decarboxylase (L-ornithine carboxylyase, EC 4.1.1.17; ODC).

The *fos* gene product is involved in the regulation of cellular growth and proliferation and in neoplastic transformation. When complexed with the c-*jun* protein, the *fos/jun* complex activates the transcription of genes whose products are necessary for proliferation.[43] C-*fos* is described as an immediate-early gene; it is expressed in quiescent cells as soon as 5 min after a growth stimulus, and it reaches peak levels of expression within 30 to 60 min.[44,45] C-*fos* is induced by a variety of extracellular stimuli.[46,47] The *fos* oncogene (v-*fos*) induces osteogenic sarcomas in FBJ virus-infected mice,[48] and the c-*fos* gene induces neoplastic transformation when transduced by retroviruses[49,50] or when inappropriately expressed.[51]

ODC is the initial enzyme in the pathway of polyamine biosynthesis. Polyamines are essential for cell proliferation; if cells are deprived of polyamines, DNA synthesis and cell proliferation cease.[52-54] ODC is induced by hormones acting on their target tissues[55] and by the tumor-promoting phorbol esters,[56] and it is generally expressed at high levels in transformed cells.[55] When quiescent cells are exposed to appropriate stimuli, ODC is expressed at peak levels 4 to 8 h later.

Serum induces a low level of c-*fos* expression in the tsSV40-transfected endometrial stromal cells under proliferative and nonproliferative conditions (permissive and nonpermissive temperatures). Exposure to 12-O-tetradecanoyl-phorbol-13-acetate (TPA) results in much higher levels of expression (Figure 4). However, there is relatively little difference in the induction of c-*fos* by either agent in the presence or absence of T-antigen activity. No c-*fos* mRNA is detected in the lanes from serum-starved, unstimulated control cells.

ODC induction in normal and tsSV40-transfected endometrial stromal cells was compared at 33, 37, and 39°C (Table 1). Serum deprivation (0.2%) reduces ODC activity to 1.0 unit in tsSV40-transfected stromal cells and below the level of detection in untransfected stromal cells. ODC activity is

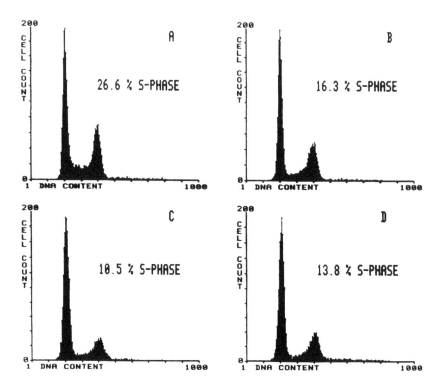

FIGURE 3. Cell cycle distribution of tsSV40-transfected cells. The cells were incubated in RPMI 1640 supplemented with 0.2% serum at either the permissive (A) or nonpermissive temperatures (B, C, and D). The cells were harvested and the DNA content was analyzed at days 1 (B), 2 (C), or 3 (D) after the temperature shift. (From Rinehart et al., *J. Virol.*, March, 1458, 1991. With permission.)

induced to high levels in the tsSV40-transfected stromal cells only at the permissive temperature. In untransfected stromal cells, the converse is true; as the temperature is increased, ODC induction is increased. Endometrial stromal cells also proliferate faster at 39 than at 33°C.[37] The ratio of ODC activity in transfected cells to that in normal control cells decreases from 52 after 8 h at 33°C to 0.4 at 39°C. This indicates a 130-fold difference in the ratio of ODC activity in the presence and absence of T-antigen activity at the two temperatures.

These results do not discriminate between the possibilities that the large T-antigen induces ODC activity autonomously, or functions in a permissive role, allowing other agents to act as inducers of ODC activity. These possibilities are analyzed in the experiment shown in Figure 5. The tsSV40-transfected cells were shifted from nonpermissive to permissive temperature in the presence or absence of serum. No ODC activity was detectable in the absence of serum (data not shown). ODC was induced to much higher levels at 33 than at 39°C (see also Table 1). Only those cells in the presence of serum express ODC, implying that the large T-antigen does not induce ODC

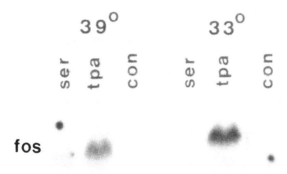

FIGURE 4. Northern blot analysis of the temperature dependence of induction of c-*fos* expression. Confluent tsSV40-trans-
fected stromal cells were maintained at the indicated temperature
for 3 d prior to analysis and serum starved for 16 h before stimulation. The additions were 10% fetal bovine serum or 10^{-7} M
TPA. Control cultures (con) were unstimulated. RNA was isolated 1 h after the stimulus. (From Rinehart et al., *J. Virol.*,
March, 1458, 1991. With permission.)

directly, but permits other agents to do so. Putrescine at 2 μM accelerated
cell growth at the permissive temperature, but had no effect at the nonpermissive temperature (data not shown), indicating that other alterations in
addition to ODC activity are responsible for the lack of growth in the absence
of T-antigen activity.

3. Large T-Antigen Effects on Actin Organization

Actin in mesenchymal cells *in vitro* often forms large polymers organized
into stress fibers which traverse the cytoplasm.[58,59] Neoplastic transformation
may result in the disruption and disorganization of stress fibers,[60,61] with a
concomitant disorganization of other cytoskeleton components which may
result in an alteration of cell shape.[60-65] We tested the ability of the SV40
large T-antigen to alter the organization of actin stress fibers in human endometrial stromal cells. Actin in normal stromal cell strains is organized into
an elaborate system of crisscrossing stress fibers which traverse the cytoplasm

TABLE 1
ODC Activity Induced by Serum in tsSV40-Transformed or Normal Cells

Time (h)	tsSV40	Normal cells	Significance	Ratio: tsSV40/control
		33°C		
0	1.0 ± 0.29	0	—	—
4	3.68 ± 0.43	0.03 ± 0.06	$p < 0.01$	123
6	9.57 ± 2.83	0.38 ± 0.19	$p < 0.01$	24
8	15.9 ± 2.33	0.33 ± 0.14	$p < 0.01$	52
		37°C		
4	2.20 ± 0.12	0.92 ± 0.06	$p < 0.01$	2.4
6	2.08 ± 0.48	1.74 ± 0.10	$p < 0.30$	1.2
8	1.05 ± 0.30	3.14 ± 0.58	$p < 0.01$	0.3
		39°C		
4	2.28 ± 0.18	3.84 ± 0.54	$p < 0.01$	0.6
6	1.55 ± 0.81	4.40 ± 0.79	$p < 0.05$	0.4
8	1.75 ± 0.05	4.17 ± 0.44	$p < 0.01$	0.4

Note: The cells were grown to confluence, incubated at the indicated temperatures for 48 h, serum deprived in 0.2% serum at the same temperature for an additional 16 h, and stimulated with 10% FBS. ODC activity is represented as nanomoles of CO_2 released per milligram of protein per hour. The numerical values are the mean ± SD of triplicate dishes from one of two experiments. The control (normal stromal) cells were at PD 6. Statistical significance was tested by the Student t test.

From Rinehart et al., *J. Virol.*, March, 1458, 1991. With permission.

and often run parallel to the plasma membrane.[66] In stromal cell strains, actin organization was the same at 33 and 39°C. In tsSV40-transfected stromal cells at the temperature-permissive conditions for large T-antigen function, the stress fibers are disrupted, and apparently unpolymerized actin is concentrated near the plasma membrane. Actin is largely absent from the cell centers.[66] Similar results have been obtained in other cell types transfected or infected with SV40.[60,67,68] Inactivation of the large T-antigen by shift to the nonpermissive temperature results in the cessation of proliferation and reassembly of stress fibers.[66] Loss of actin organization may be a marker for preneoplastic events. Suppressor-positive Syrian hamster embryo (SHE) cells display organized actin stress fibers, similar to normal SHE fibroblasts, whereas suppressor negative lines exhibit disorganized diffuse actin typical of tumorigenic cells.[69] The SV40 large T-antigen binds to several cellular proteins, including certain tumor suppressor genes (see Section IV). This is of interest, since human endometrial stromal cells expressing the SV40 large T-antigen and suppressor-negative SHE cells both contain disorganized actin stress fibers.

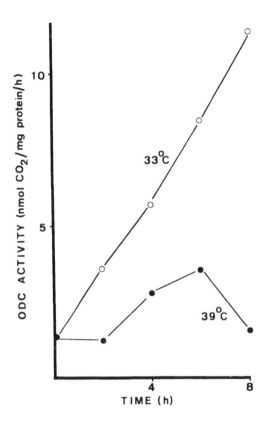

FIGURE 5. The temperature dependence of induction of ODC activity. The cells were prepared for induction as in Figure 4. The cells were stimulated with or without 10% FBS in RPMI 1640 and harvested at the indicated times. The data shown represent one of two separate experiments, each done with duplicate dishes. The variability between dishes at each point is less than 10%.

The EJ/T24 *ras* oncogene when transfected into tsSV40-transformed endometrial stromal cells does not support proliferation or disrupt actin stress fibers in the absence of the SV40 large T-antigen activity.[66] The observed alterations in actin organization were not the result of altered transcription, since neither the SV40 large T-antigen nor the *ras* oncogenes altered the level of actin mRNA.[66]

4. Increase in DNA Content Due to the SV40 Large T-Antigen

The multistep process of carcinogenesis usually involves several distinct genetic aberrations.[70-73] Flow cytometric analysis has implicated the presence of aneuploidy in the progression to clinically aggressive disease for many solid tumors.[74-82] Aneuploidy measured by flow cytometry almost invariably indicates the presence of cell populations with abnormal numbers of chromosomes.[83] In general, the diversity of the tumor cell population increases

TABLE 2
DNA Content of tsSV40-Transfected Clonal Isolates

Primary cell strain	Analyzed at PD	DNA content		
		Diploid	Mixed	Tetraploid
88-035	20—25	8	7	1
88-038	22—27	9	0	0
88-053	20—27	0	6	14
88-054	20—27	2	12	10
89-002	26—27	0	8	0
89-003	23	0	0	6
89-005	22—24	0	11	3
Total		19 (20%)	44 (45%)	34 (35%)

as neoplasms progress.[84] This tumor cell heterogeneity has been postulated to result from genetic instability, which is a characteristic of the neoplastic population.[85-87]

In order to study the changes in DNA content, a large number of clones were isolated. Approximately 6 to 8 weeks after transfection, round colonies sufficiently isolated from near neighbor to be clonal in origin were subcultured with cloning rings. These clonal cell populations subsequently were expanded for further analysis. A total of 97 clonal cell lines were obtained.

The clonal lines isolated from tsSV40-transfected endometrial stromal cells and their DNA content are described in Table 2. These determinations of DNA content were obtained as soon as sufficient cell numbers were available. At the initial characterization, approximately 20% (19/97) had a diploid DNA content, 35% (34/97) possessed a tetraploid DNA content, and the remainder consisted of a mixed population composed of varying proportions of cells with 2n or 4n DNA content.[88] Profiles of DNA content from flow cytometry illustrate these variations, as shown in Figure 6. Those cell lines lacking discernible cell populations in the region above 4n were classified as diploid (panel a). Those cell lines lacking discernible populations in the region below 4n were classified as tetraploid (panel b). Mixed populations were those that contained readily identifiable populations at 2, 4, and 8n, but in varying proportions (panels c, d, and e). The determinations of DNA content were performed using human lymphocytes as the standard. Our use of the designations 2n (diploid) and 4n (tetraploid) is not intended to imply that relatively minor changes in DNA content due to chromosomal loss or gain have not taken place. Those designations are used here to indicate the approximate net DNA content, not two or four copies of the 23 chromosomes.

The finding of endometrial stromal cell populations with tetraploid DNA content resulting from transformation implies that the SV40 large T-antigen can alter the DNA content of human endometrial cells. The diploid cells may give rise to cells with a tetraploid DNA content because the cell genome is destabilized. However, other interpretations are possible. The alterations may

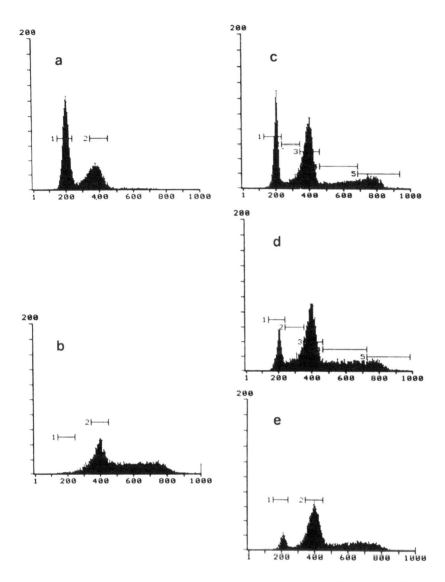

FIGURE 6. Profiles of DNA content of several clonal isolates demonstrating the variability of DNA content among the clonally derived cell lines. Cell number is plotted on the ordinate; fluorescence intensity (proportional to DNA content) is plotted on the abscissa. The units on the abscissa were standardized with human lymphocytes so that 200 equals 2n. A diploid clone is depicted in a, a tetraploid clone in b, and three clones containing mixed populations in c, d, and e.

be due to the transfection procedure. Alternatively, an increased life-span may provide the number of cell doublings that are needed for tetraploid cells to form.

Normal human endometrial stromal cells do not give rise to a tetraploid

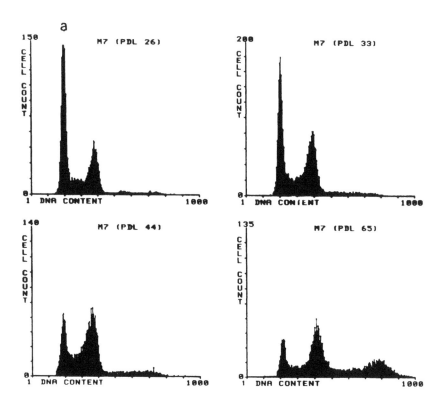

FIGURE 7. Alteration of DNA content during continuous passage. The abscissa and ordinate are as described for Figure 6. The cells were harvested and DNA content determined periodically during continuous passage of control (untransfected) stromal cells at 37°C (b). Note the absence of tetraploid cells throughout their life-span in culture (PD 20).

population over the course of their life-span in culture (Figure 7). The DNA content of the untransfected cells is shown from PD 2 until PD 20. The untransfected endometrial stromal cells became senescent at PD 20 to 24. Many populations of transfected cells were entirely tetraploid at PD levels equivalent to the 20 to 24 PDs obtained prior to the senescence of the untransfected control cell populations (Table 2).

The DNA content of several tsSV40-transfected stromal cell clones beginning in early passage and ending with entry into senescence crisis is described in Table 3. There is an ordered, unidirectional change in DNA content during the presenescence crisis phase of proliferation. All clones of tsSV40-transfected endometrial stromal cells which were 2n early in their presenescence "crisis" phase of proliferation developed a portion of the population with 4n DNA content. Those clones which were 4n remained so. The DNA contents appear to be constrained to either 2n or 4n, and the alterations are unidirectional, always from 2n to 4n.

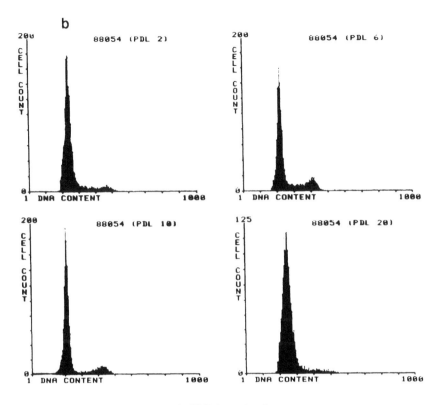

FIGURE 7 (continued)

5. Azacytidine Causes Increased Anchorage-Independent Growth

Hypomethylation of cellular DNA has been associated with neoplastic progression. DNA from both benign colon polyps and malignant carcinomas is hypomethylated compared to adjacent normal epithelium.[89] Cellular oncogenes have been found to be hypomethylated in human tumor cell lines. The third exon of the c-*myc* gene is hypomethylated in melanoma, lung carcinoma, and fibrosarcoma-derived cell lines.[90] In DNA from normal human fibroblasts, this region is completely methylated. In addition, both the c-Ha-*ras* and c-Ki-*ras* genes have been found to be hypomethylated in colonic adenocarcinomas compared to adjacent tissue.[91]

Our previous experiments indicated that 5-azacytidine could act in synergy with MNNG to induce morphological alterations in endometrial stromal cells.[23] We also investigated the ability of 5-azacytidine to further the neoplastic progression in tsSV40-transfected endometrial stromal cells.[92] Treatment with 50 μ*M* 5-azacytidine slowed the growth rate of the transfected cells in monolayer culture, but increased their anchorage-independent growth at the permissive temperature (Table 4). There were no discernible morphological alterations induced by 5-azacytidine at either permissive or nonpermissive temperature. Furthermore, 5-azacytidine did not increase the proliferative

TABLE 3
Alterations in DNA Content During Continuous Culture

Primary cell strain	Clone	Initial PD	Initial DNA content	Crisis PD	Crisis DNA content
88-035	M3	30	2n	43	2n and 4n
	M5	24	2n	44	2n and 4n
	M6	25	2n	37	2n and 4n
	M7	32	2n	59	2n and 4n
	M9	27	2n	54	2n and 4n
	M10	24	2n and 4n	33	2n and 4n
	M12	24	4n	32	4n
89-003	P2	23	4n	38	4n
	P6	23	4n	34	4n
89-002	C2	29	2n and 4n	40	2n and 4n
	C4	26	2n	49	2n and 4n

TABLE 4
5-Azacytidine-Induced Alterations

Cell type	Doubling time (days)	CFE (% ± SD) Exp. 1	CFE (% ± SD) Exp. 2
Endometrial stromal cell	1.8	0	0
M7	1.0	1.00 ± 0.26	1.03 ± 0.28
M7/5-azaC	2.3	$2.14^a \pm 0.32$	$3.16^a \pm 0.38$

Note: The tsSV40-transfected clone M7 was assayed for anchorage-independent growth between passages 36 and 43. All determinations were performed at 33°C; CFE, colony-forming efficiency.

[a] By the Student t test, $p < 0.1$ compared to untreated M7 cells.

capacity at the nonpermissive temperature. In these experiments, demethylation with 5-azacytidine appeared to further neoplastic progression in that it increased the ability of the tsSV40-transfected cells to grow in soft agar, but was ineffective in replacing the life-span-extending function of the large T-antigen. This result is in agreement with the previous finding that 5-azacytidine eliminated contact inhibition, resulting in focus formation in MNNG-treated normal endometrial stromal cells. However, 5-azacytidine did not prolong the life-span of the MNNG-treated stromal cells, nor could 5-azacytidine-induced changes replace the life-span-extending function of the SV40 large T-antigen at the nonpermissive temperature.

C. IMMORTALIZATION

Immortalization may be the rate-limiting step in carcinogenesis, and in SV40-transfected or infected human cells, immortalization appears to consist of two stages.[93] In the first stage, the cells have an extended but finite life-

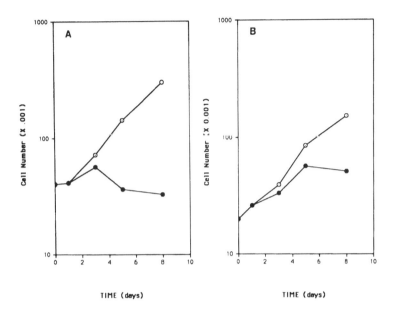

FIGURE 8.   Temperature-sensitive SV40-transfected cells (M4, A; B10T1, B) proliferate rapidly at the permissive temperature of 33°C (○), but cease proliferation when shifted to the nonpermissive temperature of 39°C (●) at day 1.

span. These cells eventually enter a period of crisis in which the cells continue to attempt to divide. However, the population becomes static and, over a period of time which may last as long as 6 months, subsequently declines. In the second stage, a few cells may emerge from this "crisis" period to become capable of indefinite life-span in culture. This second stage has been interpreted as the result of the loss of a dominantly acting growth repressor gene activity.[32] This emerging view of immortalization itself as a multistage process is consistent with and increases our understanding of carcinogenesis as a multistep process.

1. Description of Immortalized Cell Lines

We have characterized two unrelated cell lines which appear to be capable of unlimited growth. Both lines continue to require functional large T-antigen for proliferation (Figure 8). Two recent studies with human fibroblasts immortalized with controllable SV40 genes indicate a continued, postcrisis dependence upon large T-antigen for growth.[93,94]

The histories of these two cell lines, designated M4 and T1, are described in Table 5. The cell line M4 was derived as one of the original colonies which appeared following transfection and which was isolated with a cloning ring. In early passage, M4 had a 2n DNA content. About 8 weeks after M4 entered "crisis" at PD 58, many colonies appeared simultaneously, and the culture resumed proliferation. Cell line B10 was also cloned from colonies that arose

TABLE 5
Description of Immortal Cell Lines

Specimen	Precrisis line	DNA content	Crisis at PD	Immortal line	Current PD
88-035	M4	2n	58	M4	230
88-054	B10	4n	28	B10T1	125

FIGURE 9.   Karyotypes of M4 (A) at PD 200, and of B10T1 (B) at PD 100.

following transfection. B10 had a 4n DNA content at the initial determination. This clone entered crisis at PD 28. After 12 weeks in crisis, two colonies appeared in the culture dish, and were subcultured separately. One of these, B10T1, has been maintained in culture and characterized. It has now achieved PD 125.

2. Genetic Instability

Cytogenetic analysis indicates the development of aneuploidy subsequent to immortalization.[95] The M4 cell line is hypertetraploid, and the B10 T1 cell line is hypotetraploid. Both cell lines have cytogenetic abnormalities, including numerous marker chromosomes (Figure 9A and B).

3. Comparison of Precrisis and Postcrisis Cells

Escape from crisis resulted in increased anchorage-independent growth compared to parental cells, but did not consistently increase the growth rate.[96] Three clones during the precrisis period of growth (M4, M7, and B10) were compared to five immortalized daughter clones (M4, M7C1, M7C9, B10T1, and B10T2) (Table 6). There was no consistent change in doubling time in monolayer culture, but there is a significant increase in capacity for anchorage-independent growth subsequent to immortalization.

TABLE 6
Comparison of tsSV40-Transfected Endometrial Stromal Cells During the Precrisis and Postcrisis Periods of Growth

Clone	Doubling time (days)	CFE (% ± SD)
B10 (precrisis)	3.2	0
B10T1 (postcrisis)	2.0	4.29 ± 0.96[a]
B10T2 (postcrisis)	3.6	0.033 ± 0.058[a]
M4 (precrisis)	4.8	0
M4 (postcrisis)	2.0	0.30 ± 0.06[a]
M7 (precrisis)	2.0	0.30 ± 0.06
M7C1 (postcrisis)	2.3	0.59 ± 0.07[a]
M7C9 (postcrisis)	1.4	0.88 ± 0.14[a]
87-061 (uncloned, postcrisis)		0.067 ± 0.029
J15 W5 (postcrisis)		0.15 ± 0.05

Note: CFE, colony-forming efficiency.

[a] Significantly different from cells during precrisis growth (*p* <0.01 by the Student *t* test).

Seven postcrisis cell lines have been established. All continue to be dependent upon the large T-antigen for proliferation, as determined by their growth arrest at the nonpermissive temperature. The capacity of these cell lines for anchorage-independent growth is described in Table 6. These lines are remarkably diverse in their capacity for anchorage-independent growth, exhibiting a range in excess of two orders of magnitude.

4. Promotional Effects of DES on tsSV40-Immortalized Endometrial Stromal Cells

The nonsteroidal estrogen DES induced malignancies of the female reproductive tract in young women[97] and rodents[98] whose mothers had been administered the drug during pregnancy. In addition, our previous results with MNNG treatments of normal stromal cells indicated that DES could increase the frequency of morphological alterations.[22] Using the cell line M4, we demonstrated that DES can increase the anchorage-independent growth of tsSV40-immortalized endometrial stromal cells.[99] Acute treatment of the cells did not have an effect upon the doubling time in monolayer culture (Table 7). DES did enhance the anchorage-independent growth significantly at the highest dose. Chronic treatment of the cells for 6 months decreased the doubling time of the DES-treated cells, and increased their capacity for anchorage-independent growth three- to fourfold. These results indicate that DES may be capable of enhancing or promoting the appearance of the transformed phenotype in the tsSV40-immortalized stromal cells. The increase in transformation seen with increasing time may reflect the time necessary for

TABLE 7
Promotional Effects of DES on tsSV40-Immortalized Stromal Cells

| DES | Acute treatment | | Chronic treatment (6 mo.) | |
| | CFE | Doubling time | CFE | Doubling time |
(M)	*(% ± SD)*	(days)	*(% ± SD)*	(days)
0 (con)	0.60 ± 0.15	3.2	0.49 ± 0.21	2.7
10^{-10}	1.16 ± 0.38	3.2	1.73[a] ± 0.27	1.9
10^{-9}	0.96 ± 0.20	3.2	1.99[a] ± 0.20	2.1
10^{-8}	1.26[a] ± 0.32	3.2	1.75[a] ± 0.28	2.3

Note: CFE, colony-forming efficiency; average of triplicate dishes.

[a] Significantly different from control cells ($p < 0.05$).

the DES-enhanced cells to exert a selective advantage, or may reflect the cumulative effects of multiple doses.

5. Divergent Effects of *ras* Transformation

The immortalized M4 cell line was transfected with a plasmid containing the human c-Ha-*ras* gene mutated at codon 12 to code for valine at that position (the EJ or T24 Ha-*ras* oncogene), and the bacterial Tn5 gene coding for neomycin resistance (pSV$_2$neo/EJ-*ras*).[100] Transfectants were selected by their resistance to the antibiotic G418 (a neomycin analog). Two clones were selected for further study. Both contain the mutated *ras* gene as demonstrated by restriction enzyme analysis of a fragment of exon 1 containing codon 12 (Figure 10). The codon 12 mutation eliminates a recognition site for the restriction endonuclease Msp I. This was used as the basis for identifying the transfected gene.[101] First, the region around codon 12 was amplified by the polymerase chain reaction (PCR) technique.[102] The product of the PCR reaction was then digested with Msp I. The results indicate that the two transfected clones (M4C1 and M4C4) both contain the transfected gene. The biological effects of *ras* transfection are quite different in the two clones (Table 8). In M4C1, the cells are more transformed, as measured by their CFE in soft agar, while M4C4 is less transformed. The mutated *ras* gene may have divergent biological effects dependent upon the recipient cell. These results point out the importance of analyzing several clones in order to identify variable effects and to avoid potentially incorrect generalizations. Note that SV40-infected human epithelial cells have been completely transformed by secondary transfection of an activated *ras* oncogene.[103]

IV. DISCUSSION

Carcinogenesis in human tissue is a complex, multistep process which has proved to be difficult to model with human cells *in vitro*.[24] However, there are advantages in using human cells and tissues for carcinogenesis

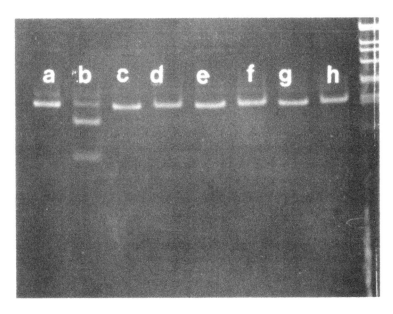

FIGURE 10.   Integration of the *ras* oncogene. Lanes a and b contain the amplification product of DNA from M4, c and d of DNA from M4C1, e and f from M4C4, and g and h from the T24 bladder carcinoma line (positive control). Lanes b, d, f, and h were cut with msp 1.

TABLE 8
Effects of Transfection with Activated Ha-*ras*

Cell line	Anchorage-independent growth (CFE ± SD)	Doubling time (days)
M4	1.52 ± 0.4	3.2
M4 EJ C1	13.8 ± 0.4 (*p* <0.0001)	2.5
M4 EJ C4	0.03 ± 0.04 (*p* <0.02)	3.4

Note: CFE, colony-forming efficiency.

studies. For example, the use of human cells eliminates the difficulties in extrapolating from data obtained with rodent systems. It also allows study of congenital dispositions for which no animal models exist.[104] The conversion of normal cells capable of a limited number of PDs to a cell line capable of indefinite life-span is a critical step, and perhaps the rate-limiting step, in the neoplastic transformation process.[24,105,106] SV40-immortalized cell lines[105,107,108] have been used to study the role of cooperating oncogenes in the malignant conversion of human epithelial cells.[106,107]

In order to create cell lines with enhanced proliferative potential for the study of the initial phases of carcinogenesis, we have transfected human endometrial stromal cells with a plasmid containing a temperature-sensitive mutant SV40. The cells eventually entered a "crisis", but did increase their proliferative potential from the 10 to 20 PDs characteristic of endometrial stromal cells obtained from adult donors to at least 30, and in some clones to as much as 70 to 80 PDs. This result is in agreement with previous studies which indicate that human cells must pass through a senescence "crisis" to become immortal.

The analysis of SV40 sequences in the transfected cells indicted that the genes had integrated at a single site, and in a low number of copies, possibly as a single copy. The transformation induced by a low copy number of transfected oncogenes may be more relevant to carcinogenesis *in vivo* than some studies in which a high multiplicity of SV40 infection is used.

The temperature-sensitive feature of the SV40 large T-antigen used in this study permits the direct comparison of cells in the presence and absence of large T-antigen activity. It also eliminates the possibility of genetic differences due to the selection of transformed subpopulations or to the genetic drift which often occurs when cells are passaged in culture for extended periods. We have used the ability to regulate the function of the large T-antigen to begin to localize its effects upon the signal transduction pathway(s). At the nonpermissive temperature, the cells cease proliferation and ODC activity is much diminished, but the induction of c-*fos* expression by serum is relatively unaffected. Membrane-associated oncogenes such as *src* and growth factor receptors possessing a tyrosine kinase activity do induce c-*fos*.[109] This implies a sequence of events in which a membrane-associated tyrosine kinase phosphorylates an intermediate which acts through an undetermined series of steps to transcriptionally activate responsive elements of the *fos* gene. Our results indicate that the large T-antigen acts downstream of these events. The SV40 large T-antigen is a nuclear-localized phosphoprotein with no known tyrosine kinase activity,[41,42] and would therefore be unlikely to act at such an early stage in the signal transduction pathway.

Inhibition of c-*fos* expression has been reported to occur in senescent compared to proliferative human fibroblasts.[110] However, SV40 can induce DNA synthesis in these senescent cells.[40] These studies differ from ours in that they compare young, proliferative cells to older, senescent cells. It is not known if *fos* or ODC expression is blocked in senescent endometrial stromal cells. In our study, we compared age-matched cells which differ in T-antigen activity. Serum induced relatively low levels of c-*fos* mRNA at both temperatures. TPA, which is not mitogenic for these cells, induced much higher levels of c-*fos* mRNA at both temperatures. These results indicate that the extended life-span conferred upon the endometrial stromal cells by SV40 is not due to an increase in c-*fos* expression, since cells at proliferative conditions (serum stimulation at permissive temperature) do not have higher levels of *fos* expression than cells at nonproliferative conditions (serum stimulation at nonpermissive temperature).

In contrast, the expression of ODC in the presence of T-antigen activity is much higher than in the absence of T-antigen activity. Large T-antigen expression itself is not sufficient to induce ODC expression, but it is necessary to permit the high levels of induction by serum or asparagine. This result is consistent with the hypothesis that the T-antigen exerts its influence, at least in part, by neutralizing the effects of an inhibitor. ODC mRNA is expressed at similar levels in senescent and proliferating fibroblasts, but ODC activity is higher in proliferating cells,[110] indicating that the decline in ODC activity observed in senescent cells may be due to regulatory events which take place posttranscriptionally.

The mechanism by which the large T-antigen permits expression of ODC activity and disrupts actin organization is at this point a matter of conjecture. The dependence upon the large T-antigen for the extended life-span of tsSV40-transfected endometrial stromal cells derives from the ability of the T-antigen to permit the transfected cells to traverse the cell cycle and complete mitosis. The large T-antigen binds to a number of cellular proteins having regulatory functions, including the AP-2 enhancer binding protein,[111] p53,[112] and the retinoblastoma susceptibility gene product.[113] It also binds to cellular proteins with undetermined functions.[114,115] The transforming activity of DNA viruses and perhaps some of the cellular oncogenes has been ascribed to their ability to complex with these proteins having growth-inhibitory or tumor suppressor activity.[116] p53[117] and p105[RB] are the two best-characterized transformation inhibitory proteins, and the SV40 large T-antigen binds to both. The binding of the large T-antigen to the dephospho-RB present during the G_0/G_1 portions of the cell cycle may be responsible for the growth-promoting effects.[118] Our data indicate that alleviating the suppression of ODC activity and altering the cytoskeleton might be important aspects of transformation by SV40.

All of the clones studied eventually entered a senescence crisis. Several cell lines have emerged from crisis, and appear to be capable of indefinite life-span. This immortal phenotype is conditional, however. Depleting the immortal endometrial stromal cells of functional large T-antigen results in senescence within 2 to 3 PDs after shifting the transfected stromal cells to the nonpermissive temperature.

Immortalization also results in cell behaviors characteristic of an increasingly transformed phenotype. The immortalized cell populations have an increased capacity for anchorage-independent growth compared to their parental populations prior to senescence crisis. The immortalized clones also are more genetically unstable. There appears to be a progression of genetic instability. Endometrial stromal cells retain a 2n DNA complement throughout their life-span. TsSV40-transfected stromal cells during the precrisis, extended life-span phase of proliferation undergo a 2n to 4n genetic progression. This increase in DNA content takes place in every clone studied. The genetic alteration is directional, always 2n to 4n, and constrained to two allowable genetic states, 2n and 4n. In tsSV40-immortalized endometrial stromal cells, increasing genetic instability is manifest, resulting in aneuploidy. These results may indicate that escape from "crisis" is a major factor in the genetic instability. The genetic losses or gains intrinsic to this process may be responsible for the increase in genetic instability.

The tsSV40-immortalized endometrial stromal cells have been used to study the transforming effects of DES and the mutated Ha-*ras* oncogene. Chronic treatment of the transfected cells with DES results in cell lines with a higher capacity for anchorage-independent growth and increased growth rate. Transfection with the mutated Ha-*ras* gene can be transforming or transformation-inhibiting, depending upon as yet undetermined components of the recipient cell.

ACKNOWLEDGMENTS

The authors acknowledge the expert technical assistance of Mr. Larry Wolfe and Mr. John P. Mayben. This work was supported by National Institutes of Health (NIH) Grants CA31733 and ES07017 and American Cancer Society Grant IN-15-30.

REFERENCES

1. **Bishop, J. M.,** The molecular genetics of cancer, *Science,* 235, 305, 1987.
2. **Weinberg, R. A.,** The action of oncogenes in the cytoplasm and nucleus, *Science,* 230, 720, 1985.
3. **Spandidos, D. A. and Anderson, M. L. M.,** Oncogenes and oncosuppressor genes: their involvement in cancer, *J. Pathol.,* 157, 1, 1989.
4. **DiPaolo, J. A.,** The relative difficulties in transforming human and animal cells in vitro, *J. Natl. Cancer Inst.,* 70, 3, 1983.
5. **Cerutti, P. A.,** Response modification creates promotability in multistage carcinogenesis, *Carcinogenesis,* 9, 519, 1988.
6. **Dallenbach-Helleg, G.,** *Histopathology of the Endometrium,* Springer-Verlag, New York, 1975, 22.
7. **Wynnn, R. M.,** in *Biology of the Uterus,* Wynn, R. M., Ed., Plenum Press, New York, 1977, 341.
8. **Finn, C. A.,** in *Biology of the Uterus,* Wynn, R. M., Ed., Plenum Press, New York, 1977, 245.
9. **Rinehart, C. A., Lyn-Cook, B. D., and Kaufman, D. G.,** Gland formation from human endometrial epithelial cells *in vitro, In Vitro Cell. Dev. Biol.,* 24, 1037, 1988.
10. **Gluzman, Y., Frisque, R. J., and Sambrook, J. F.,** Origin defective mutants of SV40, *Cold Spring Harbor Symp. Quant. Biol.,* 44, 293, 1983.
11. **Chou, J. Y.,** Human placental cells transformed by ts A mutants of simian virus 40: a model system of placental function, *Proc. Natl. Acad. Sci. U.S.A.,* 75, 1409, 1975.
12. **Martin, R. G. and Chou, J. Y.,** Simian virus 40 functions required for the establishment and maintenance of malignant transformation, *J. Virol.,* 15, 599, 1975.
13. **Nobile, C. and Martin, R. G.,** Stable stem-loop and cruciform DNA structures: isolation of mutants with rearrangements of the palindromic sequence at the simian virus 40 replication origin, *Intervirology,* 25, 158, 1986.
14. **Southern, E.,** Detection of specific sequences among DNA fragments separated by gel electrophoresis, *J. Mol. Biol.,* 98, 503, 1975.
15. **Thomas, P. S.,** Hybridization of denatured RNA and small DNA fragments transferred to nitrocellulose, *Proc. Natl. Acad. Sci. U.S.A.,* 77, 5201, 1980.

16. Curran, T., Peters, G., Van Beveren, C., Teich, N. Z. M., and Verma, I. M., FBJ murine osteosarcoma virus: identification and molecular cloning of biologically active proviral DNA, *J. Virol.*, 44, 674, 1982.

17. Chen, K. Y., Heller, J. S., and Canellakis, E. S., Studies on the regulation of ODC activity by the microtubules: the effect of colchicine and vinblastine, *Biochem. Biophys. Res. Commun.*, 68, 401, 1976.

18. Rinehart, C. A., Viceps-Madore, D., Fong, W.-F., Ortiz, J. G., and Canellakis, E. S., The effect of transport system A and N amino acids and of nerve and epidermal growth factors on the induction of ornithine decarboxylase activity, *J. Cell. Physiol.*, 123, 435, 1985.

19. Dorman, B. H., Siegfried, J. M., and Kaufman, D. G., Alterations of human endometrial stromal cells produced by *N*-methyl-*N*-nitro-*N*-nitrosoguanidine, *Cancer Res.*, 43, 3348, 1983.

20. Kaufman, D. G., Siegfried, J. M., Dorman, B. H., and Nelson, K. G., Carcinogen induced changes in human endometrial cells, in *Human Carcinogenesis*, Harris, C. C. and Autrup, H., Eds., Academic Press, New York, 1983, 469.

21. Siegfried, J. M. and Kaufman, D. G., Enhancement by TPA of phenotypes associated with transformation in carcinogen treated human cells; evidence for a selective mechanism, *Int. J. Cancer*, 32, 423, 1983.

22. Siegfried, J. M., Nelson, K. G., Martin, J. L., and Kaufman, D. G., Promotional effect of diethylstilbestrol on human endometrial stromal cells pretreated with a direct-acting carcinogen, *Carcinogenesis*, 5, 641, 1984.

23. Barrett, L. L., Rinehart, C. A., and Kaufman, D. G., Synergistic effect of 5-aza-cytidine in cell-cycle dependent transformation of human endometrial stromal cells by MNNG, *Proc. Am. Assoc. Cancer Res.*, 29, 452a, 1988.

24. McCormick, J. J. and Maher, V. M., Towards an understanding of the malignant transformation of diploid human fibroblasts, *Mutat. Res.*, 199, 273, 1988.

25. Hayflick L. and Moorhead, P. S., The serial cultivation of human diploid cell strains, *Exp. Cell Res.*, 25, 585, 1961.

26. Matsumura, T., Zerrudo, Z., and Hayflick, L., Senescent human diploid cells in culture: survival, DNA synthesis and morphology, *J. Gerontol.*, 34, 328, 1979.

27. Chang, S. E., *In vitro* transformation of human epithelial cells, *Biochim. Biophys. Acta*, 823, 161, 1986.

28. Sack, G. H., Jr.,, Human cell transformation by simian virus 40: a review. *In Vitro*, 17, 1, 1981.

29. Meisner, L. F., Wu, S.-Q., Christian, B. J., and Reznikoff, C. A., Cytogenetic instability with balanced chromosome changes in an SV40 transformed human uroepithelial cell line, *Cancer Res.*, 48, 3215, 1988.

30. Gotoh, S., Gelb, L., and Schlessinger, D., SV40-transformed human diploid fibroblasts that remain transformed throughout their limited lifespan, *J. Gen. Virol.*, 42, 409, 1979.

31. Huschtscha, L. I. and Holliday, R., Limited and unlimited growth of SV40-transformed cells for human diploid MRC-5 fibroblasts, *J. Cell Sci.*, 63, 77, 1983.

32. Pereira-Smith, O. M. and Smith, J. R., Functional SV40 T antigen is expressed in hybrid cells having finite proliferative potential, *Mol. Cell. Biol.*, 7, 1541, 1987.

34. Nagata, Y., Diamond, B., and Bloom, B. R., The generation of human monocyte/macrophage cell lines, *Nature*, 306, 597, 1983.

35. Small, M. B., Gluzman, Y., and Ozer, H. L., Enhanced transformation of human fibroblasts by origin-defective simian virus 40, *Nature*, 296, 671, 1982.

36. Neufeld, D. S., Ripley, S., Henderson, A., and Ozer, H. L., Immortalization of human fibroblasts transformed by origin-defective simian virus 40, *Mol. Cell. Biol.*, 7, 2794, 1987.

37. Rinehart, C. A., Haskill, J. S., Morris, J. S., Butler, T. D., and Kaufman, D. G., Extended lifespan of human endometrial stromal cells transfected with a cloned origin defective, temperature sensitive simian virus 40, *J. Virol.*, 65, 1458, 1991.

38. **Jat, P. S. and Sharp, P. A.,** Cell lines established by a temperature-sensitive simian virus 40 large-T-antigen gene are growth restricted at the nonpermissive temperature, *Mol. Cell. Biol.,* 9, 1672, 1989.

39. **Petit, C. A., Gardes, G., and Feunteun, J.,** Immortalization of rodent embryo fibroblasts by SV40 is maintained by the A gene, *Virology,* 127, 74, 1983.

40. **Gorman, S. D. and Cristofalo, V. J.,** Reinitiation of cellular DNA synthesis in BrdU-selected nondividing senescent Wi-38 cells by Simian Virus 40 infection, *J. Cell. Physiol.,* 125, 122, 1985.

41. **Livingston, D. M. and Bradley, M. K.,** The simian virus 40 large T antigen: a lot packed into a little, *Mol. Biol. Med.,* 4, 63, 1987.

42. **Stahl, H. and Knippers, R.,** The simian virus 40 large T antigen, *Biochim. Biophys. Acta,* 910, 1, 1987.

43. **Sassone-Corsi, P., Ransone, L. J., Lamph, W. W., and Verma, I. M.,** Direct interaction between *fos* and *jun* nuclear oncoproteins: role of the leucine zipper domain, *Nature,* 336, 692, 1988.

44. **Greenberg, M. E. and Ziff, E. B.,** Stimulation of 3T3 Cells induces transcription of the c-*fos* proto-oncogene, *Nature,* 311, 433, 1984.

45. **Kruijer, W., Cooper, J. A., Hunter, T., and Verma, I. M.,** Platelet-derived growth factor induces rapid but transient expression of the c-*fos* gene and protein, *Nature,* 312, 711, 1984.

46. **Cohen, D. R. and Curran, T.,** Fra-1: a serum-inducible, cellular immediate-early gene that encodes a *fos*-related antigen, *Mol. Cell. Biol.,* 8, 2063, 1988.

47. **Franza, B. R., Sambucetti, L. C., Cohen, D. R., and Curran, T.,** Analysis of *fos* protein complex and *fos*-related antigens by high-resolution two dimensional gel electrophoresis, *Oncogene,* 1, 213, 1987.

48. **Curran, T. and Teich, N. M.,** Candidate product of the FBJ murine osteosarcoma virus oncogene: characterization of a 55,000 dalton phosphoprotein, *J. Virol.,* 42, 114, 1982.

49. **Finkel, M. P., Biskis, B. P., and Jenkins, P. B.,** Virus induction of osteosarcomas in mice, *Science,* 151, 698, 1966.

50. **Finkel, M. P., Reilly, C. A., and Biskis, B. O.,** Viral etiology of bone cancer, *Front. Radiat. Ther. Oncol.,* 10, 28, 1975.

51. **Verma, I. M.,** Proto-oncogene fos: a multifaceted gene, *Trends Genet.,* 2, 93, 1986.

52. **Bachrach, U.,** *Function of the Naturally Occurring Polyamines,* Academic Press, New York, 1974, 1.

53. **Cohen, S. S.,** *Introduction to the Polyamines,* Prentice-Hall, Englewood Cliffs, NJ, 1971, 1.

54. **Tabor, C. W. and Tabor, H.,** 1,4-diaminobutane (putrescine), spermidine, and spermine, *Annu. Rev. Biochem.,* 45, 285, 1976.

55. **Russell, D. H. and Durie, B. G. M.,** *Polyamines as Biochemical Markers of Normal and Malignant Growth,* Raven Press, New York, 1978, 1.

56. **O'Brien, T. G., Simsimian, R. C., and Boutwell, R. K.,** Induction of the polyamine-biosynthetic enzymes in mouse epidermis by tumor-promoting agents, *Cancer Res.,* 35, 1662, 1975.

57. **Boutwell, R. K.,** The function and mechanisms of promoters of carcinogenesis, *CRC Crit. Rev. Toxicol.,* 2, 419, 1974.

58. **Clark, M. and Spudich, J.,** Nonmuscle contractile proteins: the role of actin and myosin in cell motility and shape determination, *Annu. Rev. Biochem.,* 46, 797, 1977.

59. **Stossel, T. P.,** Contribution of actin to the structure of the cytoplasmic matrix, *J. Cell Biol.,* 99, 15s, 1984.

60. **Pollack, R., Osborn, M., and Weber, K.,** Patterns of organization of actin and myosin in normal and transformed cultured cells, *Proc. Natl. Acad. Sci. U.S.A.,* 72, 994, 1975.

61. **Verderame, M., Alcorta, D., Egnor, M., Smith, K., and Pollack, R.,** Cytoskeletal F-actin patterns quantitated with flurosein isothiocyanate-phalloidin in normal and transformed cells, *Proc. Natl. Acad. Sci. U.S.A.,* 77, 6624, 1980.

62. **Shapland, C., Lowings, P., and Lawson, D.,** Identification of new actin-associated polypeptides that are modified by viral transformation and changes in cell shape, *J. Cell Biol.,* 107, 153, 1988.

63. **Boschek, C. B.,, Jockusch, B. M., Friis, R. R., Back, R., Grundmann, E., and Bauer, H.,** Early changes in the distribution and organization of microfilament proteins during cell transformation, *Cell,* 24, 175, 1981.

64. **Holme, T. C., Koestler, T. P., and Crawford, N.,** Actin in B16 melanoma cells of differing metastatic potential: effects of trypsin and serum, *Exp. Cell Res.,* 169, 442, 1987.

65. **Pienta, K. J., Partin, A. W., and Coffey, D. S.,** Cancer as a disease of DNA organization and dynamic cell structure, *Cancer Res.,* 49, 2525, 1989.

66. **Carter, C. A., Rinehart, C. A., Bagnel, C. R., and Kaufman, D. G.,** Fluorescent laser scanning microscopy of F-actin disruption in human endometrial stromal cells expressing the SV40 large T antigen and the EJ *ras* oncogene, *Pathobiology,* 59, 36, 1991.

67. **Katsumoto, T., Takenouchi, N., Kamahora, T., and Kurimura, T.,** Comparison of cytoskeletons of SV40-transformed cells and their revertants by electron microscopy, *J. Electron Microsc.,* 33, 68, 1984.

68. **Osborn, M. and Weber, K.,** Simian virus 40 A gene function and maintenance of transformation, *J. Virol.,* 15, 636, 1975.

69. **Boyd, J. A., Jones, C. A., and Barrett, J. C.,** Cytoskeletal changes and alterations in signal transduction which correlate with loss of the tumor suppressor phenotype in Syrian hamster embryo (SHE) cells, *Proc. Am. Assoc. Cancer Res.,* 30, 442, 1989.

70. **Pitot, H. C., Goldsworthy, T., and Moran, S.,** The natural history of carcinogenesis: implications of experimental carcinogenesis in the genesis of human cancer, *J. Supramol. Struct. Cell Biochem.,* 17, 133, 1981.

71. **Weinstein, I. B., Gattoni-Celli, S., Kirschmeier, P., Lambert, M., Hsiao, W., Backer, J., and Jeffrey, A.,** Multistage carcinogenesis involves multiple genes and multiple mechanisms, *J. Cell Physiol.,* 3, 127, 1984.

72. **Nowell, P. C.,** The clonal evolution of tumor cell populations, *Science,* 194, 23, 1976.

73. **Wolley, R. C., Schrieber, K., Koss, L. G., Karas, M., and Sherman, A.,** DNA distribution in human colon carcinomas and its relationship to clinical behavior, *J. Natl. Cancer Inst.,* 69, 15, 1982.

74. **Armitage, N. C., Robins, R. A., Evans, D. F., Turner, D. R., Baldwin, R. W., and Hardcastle, J. D.,** The influence of tumor cell abnormalities on survival in colorectal cancer, *Br. J. Surg.,* 72, 828, 1985.

75. **Kokal, W., Sheibani, K., Terz, J., and Harada, J. R.,** Tumor DNA content in the prognosis of colorectal carcinoma, *J. Am. Med. Assoc.,* 255, 3123, 1986.

76. **Emdin, S. O., Stenling, R., and Roos, G.,** Prognostic value of DNA content in colorectal carcinoma. A flow cytometric study with some methodologic aspects, *Cancer,* 60, 1282, 1987.

77. **Hedley, D. W., Rugg, C. A., Ng, A. B. P., and Taylor, I. W.,** Influence of cell DNA content on disease free survival of Stage II breast cancer patients, *Cancer Res.,* 44, 5395, 1984.

78. **Volm, M., Hahn, E. W., Mattern, J., Muller, T., Vogt-Moykopf, I., and Weber, E.,** Five-year follow-up study of independent clinical and flow cytometric prognostic factors for the survival of patients with non-small cell lung carcinoma, *Cancer Res.,* 48, 2923, 1988.

79. **Zimmerman, P. V., Hawson, G. A. T., Bint, M. H., and Parsons, P. G.,** Ploidy as a prognostic determinant in surgically treated lung cancer, *Lancet,* 2, 530, 1987.

80. **Stephanson, R. A., Jamies, B. C., Gay, H., Fair, W. R., Whitmore, W. F., Jr., and Melamed, M. R.,** Flow cytometry of prostate cancer: relationship of DNA content to survival, *Cancer Res.,* 471, 2504, 1987.

81. **Iverson, D. E.,** Prognostic value of flow cytometric DNA index in human ovarian carcinoma, *Cancer,* 61, 971, 1988.

82. **Friedlander, M. L., Hedley, D. W., Swanson, C., and Russel, P.,** Prediction of long term survival by flow cytometric analysis of cellular DNA content in patients with advanced ovarian cancer, *J. Clin. Oncol.,* 6, 282, 1988.

83. **Barlogie, B., Raber, M. N., Schumann, J., Johnson, T. S., Drewinko, B., Swartz-endruber, D. E., Gohde, W., Andreef, M., and Freireich, E. J.,** Flow cytometry in clinical cancer research, *Cancer Res.,* 43, 3982, 1983.

84. **Nicolson, G. L.,** Tumor cell instability, diversification, and progression to the metastatic phenotype: from oncogene to oncofetal expression, *Cancer Res.,* 47, 1473, 1987.

85. **Fidler, I. J. and Hart, I. R.,** The development of biological diversity and metastatic potential in malignant neoplasms, *Oncodev. Biol. Med.,* 4, 162, 1982.

86. **Ohno, S.,** Genetic implication of karyological instability of malignant somatic cells, *Physiol. Rev.,* 51, 496, 1971.

87. **Shackney, S. E., Smith, C. A., Miller, B. W., Burholt, D. R., Murtha, K., Giles, H. R., Ketterer, D. M., and Police, A. A.,** Model for the genetic evolution of human solid tumors, *Cancer Res.,* 49, 3344, 1989.

88. **Rinehart, C. A., Mayben, J. P., Butler, T. D., and Kaufman, D. G.,** Alterations in DNA content associated with transformation of human endometrial stromal cells with an SV40 containing plasmid, *J. Cell Biol.,* 109, 182a, 1989.

89. **Golez, S. and Vogelstein, B.,** Hypomethylation of DNA from benign and malignant colon neoplasms, *Science,* 228, 187, 1985.

90. **Cheah, M. S. C., Wallace, C. D., and Hoffman, R.,** Hypomethylation of DNA in human cancer cells: a site-specific change in the c-*myc* oncogene, *J. Natl. Cancer Inst.,* 73, 1051, 1984.

91. **Feinberg, A. and Vogelstein, B.,** Hypomethylation distinguishes genes of some human cancers from their normal counterparts, *Nature,* 301, 89, 1983.

92. **Barrett, L. L.,** 5-Azacytidine Increases Anchorage Independent Growth of Human Endometrial Stromal Cells Expressing the SV40 Large T Antigen, Master's thesis, University of North Carolina, Chapel Hill, 1990.

93. **Wright, W. E., Pereira-Smith, O. M., and Shay, J. W.,** Reversible cellular senescence: implications for immortalization of normal human diploid fibroblasts, *Mol. Cell. Biol.,* 9, 3088, 1989.

94. **Radna, R. L., Caton, Y., Jha, K. K., Kaplan, P., Li, G., Traganos, F., and Ozer, H. L.,** Growth of immortal simian virus 40 tsA-transformed human fibroblasts is temperature dependent, *Mol. Cell. Biol.,* 9, 3093, 1989.

95. **Rinehart, C. A., Laundon, C. H., Mayben, J. P., Butler, T. D., and Kaufman, D. G.,** Immortalization of human endometrial stromal cells with a temperature-sensitive simian virus 40, *J. Cell Biol.,* 111, 477a, 1990.

96. **Butler, T. D., Rinehart, C. A., and Kaufman, D. G.,** Escape from senescence crisis by SV40 transformed human endometrial stromal cells results in increased anchorage independent growth, *Proc. Am. Assoc. Cancer Res.,* 31, 638a, 1990.

97. **Herbst, A. L., Ulfelder, H., and Paskanzer, D. C.,** Adenocarcinoma of the vagina; association of the maternal stilbestrol therapy with tumor appearance in young women, *N. Engl. J. Med.,* 284, 878, 1971.

98. **McLachlan, J. A., Newbold, R. R., and Bullock, B. C.,** Long-term effects on the female mouse genital tract associated with prenatal exposure to diethylstilbestrol, *Cancer Res.,* 40, 3988, 1990.

99. **Xu, L. H., Rinehart, C. A., and Kaufman, D. G.,** The promotion of neoplastic transformation by diethylstilbestrol in human endometrial stromal cells transfected with a temperature-sensitive simian virus 40 large T antigen, *FASEB J.,* 5, A706, 1991.

100. **Rinehart, C. A., Van Le, L., Mayben, J. P., and Kaufman, D. G.,** Effects of *ras* transformation on temperature sensitive SV40 immortalized human endometrial stromal cells, *FASEB J.,* 5, A1442, 1991.

101. **Kahn, S. M., Jiang, W., and Weinstein, I. B.,** Rapid nonradioactive detection of *ras* oncogenes in human tumors, *Amplifications,* 4, 22, 1990.

102. **Mullis, K. B. and Faloona, F. A.**, Specific synthesis of DNA *in vitro* via a polymerase-catalyzed chain reaction, in *Methods in Enzymology*, Wu, R., Ed., Academic Press, San Diego, 1987, 155, 335.

103. **Reznikoff, C. A., Loretz, L. J., Christian, B. J., Wu, S.-Q., and Meisner, L. F.**, Neoplastic transformation of SV40-immortalized human urinary tract epithelial cells by *in vitro* exposure to 3-methylcholanthrene, *Carcinogenesis*, 9, 1427, 1988.

104. **Harris, C. C.**, Human tissues and cells in carcinogenesis research, *Cancer Res.*, 47, 1, 1987.

105. **Christian, B. J., Loretz, L. J., Oberly, T. D., and Reznikoff, C. A.**, Characterization of human uroepithelial cells immortalized in vitro by simian virus 40, *Cancer Res.*, 47, 6066, 1987.

106. **Rhim, J. S., Jay, G., Arnstein, P., Price, F. M., Sanford, K. K., and Aaronson, S. A.**, Neoplastic transformation of human epidermal keratinocytes by AD12-SV40 and Kirstein sarcoma viruses, *Science*, 227, 1250, 1985.

107. **Clark, R., Stampfer, M. R., Milley, R., O'Rourke, E., Walen, K. H., Krieger, M., Kopplin, J., and McCormick, F.**, Transformation of human mammary epithelial cells by oncogenic retroviruses, *Cancer Res.*, 48, 4689, 1988.

108. **Reddel, R. R., Ke, Y., Gerwin, B. I., McMenemin, M. G., Lechner, J. F., Su, R. T., Brash, D. E., Park, J.-B., Rhim, J. S., and Harris, C. C.**, Transformation of human bronchial epithelial cells by infection with SV40 or adenovirus-12SV40 hybrid virus, or transfection via strontium phosphate coprecipitation with a plasmid containing SV40 early region genes, *Cancer Res.*, 48, 1904, 1988.

109. **Fujii, D., Shalloway, D., and Verma, I. M.**, Gene regulation by tyrosine kinases: *src* protein activates various promoters, including c-*fos*, *Mol. Cell. Biol.*, 9, 2493, 1989.

110. **Seshadri, T. and Campesi, J.**, Repression of c-*fos* transcription and an altered genetic program in senescent human fibroblasts, *Science*, 247, 205, 1990.

111. **Mitchell, P. J., Wang, C., and Tijan, R.**, Positive and negative regulation of transcription *in vitro*: enhancer-binding protein AP-2 is inhibited by SV40 T antigen, *Cell*, 50, 847, 1987.

112. **Gannon, J. V. and Lane, D. P.**, p53 and DNA polymerase-α compete for binding to SV40 T antigen, *Nature*, 329, 456, 1987.

113. **DeCaprio, J. A., Ludlow, J. W., Figge, J., Shew, J. W., Huang, C. M., Lee, W.-H., Marsilo, E., Paucha, E., and Livingstone, D. M.**, SV40 large tumor antigen forms a specific complex with the product of the retinoblastoma susceptibility gene, *Cell*, 54, 275, 1988.

114. **Dyson, N., Buchkovich, K., Whyte, P., and Harlow, E.**, The cellular 107K protein that binds to adenovirus E1A also associates with the large T antigens of SV40 and JC virus, *Cell*, 58, 249, 1989.

115. **Ewen, M. E., Ludlow, J. W., Marsilio, E., Decaprio, J. A., Millikan, R. C., Cheng, S. H., Paucha, E., and Livingston, D. M.**, An N-terminal transformation-governing sequence of SV40 large T antigen contributes to the binding of both p110^{RB} and a second cellular protein, p120, *Cell*, 58, 257, 1989.

116. **Green, M. R.**, When the products of oncogenes and anti-oncogenes meet, *Cell*, 56, 1, 1989.

117. **Finlay, C. A., Hinds, P. W., and Levine, A. J.**, The p53 proto-oncogene can act as a suppressor of transformation, *Cell*, 57, 1083, 1989.

118. **Cooper, C. A. and Whyte, P.**, RB and the cell cycle, *Cell*, 58, 1009, 1989.

119. **Lee, D.**, personal communication.

Chapter 6

FACTORS INFLUENCING GROWTH AND DIFFERENTIATION OF NORMAL AND TRANSFORMED HUMAN MAMMARY EPITHELIAL CELLS IN CULTURE

Martha R. Stampfer and Paul Yaswen

TABLE OF CONTENTS

I. INTRODUCTION

The mammary gland is a very relevant organ system for carcinogenesis studies, since approximately one out of every nine women in the U.S. will develop breast cancer in her lifetime. The reasons for this high incidence of breast cancer are still unknown, necessitating a thorough analysis of the normal and abnormal processes of growth and differentiation which occur in this organ. The use of the human mammary gland as a model system for human carcinogenesis is facilitated by the abundant quantities of both normal and diseased tissues which are readily available as discard material from surgical procedures. These procedures include reduction mammoplasties, which supply tissue with normal epithelial cells from women of all ages; mastectomies, which supply primary tumor and nontumor tissues from the same individual; effusions and secondary tumor sites, which supply metastatic cells; and gynecomastias, which supply benign male breast tissue. Normal mammary epithelial cells can also be obtained from lactational fluids. These cells and tissues provide valuable resources for *in vitro* studies on the factors which contribute to malignant progression.

Elucidating the interrelationship between differentiation and carcinogenesis is an important step in improving our understanding of malignant progression in epithelial cells. The mammary gland provides advantages and disadvantages for studies on this relationship. On the one hand, the variety of physiologic states of the mammary gland *in vivo*, i.e., normal cycling, pregnant, lactating, involuting, and postmenopause, each with distinct specialized functions, potentially provides many different markers of differentiation. On the other hand, the complexity inherent in such a range of phenotypes makes the mammary gland a difficult organ for delineating pathways of differentiation.

Our laboratory has developed a culture system utilizing human mammary epithelial cells (HMEC) in order to facilitate studies on the normal mechanisms controlling growth and differentiation in these cells, and to understand how these normal processes may become altered as a result of immortal and malignant transformation. Since cell culture systems provide the only means for systematic experimentation on cells of human origin, one of our goals has been to optimize the usefulness of this *in vitro* model system. We have approached this goal by (1) characterization of the cells grown in culture, particularly with reference to their relationship to cell types which exist *in vivo*, and (2) modifications of the culture system so the cells may better reflect the phenotypes and patterns of growth found *in vivo*. Underlying this work has been the assumption that carcinogenesis involves aberrations in the normal pathways of proliferation and differentiation, and that development of optimized culture systems to examine the behavior of normal HMEC will aid in our understanding of the mechanisms of carcinogenesis.

For the purposes of our characterization of HMEC in culture, we have defined two different kinds of differentiation. The first type, functional differentiation, refers to those properties of the mammary gland associated with its role in milk production. These include the capacity for rapid proliferation in response to specific hormonal stimuli during the first half of pregnancy, the preparation of the gland for milk production in the second half of pregnancy, the synthesis and secretion of a variety of milk products during pregnancy (e.g., caseins, α-lactalbumin, and medium-chain fatty acids), and the remodeling of the gland, including protease activity and reduction in epithelial content, during involution. Since human mammary tissues, unlike tissues from animal model systems, cannot readily be obtained in these functionally differentiated states, analysis of these properties in culture is extremely difficult. Maintenance of functional differentiation is difficult even in rodent mammary cultures obtained from pregnant and lactating glands; it has only recently been demonstrated in rodent tissues obtained from virgin animals.[1] However, mammary carcinoma cells do not exhibit functional differentiation. Thus, the absence of functionally differentiated cells in culture does not necessarily limit the usefulness of cultured HMEC for studies of carcinogenesis. A relationship that has been observed between functional differentiation and carcinogenesis is that women who have had a full-term pregnancy at a young age have a reduced incidence of breast cancer. This suggests that changes which occur during parity may influence the subsequent capacity of the mammary epithelial cells to become malignant. Since normal reduction mammoplasty tissues are available from women of differing ages and parity histories, experimental examination of this relationship can be made *in vitro*.

The second type of differentiation we have termed "maturation". This refers to the developmental history of a cell from a proliferative stem cell population to a cell with diminished reproductive capacity to a "terminally differentiated" cell no longer capable of division. The actual lineage of human mammary epithelial cells *in vivo* has not been fully defined. The mammary gland consists of pseudostratified epithelia, with a basal layer resting upon a basement membrane and an apical layer facing the lumen of the ducts and alveoli. The basal layer of cells does not contact the lumen, whereas the apical layer may contact the basement membrane as well as the lumen. Apical cells display a polarized morphology, with microvilli at the luminal side. The myoepithelial cells, which contain muscle-like myofilaments and which contract upon appropriate hormonal stimuli to cause expulsion of milk, lie in the basal layer of cells. Based upon examination of keratin expression and other marker antigens, it has been proposed for the rodent mammary gland that a stem cell population capable of differentiating into both myoepithelial cells and the apical glandular epithelial cells also resides in the basal cell layer.[2]

Functional differentiation and maturation are two separate, though not necessarily independent processes in the mammary gland. Thus, a woman's epithelial cells may undergo lineage development from stem cell to a nonproliferative fully mature cell without every becoming functionally differentiated, i.e., without ever undergoing the changes of pregnancy and lactation. In this sense, the mammary gland is unlike most of the other epithelial organs.

The relationship of functional differentiation to stages of maturation is not known, i.e., what cell type responds to the hormonal stimulus to proliferate during pregnancy, or what maturation stage responds to the hormonal stimuli to cease proliferation and start synthesizing milk products. These questions are of particular interest because, as will be described below, the phenotype of the cancer cells found *in vivo,* as well as that of most breast tumor cell lines *in vitro,* most closely resembles the phenotype of the normal mature apical cell *in vivo.*

Maturation, or terminal differentiation, may also follow a biological pathway distinct from cellular senescence observed *in vitro.* Normal human fibroblasts and epithelial cells in culture display a limited, fixed number of population doublings which varies with cell type and culture conditions. For example, normal HMEC grown in the serum-free medium MCDB 170 will undergo 45 to 80 population doublings, depending upon the individual specimen donor, and then show no net increase in cell number. These nondividing cells may maintain viability for months in culture. It is likely that the controls which limit the number of times a given cell may complete the cell cycle are distinct from those which lead to a mature, nondividing, and ultimately nonviable phenotype. Thus, a cell may senesce in culture without ever exhibiting the phenotype of the most mature or functionally differentiated cell type in its lineage.

In order to characterize the cell types grown *in vitro,* and to compare them to cell lineages *in vivo,* we have examined them for expression of potential markers of mammary epithelial cell maturation and differentiation. These include intermediate filaments (keratins and vimentin), the large polymorphic epithelial mucins, extracellular matrix-associated proteins (fibronectin, collagen, laminin, proteases, and protease inhibitors), and milk products (caseins and α-lactalbumin). In order to examine factors which control normal HMEC proliferation, we have measured the effects of a variety of potential growth stimulators and growth inhibitors on the growth and differentiation of cells which display long-term growth in a serum-free medium. In order to compare the properties of normal and transformed HMEC, we have utilized chemical carcinogens and oncogenes to transform normal HMEC from reduction mammoplasty tissue to cell lines displaying indefinite lifespan, reduced growth factor requirements, and tumor formation in nude mice.

II. GROWTH AND CHARACTERIZATION OF CELL TYPES IN CULTURE

A. ISOLATION AND GROWTH OF NORMAL HMEC
Our laboratory, in collaboration with other groups, has developed culture systems which support the long-term growth of HMEC derived from reduction mammoplasty tissues.[3-5] Surgical discard material is crudely dissected to separate glandular from fatty and stromal tissue. This material is then digested with collagenase and hyaluronidase to yield epithelial cell clusters (termed

organoids) free of stroma. The epithelial organoids are separated from single cells and small clumps via filtration, and both the epithelium and the filtrate (which preferentially contains mesenchymal cells) can be stored frozen in liquid nitrogen. A single-reduction mammoplasty can easily yield 10 to 60 frozen ampules containing 0.1 ml each of concentrated epithelial organoids, permitting multiple experiments utilizing cells from the same individual.

We have used two main types of medium to support growth of the HMEC — a serum-containing medium, designated MM,[6] and a scrum-frcc mcdium, designated MCDB 170.[4] Both media contain a variety of growth factors, including insulin, hydrocortisone, EGF, and a cAMP stimulator. MM contains 0.5 fresh fetal bovine serum and 30% conditioned media from other human epithelial cell lines; MCDB 170 contains 70 μg/ml bovine pituitary extract (BPE). We have also used formulations of MM that lack the conditioned media (designated MM4).

Organoids placed in primary culture show initial cell migration followed by rapid division from the edges of the outgrowth. In MM, there is active epithelial division for three to five passages at 1:10 dilutions. The cells then acquire a mixed morphology, with larger, flatter, nondividing cells mixed with smaller cells growing with a cobblestone morphology. Some cobblestone-appearing cells may maintain growth for an additional two to five passages, particularly in MM4, with overall slower growth of the cultures and the continued presence of nondividing, flatter cells. In MCDB 170, there is initial active cell division for two to three passages of cobblestone-appearing cells. These cells gradually change morphology, becoming larger, flatter, striated, with irregular edges, and reduced proliferative capacity. As these larger cells cease growth and die, a small number of cells with the cobblestone morphology maintain proliferative capacity. These smaller cells soon dominate the culture, and continue growing with a fairly uniform cobblestone appearance for an additional 7 to 24 passages, depending upon the individual reduction mammoplasty specimen. At senescence, the cells maintain the smooth-edged cobblestone appearance, but become larger and more vacuolated. We have referred to this process, whereby only a small fraction of the cells grown in MCDB 170 display long-term growth potential, as self-selection. Self-selection can also be observed in primary cultures which are subjected to repeated partial trypsinization, a process wherein approximately 50% of the cells are removed and the remaining cells allowed to regrow. After about ten partial trypsinizations, most of the cells remaining in the dish display the flat, striated, nondividing morphology. However, nearly every organoid patch also gives rise to areas of the growing cobblestone cells, indicating a widespread distribution of the cell type with long-term growth potential.

Most of our current studies on normal HMEC biology utilize the post-selection cells which display long-term growth in MCDB 170. These cells have doubling times of 18 to 24 h, and will grow clonally with 15 to 50% colony-forming efficiency. Large batches of post-selection cells can be stored frozen, permitting repetition of experiments with cells from the same frozen batch, as well as from the same individual. We have carefully followed the growth of each cell batch until senescence,[7] so that the remaining reproductive potential of a given cell batch is known. All reduction mammoplasty-derived cells thus far examined have shown a normal karyotype;[8,9] however, we have not examined cells near senescence.

B. *IN VITRO* TRANSFORMATION OF HMEC

Normal HMEC from specimen 184 have been transformed to immortality following exposure to the chemical carcinogen benzo(a)pyrene (BaP).[7,10] Primary cultures were grown in MM and exposed two to three times to 2 μg/ml BaP. Selection for transformed cells was based on the ability of BaP-treated cells to continue growing past the time that the control cells senesced. Treated cultures typically contained cells with an extended life-span compared to controls. These extended life cultures were very heterogeneous with respect to morphology and growth potential. Often, they represented the outgrowths of individual patches or colonies. However, almost all of these extended-life cells eventually ceased growth. In only two instances have we observed escape from senescence, leading to cell lines with indefinite life-span. The two resulting cell lines, 184A1 and 184B5, each show specific clonal karyotypic aberrations, indicating their independent origins from single cells.[9] Some of the karyotypic abnormalities found in 184B5, e.g., 1q22 breaks and tetrasomy for 1q, are also frequently observed in cells obtained from breast tumors.[11] Upon continued passage in culture, these two lines show some genetic drift, but it is relatively minimal compared to that observed in most human breast tumor cell lines. Thus, the vast majority of the cell population would be expected to remain karyotypically stable when studied over the course of a few passages in culture, yet the presence of some genetic drift could give rise to rare variants in the cell population. Although 184A1 and 184B5 are immortally transformed, they do not have properties associated with malignant transformation. They do not form tumors in nude mice and they show very little or no capacity for anchorage-independent growth (AIG).

Malignant derivatives of 184A1 and 184B5 have been obtained with the use of oncogene-containing retroviral vectors and viruses. In the case of 184A1, A1N4, a clonal derivative with reduced nutritional requirements (i.e., growth in MM4), was exposed to the genes for SV40 large T-antigen, v-H-*ras*, and v-*mos*, singly and in combination.[12] The combination of H-*ras* and SV40-T led to cells (designated A1N4-TH) which formed progressively growing tumors in nude mice and showed AIG. Exposure to v-H-*ras* or v-*mos* alone led to cells that produced tumors with reduced frequency and longer latency. SV40-T alone did not yield tumorigenic cells, but did affect the growth factor requirements for anchorage-dependent and independent growth.[13] In all cases of oncogene exposure, the resultant cells were capable of proliferation in media that did not support the growth of the parental A1N4 cells. The karyotype of the A1N4 cells is aneuploid, near triploid, with one additional clonal chromosomal aberration beyond the three present in the parental

184A1 cells. A1N4-TH has a near-tetraploid karyotype, which is missing the A1N4 chromosomal marker and contains only one additional clonal chromosomal aberration relative to 184A1. Thus, the malignantly transformed cell line, containing, v-H-*ras*, does not show an unstable karyotype in terms of chromosomal aberrations.

The 184B5 cell line has been exposed to v-K-*ras* (cell line designated B5-K). In 184B5, the k-*ras* gene alone was capable of producing cells which were 100% tumorigenic in nude mice, with short latency. However, these tumors did not grow beyond approximately 5 cm in diameter.[7] Most of our studies have utilized the culture designated B5KTu, a cell line originating from a tumor resected from a nude mouse and placed in culture. B5-K and B5KTu do not display AIG.

We have also conducted a series of experiments to attempt to obtain malignantly transformed derivatives of 184A1 and 184B5 following additional exposure to chemical carcinogens. The design of these experiments was based on the observation that the oncogene derivatives of A1N4 showed reduced nutritional requirements, as well as on the existing literature on the reduced nutritional requirements of many transformed cells. To perform these experiments, we first determined the requirements of 184, 184A1, and 184B5 for the various growth factors present in MCDB 170 for short-term culture (Table 1) and for continued passage in culture. In the short-term experiments, 184A1 and 184B5 showed a few differences from each other and normal HMEC in their growth requirements. Both were more dependent upon EGF for growth in mass culture than the normal cells. 184A1 showed little effect upon removal of hydrocortisone (HC). In the long-term experiments, removal of HC or BPE from mass cultures of normal HMEC led to cessation of growth over the course of one to three passages. Removal of insulin (I) did not prevent continued proliferation, but led to slower growth, a less healthy appearing culture, and earlier senescence. Removal of I from 184A1 and 184B5 also did not prevent continued growth. There was an initial reduction in growth rate, but the cell populations resumed active growth within two passages. However, removal of BPE or EGF led to cessation of growth for the vast majority of 184A1 and 184B5 cells. Nonetheless, a few cells could be observed which maintained growth without addition of BPE or EGF. The growth rates of these variants slowly increased upon continued subculture for four to six passages, leading to selected subpopulations which maintained active growth in the absence of BPE or EGF.

We next examined the effect of removal of multiple growth factors, and were able to define conditions which did not support the continued growth of 184A1 or 184B5, e.g., removal of I and EGF, I and BPE, or EGF and BPE for 184A1, and removal of I and EGF, or I and BPE for 184B5. Populations of 184A1 and 184B5 were then tested for colony-forming ability in the presence of the direct-acting carcinogen, *N*-nitroso-ethyl-urea (ENU). Concentrations of ENU that yielded 80% inhibition were chosen for further experiments, i.e., 1500 μg/ml for 184A1 and 750 μg/ml for 184B5. Two

TABLE 1
Growth Factor Requirement of Normal and Transformed HMEC in MCDB 170

| | Percentage of control cell grwoth | | | | | |
| | 184 | | 184A1 | | 184B5 | |
Medium	MC	CFE	MC	CFE	MC	CFE
Complete MCDB 170 + IP	100	100	100	100	100	100
minus I	49	47	11	18	26	73
minus HC	36	32	84	88	18	61
minus EGF	86	2	20	0	12	0
minus BPE	15	21	21	24	16	75

Note: Abbreviations used: I, insulin; HC, hydrocortisone; EGF, epidermal growth factor; BPE, bovine pituitary extract; IP, isoproterenol; MC, mass culture growth; CFE, colony-forming efficiency. Cells from specimen 184 (passage 11), and cell lines 184A1 and 184B5 (passages 17 to 20) were grown in complete MCDB 170 with isoproterenol. For mass culture, cells were subcultured into duplicate 35-mm dishes (5 × 10⁴ per dish) in the indicated media. When control cultures were subconfluent or just confluent, all the cultures were trypsinized and the cells counted by hemocytometer. For clonal cultures, single cells (100 to 1000) were seeded into triplicate 100-mm dishes. After 10 to 14 d, cells were stained with Giemsa and colonies greater than 30 cells counted.

T-75 flasks each of treated and control cells were exposed to ENU or solvent alone for two or three consecutive passages. The resulting cell populations were then tested for their ability to grow in the restrictive media and for AIG. Under some conditions, the ENU-treated cells were capable of sustained growth, whereas the untreated cell lines quickly ceased growth. The resulting growth factor-independent variants, such as A1ZNEB, which does not require either EGF or BPE, and B5ZNEI, which does not require EGF or I, may represent a further step in malignant progression. However, none of the variants showed AIG, nor did they form tumors in nude mice. Thus, we have not been able to derive cells that showed tumorigenic properties following use of chemical carcinogens alone.

C. CHARACTERIZATION OF NORMAL HMEC GROWN IN CULTURE

In order to relate the HMEC which maintain growth *in vitro* to the different cell types identified *in vivo,* we have examined the normal and transformed cells for phenotypes which have been characterized using sectioned human breast tissues. A variety of studies have defined properties which can be used to distinguish basal vs. luminal breast cells, and which change during the course of lactation.[14-16] In general, mammary basal cells, similar to basal cells in stratified tissues such as the skin, express keratins 5 and 14. Some reports have indicated that a subpopulation expresses the common mesenchymal intermediate filament, vimentin.[15,17] Expression of α-actin has also been localized to the basal cell layer. Luminal cells express the keratins 8 and 18 found in simple epithelia such as the lung. Keratin 19 shows variable expression in the luminal cells. The keratin 19-positive cells display low proliferative potential in culture, suggesting that they may represent the least proliferative, or most mature luminal cell type *in vivo*.[18,19] Expression of specific epitopes of a polymorphic epithelial mucin (PEM) has been localized to luminal cells *in vivo*. Cells in the resting gland are weakly mucin positive, whereas cells from lactating glands may express higher levels of specific mucin epitopes.[14] High expression of specific PEM epitopes has also been correlated with a low proliferative potential in normal HMEC *in vitro*.[20] Only a small fraction of normal mammary epithelial cells *in vivo* are estrogen receptor positive, and this positive population is preferentially localized in the nonbasal layer.[21,22] It is not clear whether the mammary gland contains cells which are terminally differentiated such as those in the most mature layers of stratified epithelium, since even keratin 19, mucin-positive cells can show a limited capacity for cell division *in vitro*.

In collaboration with others,[16] we have examined the HMEC grown under our culture conditions for expression of certain phenotypic markers by Northern blot and by immunohistochemical analysis. Figure 1 shows some of the data on antigen expression; Figures 2 and 3 include some of the data on mRNA expression. Primary cultures of normal HMEC grown in MCDB 170 and early-passage cultures grown in MM are heterogeneous. Some cells have the basal phenotype — keratin 5/14 positive, mucin negative, and α-actin positive; other cells show the luminal phenotype — keratin 5/14 negative, keratin 8/18/19 positive, mucin positive (Figure 1). The cells which initially proliferate in the serum-free MCDB 170 medium are those with the basal phenotype. However, post-selection cells begin to express some properties associated with the luminal cell type, i.e., keratins 8 and 18 and some mucin epitopes. Expression of these luminal properties increases with continued passage in culture, such that the senescent cells uniformly express these markers. At the same time, expression of the basal keratins 5 and 14 is not lost. We have not been able to detect keratin 19 in any postselection population. Vimentin expression is found in both the pre- and postselection cells which proliferate in MCDB 170.

The above results led us to propose that the cells which display long-term growth in the serum-free medium represent a multipotent stem cell population initially present in the basal layer of the gland. With increasing time in culture, these cells show a partial differentiation toward the luminal phenotype. Based on phenotypic expression of keratins and other markers, a multipotent basal, stem cell population has also been proposed for the rodent mammary gland.[23] However, we cannot be sure that culture conditions have not induced some artifactual phenotypic expression. In particular, growth of cells on impermeable plastic substrates prevents the normal cell-extracellular matrix contacts and precludes the normal development of cellular polarity. The lack of normal cell polarity may in turn affect the cells' phenotypic expression in culture. Cells which are positive for keratins 5, 14, 8, and 18

FIGURE 1. Keratin staining of normal HMEC. HMEC specimen 184 was grown in MM as described,[7] and second-passage cultures stained by indirect immunoperoxidase assay,[16] using anti-keratin antibodies monospecific for (A) keratin 14, (B) keratin 18, and (C) keratin 19. The results demonstrate the heterogeneous populations present in MM medium. Cells growing in tight patches show the luminal phenotype of keratin 14 negative, keratins 18 and 19 positive, whereas the more loosely growing cells show the basal phenotype of keratin 14 positive, keratins 18 and 19 negative.

are not commonly observed *in vivo,* although coexpression of keratin 8 and vimentin has been reported.[17]

We consider it important to continue to characterize the populations growing *in vitro* and to optimize the culture conditions to permit better approximation of *in vivo*-like conditions. However, we also believe that the HMEC cultured under the current conditions are useful for certain studies in molecular biology, biochemistry, and carcinogenesis that require large number of cells and may not require stringent fidelity to the *in vivo* situation. The HMEC grown in MCDB 170 provide large standardized pools of normal human epithelial cells for such experiments.

D. CHARACTERIZATION OF HMEC TRANSFORMED *IN VITRO*

When cells from breast tumor tissues or tumor cell lines are examined, they display the maturation phenotype found in the mature luminal cell population. They rarely express keratins 5 and 14, and nearly uniformly express keratins 8, 18, and 19.[16] Most tumor cells have high levels of expression of PEM, including the epitopes found in the differentiated lactating cells.[14] Around 70% of breast tumor tissues also display high levels of the estrogen receptor. However, vimentin expression, normally confined to basal cells *in vivo,* has been observed in a subset of estrogen receptor-negative breast tumor cell lines and tissues.[24] Thus, breast tumor cells *in vivo* and *in vitro* display a phenotype which, in normal HMEC, is associated with low proliferative potential *in vitro.*

Some of our cell lines which have been transformed *in vitro* to immortality and tumorigenicity have been examined for their expression of differentiation markers, including intermediate filaments, mucins, the extracellular matrix protein fibronectin, and a newly isolated gene designated NB-1 (see below) (Figures 2 and 3). The immortalized cell lines 184A1 and 184B5 maintain some expression of keratins 5 and 14, but Northern blot analysis shows that the level of keratin 5 mRNA is decreased in 184B5, and even further decreased in 184A1, while expression of keratin 18 mRNA is increased relative to normal HMEC. These lines also differ from the normal cells in their barely detectable levels of vimentin mRNA. 184B5 expresses the luminal PEM antigens, including one epitope, recognized by the antibody HMFG-2, which is found in tumor cells (data not shown). The tumorigenic transformants A1N4-TH and B5KTu have very low levels of keratin 5 and increased levels of keratin 18 mRNA. While B5KTu remains vimentin negative, the A1N4-TH cells show reexpression of vimentin. We have not been able to detect keratin 19 mRNA in any of these lines. These results suggest that the transformed cells, particularly 184B5, have a phenotype which is closer to the luminal phenotype than that seen in the normal HMEC, but do not fully resemble breast tumor cells.

Fibronectin represents one of the major proteins secreted by normal HMEC in culture.[25] In many transformed cells, the level of fibronectin mRNA and protein synthesis is decreased. Our transformed cell model system is similar

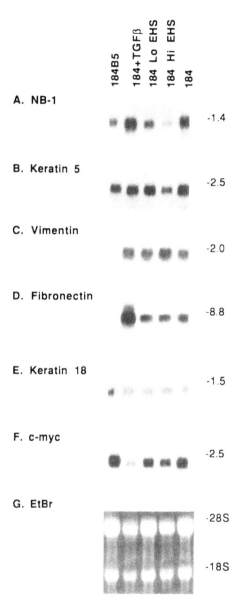

FIGURE 2. Modulation of mRNA expression in normal human mammary epithelial cells by EHS extracellular matrix and by TGF-β. (A) Normal mammary epithelial strain 184 (passage 14) was plated at 5.0 × 10⁶ (Hi EHS) and 1.0 × 10⁶ (Lo EHS) cells per 100-mm dish on EHS-derived extracellular matrix material. Control (184 and 184B5 passage 26) and TGF-β-treated cultures were plated at 0.5 × 10⁶ cells per 100-mm dish directly on tissue culture-grade plastic dishes. The cells were harvested for RNA at 96 h (EHS) or at subconfluence (control and TGF-β). 10 μg of total cellular RNA was fractionated on 1.3% agarose/formaldehyde gels and transferred to nylon filters as described.[26] The filters were sequentially probed with cDNA to (A) NB-1, (B) keratin 5, (C) vimentin, (D) fibronectin, (E) keratin 18, and (F) c-*myc*. (G) shows total RNA in the original gel stained with ethidium bromide.

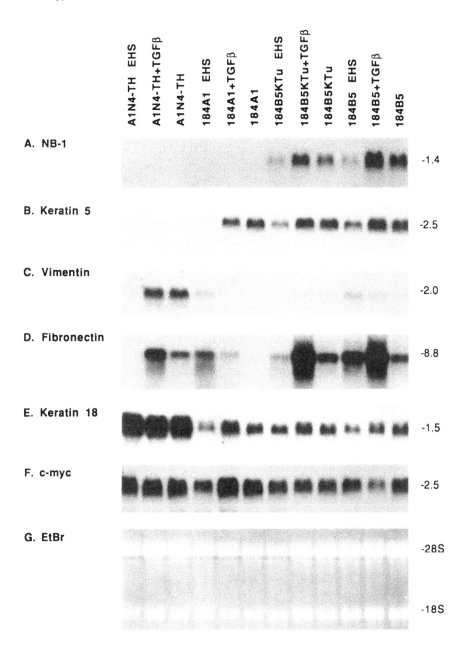

FIGURE 3. Modulation of mRNA expression in transformed HMEC lines by EHS extracellular matrix and by TGF-β. Cells (184B5 passage 26, B5KTu passage 18, 184A1 passage 47, and A1N4-TH) were plated and examined as described in Figure 2.

in that expression of fibronectin is reduced in the transformed cells (Figure 3, unpublished results). Nonetheless, even where fibronectin synthesis is low or barely detectable, as in the case of the 184A1 cell line, its expression can be upregulated by exposure of the cells to TGFβ.

The normal and transformed HMEC have also been characterized with respect to both their growth patterns and their gene expression when placed on reconstituted basement membrane material derived from the Englebreth-Holm-Swarm (EHS) murine sarcoma. Growth on EHS has previously been shown to support increased differentiated functions of a variety of cell types. Normal HMEC are capable of forming three-dimensional structures with striking resemblance to endbuds in intact mammary gland tissue (Figure 4, A and B) whereas 184A1 displays only less developed structures and 184B5 forms only small clusters (Figure 4, C and D). The A1N4-TH cells show even less structure formation than 184A1 (Figure 4, E), whereas the B5KTu cells resemble 184B5 (data not shown). Although we do not know the underlying basis for these differences in growth patterns, transformation appears to be correlated with a loss of the capacity to form glandular-like three-dimensional structures on EHS. We have examined several genes to determine if their expression is affected by culture on EHS (Figures 2 and 3). The most consistent effect thus far observed is a downregulation of keratin 5 and NB-1 mRNA.

Another approach we have taken to characterize differences between our normal and transformed HMEC cultures has been to identify genes which are expressed in the normal HMEC, but which are downregulated in the immortal and malignantly transformed cells. Toward this end, selected normal HMEC cDNAs were identified and cloned using probes enriched by subtractive hybridization between the normal 184 cell cDNA and both the B5KTu and the 184B5 cell mRNA.[26] Several genes preferentially expressed in normal 184 cells were isolated by this method, including those for fibronectin, keratin 5, and vimentin. Additionally, one 350-bp cDNA fragment was isolated which initially showed no similarity to any sequence reported in GenBank. This cDNA hybridized specifically to a 1.4-kb mRNA, designated NB-1, which was expressed in the normal HMEC, but was downregulated or undetectable in the transformed cell lines (Figures 2 and 3). Sequence analysis of a full-length NB-1 clone revealed a 447-bp open reading frame with extensive similarity (70, 71, and 80%) at the nucleic acid level to the three known human genes coding for the ubiquitous calcium binding protein, calmodulin. The similarity between the translated amino acid sequence of NB-1 and human calmodulin was 85% over the length of the entire protein.

Using Northern and PCR analysis, NB-1 mRNA has been thus far found only in normal epithelial cells and tissues from human breast, prostate, cervix, and skin. It has not been found in normal epithelial cells other than those from stratified or pseudostratified tissues, nor is it detectable in nonepithelial cells and tissues, or in epithelial tumor cell lines. Human breast cells obtained from lactational fluids were also negative for NB-1 expression by PCR analysis.

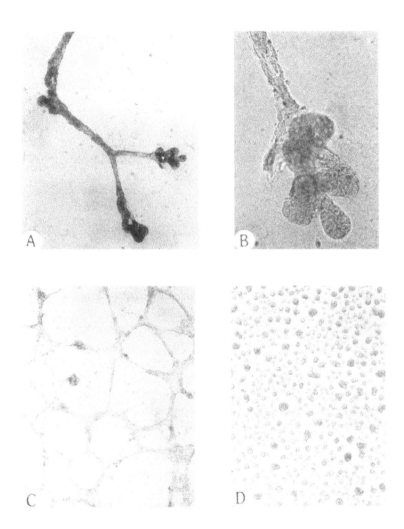

FIGURE 4. Normal and transformed HMEC grown on EHS matrix. Normal human mammary epithelial cell strain 184 (passage 14) was plated at high density (5 × 10⁶/100-mm dish) on EHS. The cells organized into structures resembling ducts and endbuds normally present *in vivo* (A and B). In contrast, 184A1 (passage 36) and 184B5 (passage 26), when plated at the same cell density on this matrix material, made poorly organized structures (C and D, respectively), while A1N4-TH was incapable of structure formation (E). Original magnifications: A, × 32; B, × 128; C, D, and E, × 32.

Although NB-1 mRNA is easily detectable by Northern analysis in total RNA from cultured normal HMEC, it is less abundant in total RNA from organoids and unprocessed reduction mammoplasty tissue. Such differences are unlikely to be due to variations in proliferative state, since expression of NB-1 mRNA is not significantly decreased when cells are growth arrested by exposure to anti-EGF receptor antibodies or in senescing cells where

FIGURE 4 (continued)

proliferation is minimal, and it is increased in cell growth arrested by TGF-β. One possible explanation is that, unlike calmodulin, NB-1 expression may be limited to a particular state of epithelial cell maturation, and thus be confined to certain subpopulations of epithelial cells *in vivo*. Since NB-1 mRNA levels are high in the postselection normal HMEC population which displays active long-term growth in MCDB 170, and which has attributes of multipotent stem cells, it is possible that expression *in vivo* may be limited to a stem cell population in the basal layer of the gland. The findings that NB-1 expression is reduced when HMEC are grown on EHS, and is absent in the nonproliferative differentiated luminal cells sloughed off into milk, are consistent with the hypothesis that NB-1 is only expressed during certain stages of epithelial differentiation.

Using full-length recombinant NB-1 protein as an immunogen, we have recently produced polyclonal antisera which can distinguish the NB-1 protein from vertebrate calmodulin.[43] Initial studies have indicated that the relative abundance of the corresponding 16-kDa protein reflects relative NB-1 mRNA levels in various cell types, being most highly expressed in normal HMEC, lower or undetectable in the immortally transformed cell lines, and virtually undetectable in tumorigenic breast and prostate cell lines as well as in normal breast fibroblasts.

The initial characterization of genomic DNA corresponding to the NB-1 transcript indicated the unexpected absence of introns. All three vertebrate calmodulin genes contain five similarly placed introns.[27] A literature search revealed the existence of a previously reported human calmodulin "pseudogene", hGH6, which shared identity with NB-1 cDNA.[28] This gene was designated a pseudogene since the authors were unable to demonstrate the existence of a corresponding mRNA. Our evidence of the expression of NB-1 at both the mRNA and protein levels suggests that NB-1 may be a rare example of an expressed retroposon.[29]

External calcium concentration has been shown to affect the proliferative potential and differentiated states of some cultured epithelial cells, including keratinocytes and mammary epithelial cells.[30-32] Loss of response to the calcium-induced differentiation signal has been shown to correlate with the early stages of transformation in keratinocyte cultures.[33] The downregulation of NB-1 expression observed after *in vitro* transformation of HMEC may reflect the fact that a particular state of differentiation is required for transformation, or that the transformed state is incompatible with high expression of NB-1.

III. GROWTH FACTORS CONTROLLING HMEC PROLIFERATION AND DIFFERENTIATION

A main focus of our research has been to study the effect of growth factors on normal HMEC proliferation, and to compare these data with growth control of the transformed HMEC. In particular, we have examined the effects of TGF-β and EGF/TGF-α. One long-term objective of these studies is to understand the parameters influencing cell cycle progression in normal, finite life-span human epithelial cells in the hope that this information may facilitate elucidating the aberrations which occur when cells attain immortality and malignancy. As part of this objective, we have tried to obtain conditions that would permit an efficient and readily reversible cell-cycle synchronization of normal HMEC.

Initial studies examined the effect of TGF-β on normal and transformed HMEC. We have demonstrated that normal HMEC are growth inhibited by TGF-β, with the extent of inhibition increasing as cells are subcultured *in vitro.*[34] All normal HMEC are ultimately growth arrested by TGF-β. Analysis by flow-activated cell sorting indicates that cells are arrested in the G_1 phase of the cell cycle (data not shown). This growth inhibition is at least partially reversible, although the extent of reversibility decreases with cell passage *in vitro*, and is relatively asynchronous. Normal HMEC show distinctive morphologic changes in the presence of TGF-β, characterized by an elongated, flattened appearance. HMEC which have been transformed to immortality or malignancy have altered growth responses to TGF-β. Although varying degrees of growth inhibition and morphologic changes are observed in 184A1 and 184B5, both lines contain populations that maintain active growth in the presence of TGF-β.[34] However, even though TGF-β may not inhibit their growth, the immortalized HMEC lines retain receptors for TGF-β[35] and, like the normal HMEC, express specific differentiated responses such as increased synthesis of the extracellular matrix-associated proteins fibronectin, collagen IV, and plasminogen activator inhibitor 1 (manuscript in preparation). Thus, TGF-β effects on cell growth can be dissociated from the effects on differentiated cell properties. The tumorigenic cell lines A1N4-TH and B5KTu maintain growth in the presence of TGF-β. This HMEC model system therefore resembles the situation observed with other normal and transformed epithelial cells in that loss of TGF-β-induced growth inhibition accompanies the carcinogenic progression.

While these studies using TGF-β have illustrated differences in growth control between normal and transformed HMEC, TGF-β growth inhibition does not appear to be a useful method for studying cell cycle effects in HMEC. The degree of reversal of growth inhibition is variable and the cells are not arrested in a resting state. Indeed, the pronounced changes in normal HMEC protein synthesis and secretion, which result in an increased level of protein synthesis per cell, suggest that TGF-β induces a particular state of cell differentiation. The relationship of this state to normal epithelial cell homeostasis or wound healing remains to be determined.

Studies on the effects of EGF/TGF-α on normal HMEC have indicated a stringent requirement for this growth factor class for clonal growth. However, growth in mass culture proceeds without addition of exogenous EGF due to the significant level of endogenous production of TGF-α.[36] As mentioned earlier, the transformed cell lines may display a progressive loss of this EGF/TGF-α requirement. Since both normal and transformed cells are capable of TGF-α production and synthesis, this aspect of the autocrine loop cannot, by itself, account for growth control differences between normal and tumor cells. It is possible that other changes, such as production of other EGF/TGF-α-related ligands, changes in the intracellular signal transduction pathway, or alterations in the cells' normally polarized pathways for secretion, may be responsible for the altered behavior of mammary tumor cells.

Addition of monoclonal antibody 225 IgG to the EGF receptor (MAb 225) prevents HMEC growth.[37] Recent experiments (manuscript in preparation) have shown that MAb 225 produces a rapid, efficient, and reversible growth arrest in an early G_1 phase of the cell cycle. Protein synthesis remains depressed in the presence of the antibody, and DNA synthesis is sharply decreased by 24 h. Removal of MAb 225 leads to a rapid increase in protein synthesis. DNA synthesis increases only after 10 h and peaks around 18 h. A 1-h exposure to EGF after MAb 225 removal is sufficient to allow the majority of cells capable of cycling to subsequently enter the S-phase. High levels of synthesis of mRNA for the early response genes c-*myc*, c-*fos*, and c-*jun* are observed within 1 h of antibody removal. Synthesis of TGF-α mRNA, which is inhibited in the presence of MAb 225, is detected by 2 h after antibody removal. It thus appears that blockage of EGF receptor signal transduction is sufficient by itself to cause normal HMEC to enter a Go-like resting state similar to the Go state described in fibroblasts. Future studies will now be able to address possible differences between normal HMEC of finite life-span and the immortally transformed HMEC cell lines with respect to their response to MAb 225 and their cell cycle controls.

IV. DISCUSSION

A. VALUES AND LIMITATIONS OF HUMAN MAMMARY CELL CULTURE SYSTEMS

Although much valuable information on the carcinogenic progress has been obtained using animal model systems, ultimately it is necessary to ascertain the nature of these processes in the cells of greatest interest, i.e., human cells. While the rodent model systems commonly employed in mammary cancer research may clarify fruitful areas of investigation, the existence of known differences between humans and rodents in mammary physiology, in response to etiologic agents, and in properties of transformation emphasizes the uncertainty in transfer of information gained in model systems to the human situation. Direct study of human cells is the only way to determine the parameters of normal and aberrant human mammary cell biology. Yet, experimentation with human cells entails problems not encountered with rodent model systems. *In vivo* experimentation is not possible, and, in the case of the mammary gland, it is nearly impossible to obtain abundant quantities of cells in functionally differentiated states.

The advantages and limitations of human cell experimentation underscore the importance of developing human culture systems that can reflect, as closely as possible, the *in vivo* situation. Obviously, this goal represents an ideal; the achievable reality is to develop ever closer approximations. Our laboratory has been working to develop an "approximate" culture system for normal HMEC biology as well as a model system for the carcinogenic process. At this stage, the normal HMEC system can provide large, standardized quantities of actively proliferating cells from individual specimen donors. The short doubling time, the large number of doublings possible before senescence, the serum-free medium, the clonal growth capability, and the ease of large-scale growth offer significant advantages for certain kinds of studies, e.g., molecular and biochemical analyses requiring large, uniform, proliferating cell populations. On the other hand, these cell populations represent only a limited range of the spectrum of mammary epithelial cell types found *in vivo*. To observe a full range of normal HMEC behavior, different culture condition are required. For example, early-passage cells, cells from lactational fluids, cells grown in different medium (varying in serum and growth factor content, calcium concentration, and presence of differentiation inducing agents), cells grown on substrates other than plastic (on or in collagen gels, extracellular matrix material, permeable membranes, and hollow fibers) will display varying phenotypes. The extensive studies with rodent mammary epithelial cells[1,38] strongly suggest that these differing conditions will prove necessary in order to obtain the variety of cell lineages and differentiated phenotypes observed *in vivo*. The crucial importance of achieving truly representative human cell culture systems will hopefully provide continued momentum for studies on optimizing these culture systems.

The model system for human mammary carcinogenesis which we have developed likewise has its values and limitations. We have available cells from one individual which display a progression of changes which correlate with changes observed during carcinogenic progression — extended life-span, immortality, growth factor independence, and tumorigenicity. The immortal cell populations retain most of the growth factor requirements of the normal HMEC, display a more luminal phenotype than the normal HMEC grown in

MCDB 170, and show minimal genetic instability. While any immortalized cell line cannot be considered to represent normal cells, lines of indefinite life-span are in many instances more convenient to use than finite life-span cells. The retention by 184A1 and 184B5 of many normal characteristics makes them useful substrates for some areas of experimentation in normal cell physiology. On the other hand, the fact that they have acquired some aberrant properties relative to normal HMEC, especially their indefinite life-span, makes them useful substrates for determining the potential capacity of additional factors (e.g., chemical and physical carcinogens, oncogenic viruses, and transfected genes) to induce malignant transformation.

Cell lines immortalized by chemical carcinogens, like the rare cell lines which have immortalized spontaneously,[39,40] may be more appropriate for some uses than lines which have been immortalized by the use of specific viral oncogenes which are not associated with mammary transformation *in vivo* or which commonly result in lines with gross genetic instability. However, the rarity of spontaneous and chemically induced transformation of human epithelial cells has meant that few immortal cell lines, reflecting a limited range of phenotypes, currently exist. Immortal transformation using SV40-T or papilloma virus-transforming genes occurs more frequently, allowing one to target specific cell phenotypes, such as in the recent report of the immortal transformation of milk-derived cells expressing a mature luminal phenotype.[41] An unanswered question in the use of immortalized lines in a model of tumor progression relates to whether or not immortality is truly a requirement for malignancy. While it is clear that only cells from malignant tissues reproducibly yield immortal cell lines, this does not mean that immortalization is a necessary step in carcinogenesis. In fact, only rare cells from a small percentage of human breast tumors show indefinite life-span in culture. It is possible that the extended life-span seen when many epithelial cells are exposed to carcinogenic agents and viruses may be a more accurate reflection of the growth control derangements present in the majority of primary breast tumor cells.

The use of *ras*-containing retroviruses for induction of malignant transformation raises issues similar to the use of SV40-T and papilloma virus oncogenes for immortal transformation. None of these viruses are known etiologic agents for human breast cancer. There is, in fact, considerable data showing the absence of *ras* mutations in human breast cancer. Yet, these oncogenes have thus far provided the only consistent means of obtaining malignant transformation of human mammary epithelial cells. Our efforts to achieve malignant transformation by chemical carcinogens alone have not been successful. Possibly, future studies on the effect of genes known to be involved in breast cancer, such as c-*erb*B-2, p53, and the retinoblastoma genes, may provide insights that will enable development of more efficient and relevant malignant model systems. An alternative method for achieving a model system of malignant progression would be to obtain cell strains and immortal lines from the nontumor and tumor tissues of one patient, although cells from the nontumor tissues could not be assumed to be fully normal. A model system has been described which provides some of the steps of malignant progression through the development of cell lines from primary and metastatic tumor tissues of one individual.[42]

B. RELATIONSHIP OF TRANSFORMATION AND DIFFERENTIATION

A relationship between transformation and differentiation is suggested by the fact that cancer cells are often found to reflect specific stages in the differentiation pathway of the organ system from which they arise, and that loss of response to differentiation-inducing agents is one of the earliest observed growth control aberrations in epithelial cell transformation. In order to understand the nature of this relationship, we need to know more about the pathways of functional differentiation and of maturation in epithelial organ systems. In addition to performing organ-specific specialized functions, epithelial cells display a maturation lineage starting from a proliferative population located next to the basement membrane leading to a more mature population with little or no proliferative potential. In some organ systems, such as the epidermis, the pathway of maturation coincides with that of functional differentiation. In simple or pseudostratified epithelia, the maturation lineage may be more difficult to define since it is not delineated by obvious positional information. In these tissues, the pathways of functional differentiation and maturation do not necessarily coincide. In the mammary gland, the situation becomes even more complex because the gland is not usually in a functionally differentiated state. This variety in the physiologies of the different epithelial organ systems suggests that there may also be variety in the specifics of the relationship between transformation and differentiation among the different epithelial tissues.

In the case of the mammary cells, the somewhat surprising observation is that the tumor cells nearly uniformly express a phenotype which most closely resembles that of the normal mature luminal cell — the cell type which shows the least proliferative potential in culture. Clearly, the tumor cells have acquired some derangement in normal growth control since they readily proliferate even though displaying this "mature" phenotype. We have no definitive explanation of this phenomenon. It is possible that cells in a particular state of maturation are more susceptible to carcinogenic transformation. On the other hand, it is possible that the transformed state is either incompatible with the basal cell phenotype or requires some aspect of the mature luminal cell phenotype, resulting in changes subsequent to transformation. We are particularly interested now in examining whether the presence or absence of the NB-1 protein plays a causal role in affecting the mammary cell's capacity to transform or to express a transformed phenotype.

The explanation for the distinctive phenotype of breast tumor cells will require a more complete understanding of both the normal pathways of growth and differentiation in this cell type and how the state of differentiation affects the cells' capacity to acquire and maintain a transformed state. This information, in turn, may allow us to develop more efficient protocols for *in vitro* transformation of HMEC. For example, it may be possible to define and develop specific culture conditions which permit cells to be in a differentiated state in which immortal or malignant transformation is more likely to occur. Conversely, this information may allow definition of conditions which will interfere with the maintenance of the transformed phenotype, opening up possibilities for novel methods of clinical intervention.

REFERENCES

1. **Hahm, H. A., Ip, M. M., Darcy, K., Black, J. D., Shea, W. K., Forczek, S., Yoshimura, M., and Oka, T.,** Primary culture of normal rat mammary epithelial cells within a basement membrane matrix. II. Functional differentiation under serum-free conditions, *In Vitro Cell Dev. Biol.,* 26, 803, 1990.

2. **Dulbecco, R., Allen, R. A., Bologna, M., and Bowman, M.,** Marker evolution during development of the rat mammary gland: stem cells identified by markers and the role of myoepithelial cells, *Cancer Res.,* 46, 2449, 1986.

3. **Stampfer, M. R., Hallowes, R., and Hackett, A. J.,** Growth of normal human mammary epithelial cells in culture, *In Vitro,* 16, 415, 1980.

4. **Hammond, S. L., Ham, R. G., and Stampfer, M. R.,** Serum-free growth of human mammary epithelial cells: rapid clonal growth in defined medium and extended serial passage with pituitary extract, *Proc. Natl. Acad. Sci. U.S.A.,* 81, 5435, 1984.

5. **Stampfer, M. R.,** Isolation and growth of human mammary epithelial cells, *J. Tissue Cult. Methods,* 9, 107, 1985.

6. **Stampfer, M. R.,** Cholera toxin stimulation of human mammary epithelial cells in culture, *In Vitro,* 18, 531, 1982.

7. **Stampfer, M. R. and Bartley, J. C.,** Human mammary epithelial cells in culture: differentiation and transformation, in *Breast Cancer: Cellular and Molecular Biology,* Dickson, R. and Lippman, M., Eds., Kluwer, Boston, 1988, 1.

8. **Wolman, S. R., Smith, H. S., Stampfer, M., and Hackett, A. J.,** Growth of diploid cells from breast cancer, *Cancer Genet. Cytogen.,* 16, 49, 1985.

9. **Walen, K. and Stampfer, M. R.,** Chromosome analyses of human mammary epithelial cells at stages of chemically-induced transformation progression to immortality, *Cancer Genet. Cytogenet.,* 37, 249, 1989.

10. **Stampfer, M. R. and Bartley, J. C.,** Induction of transformation and continuous cell lines from normal human mammary epithelial cells after exposure to benzo(a)pyrene, *Proc. Natl. Acad. Sci. U.S.A.,* 82, 2394, 1985.

11. **Dutrillaux, B., Gerbault-Seureau, M., and Zafrani, B.,** Characterization of chromosomal anomalies in human breast cancer, *Cancer Genet. Cytogenet.,* 49, 203, 1990.

12. **Clark, R., Stampfer, M., Milley, B., O'Rourke, E., Walen, K., Kriegler, M., and Kopplin, J.,** Transformation of human mammary epithelial cells by oncogenic retroviruses, *Cancer Res.,* 48, 4689, 1988.

13. **Valverius, E. M., Ciardiello, F., Heldin, N., Blondel, B., Merlo, G., Smith, G., Stampfer, M. R., Lippman, M. E., Dickson, R. B., and Salomon, D. S.,** Stromal influences on transformation of human mammary epithelial cells overexpressing c-*myc* and SV40T, *J. Cell. Physiol.,* 145, 207, 1990.

14. **Taylor-Papadimitriou, J., Millis, R., Burchell, J., Nash, R., Pang, L., and Gilbert, J.,** Patterns of reaction of monoclonal antibodies HMFG-1 and -2 with benign breast tissues and breast carcinomas, *J. Exp. Pathol.,* 2, 247, 1986.

15. **Rudland, P. S. and Hughes, C. M.**, Immunocytochemical identification of cell types in human mammary gland: variations in cellular markers are dependent on glandular topography and differentiation, *J. Histochem. Cytochem.*, 37, 1087, 1989.

16. **Taylor-Papadimitriou, J., Stampfer, M. R., Bartek, J., Lane, E. B., and Lewis, A.**, Keratin expression in human mammary epithelial cells cultured from normal and malignant tissue: relation to *in vivo* phenotypes and influence of medium, *J. Cell Sci.*, 94, 403, 1989.

17. **Guelstein, V. I., Tchypysheva, T. A., Ermilova, V. D., Litvinova, L. V., Troyanovsky, S. M., and Bannikov, G. A.**, Monoclonal antibody mapping of keratins 8 and 17 and of vimentin in normal human mammary gland, benign tumors, dysplasias and breast cancer, *Int. J. Cancer*, 42, 147, 1988.

18. **Bartek, J., Taylor-Papadimitriou, J., Miller, N., and Millis, R.**, Pattern of expression of keratin 19 as detected with monoclonal antibodies to human breast tumors and tissues, *Int. J. Cancer*, 36, 299, 1985.

19. **Bartek, J., Durban, E. M., Hallowes, R. C., and Taylor-Papadimitriou, J.**, A subclass of luminal epithelial cells in the human mammary gland, defined by antibodies to cytokeratins, *J. Cell Sci.*, 75, 17, 1985.

20. **Chang, S. E. and Taylor-Papadimitriou, J.**, Modulation of phenotype in cultures of human milk epithelial cells and its relation to the expression of a membrane antigen, *Cell Differ.*, 12, 143, 1983.

21. **Petersen, O. W., Hoyer, P. E., and van Deurs, B.**, Frequency and distribution of estrogen receptor-positive cells in normal, non-lactating human breast tissue, *Cancer Res.*, 47, 5748, 1987.

22. **Ricketts, D., Turnbull, L., Ryall, G., Bakhshi, R., Rawson, N. S. B., Gazet, J.-C., Nolan, C., and Coombes, R. C.**, Estrogen and progesterone receptors in the normal human breast, *Cancer Res.*, 51, 1817, 1991.

23. **Smith, G. H., Mehrel, T., and Roop, D. R.**, Differential keratin gene expression in developing, differentiating, preneoplastic, and neoplastic mouse mammary epithelium, *Cell Growth Differ.*, 1, 161, 1990.

24. **Sommers, C. L., Walker-Jones, D., Heckford, S. E., Worland, P., Valverius, E., Clark, R., McCormick, F., Stampfer, M., Abularach, S., and Gelmann, E. P.**, Vimentin rather than keratin expression in some hormone-independent breast cancer cell lines and in oncogene-transformed mammary epithelial cells, *Cancer Res.*, 49, 4258, 1989.

25. **Stampfer, M. R., Vlodavsky, I., Smith, H. S., Ford, R., Becker, F. F., and Riggs, J.**, Fibronectin production by human mammary cells, *J. Natl. Cancer Inst.*, 67, 253, 1981.

26. **Yaswen, P., Smoll, A., Peehl, D. M., Trask, D. K., Sager, R., and Stampfer, M. R.**, Down-regulation of a calmodulin-related gene during transformation of human mammary epithelial cells, *Proc. Natl. Acad. Sci. U.S.A.*, 87, 7360, 1990.

27. **Koller, M., Schnyder, B., and Strehler, E. E.**, Structural organization of the human CaMIII calmodulin gene, *Biochim. Biophys. Acta*, 1087, 180, 1990.

28. **Koller, M. and Strehler, E. E.**, Characterization of an intronless human calmodulin-like pseudogene, *FEBS Lett.*, 239, 121, 1988.

29. **Brosius, J.**, Retroposons — seeds of evolution, *Science*, 251, 753, 1991.

30. **Yuspa, S. H., Kilkenny, A. E., Steinert, P. M., and Roop, D. R.**, Expression of murine epidermal differentiation markers is tightly regulated by restricted extracellular calcium concentrations in vitro, *J. Cell Biol.*, 109, 1207, 1989.

31. **Boyce, S. T. and Ham, R. G.**, Calcium regulated differentiation of normal human epidermal keratinocytes in chemically defined clonal culture and serum-free serial culture, *J. Invest. Dermatol.*, 81, 33, 1983.

32. **Soule, H. D. and McGrath, C. M.**, A simplified method for passage and long-term growth of human mammary epithelial cells, *In Vitro Cell Dev. Biol.*, 22, 6, 1986.

33. **Yuspa, S. H. and Morgan, D. L.**, Mouse skin cells resistant to terminal differentiation associated with initiation of carcinogenesis, *Nature*, 293, 72, 1981.

34. **Hosobuchi, M. and Stampfer, M. R.**, Effects of transforming growth factor-β on growth of human mammary epithelial cells in culture, *In Vitro Cell Dev. Biol.*, 25, 705, 1989.

35. **Valverius, E. M., Walker-Jones, D., Bates, S. E., Stampfer, M. R., Clark, R., McCormick, F., Dickson, R. B., and Lippman, M. E.**, Production of and responsiveness to transforming growth factor β in normal and oncogene transformed human mammary epithelial cells, *Cancer Res.*, 49, 6407, 1989.

36. **Valverius, E., Bates, S. E., Stampfer, M. R., Clark, R., McCormick, F., Salomon, D. S., Lippman, M. E., and Dickson, R. B.**, Transforming growth factor alpha production and EGF receptor expression in normal and oncogene transformed human mammary epithelial cells, *Mol. Endocrinol.*, 3, 203, 1989.

37. **Bates, S. E., Valverius, E., Ennis, B. W., Bronzert, D. A., Sheridan, J. P., Stampfer, M., Mendelsohn, J., Lippman, M. E., and Dickson, R. B.**, Expression of the TGFα/EGF receptor pathway in normal human breast epithelial cells, *Endocrinology*, 126, 596, 1990.

38. **Barcellos-Hoff, M. H., Aggler, J., Ram, T. G., and Bissell, M. J.**, Functional differentiation and alveolar morphogenesis of primary mammary cultures on reconstituted basement membrane, *Development*, 105, 223, 1989.

39. **Soule, H. D., Maloney, T. M., Wolman, S. R., Peterson, W. D., Brenz, R., McGrath, C. M., Russo, J., Pauley, R. J., Jones, R. F., and Brooks, S. C.**, Isolation and characterizaton of a spontaneously immortalized human breast epithelial cell line, MCF-10, *Cancer Res.*, 50, 6075, 1990.

40. **Briand, P., Petersen, O. W., and Van Deurs, B.**, A new diploid nontumorigenic human breast epithelial cell line isolated and propagated in chemically defined medium, *In Vitro Cell Dev. Biol.*, 23, 181, 1987.

41. **Bartek, J., Bartkova, J., Kyprianou, N., Lalani, E.-N., Staskova, Z., Shearer, M., Chang, S., and Taylor-Papadimitriou, J.**, Efficient immortalization of luminal epithelial cells from human mammary gland by introduction of simian virus 40 large tumor antigen with a recombinant retrovirus, *Proc. Natl. Acad. Sci. U.S.A.*, 88, 3520, 1991.

42. **Band, V., Zajchowski, D., Swisshelm, K., Trask, D., Kulesa, V., Cohen, C., Connolly, J., and Sager, R.**, Tumor progression in four mammary epithelial cell lines derived from the same patient, *Cancer Res.*, 50, 7351, 1990.

43. **Yaswen, P., Small, A., Hosoda, J., Parry, G., and Stampfer, M. R.**, Protein product of a human intronless calmodulin-like gene shows tissue-specific expression and reduced abundance in transformed cells, submitted.

Chapter 7

TRANSFORMATION OF COLON EPITHELIAL CELLS

Dharam P. Chopra

TABLE OF CONTENTS

I. INTRODUCTION

Colorectal neoplasia is a leading cause of cancer deaths in the U.S., but its etiology remains unknown. Diet and environmental factors have been implicated since high-fat diet and standard of living exhibit a significant correlation with the geographic incidence of this disease.[1-10] For instance, patients with colon cancer and high-risk populations have high levels of fecal bile acids and cholesterol,[11-14] which reportedly act as cocarcinogens and/or tumor promoters.[15-17] Also, several carcinogens specifically induce neoplasia in the colon of experimental animals.[18,19] However, the role of these factors and cellular, biochemical, and genetic events involved in the initiation and progression of colorectal neoplasia are not clearly understood, mainly because an appropriate experimental model for the disease is not yet available. Although *in vitro* propagation of normal human colon epithelial cells has been described,[20-22] attempts to achieve typical phenotypes of neoplastic transformation have not been successful. Human colorectal carcinoma, however, provides excellent opportunities to elucidate molecular and genetic events associated with various stages of tumor progression because the development of the neoplasia apparently occurs in relatively well-defined stages of hyperplasia, dysplasia, adenoma, and carcinoma which can be easily procured for analysis. Consequently, with the advent of new molecular biology techniques, major strides have recently been made in the elucidation of genetic events associated with the progression of colorectal neoplasia. In this respect, a major hypothesis describing two types of genetic alterations has evolved from studies of various colorectal lesions.[23] One type of alteration includes activation of cellular protooncogenes through amplification, rearrangements, and/or point mutations. In particular, mutations in the Ki-*ras* protooncogene have been reported in approximately 50% of the large adenomas and carcinomas.[24-26] Additionally, amplification of c-*myc* and c-*erb*B-2 oncogenes occurs frequently in colon carcinomas.[27] The second type of genetic alteration involves allelic deletion in certain chromosomes assumed to contain tumor suppressor genes whose products regulate normal growth and thus suppress neoplastic transformation.[23] The most common allelic deletions in colorectal tumors reportedly occur in chromosomes 5, 17, and 18.[28-30] It is also postulated that such sequences involved tumor suppressor genes; in fact, recent evidence has reported that the loss of chromosome 17p in colon neoplasia is associated with the p53 gene.[31,32] This chapter reviews studies on current concepts of the multistep nature of human colorectal neoplasia and the genetic alterations that may be involved in the tumorigenesis. Additionally, *in vitro* transformation of human and rat colon epithelial cells is discussed.

II. DEVELOPMENT AND PROGRESSION OF COLORECTAL NEOPLASIA

In spite of detailed observations by numerous investigators on hundreds of specimens of human colorectal lesions, controversy still exists as to the true origin of the cancer. There have been three major viewpoints, i.e., the carcinomas originate from hyperplastic polyps, adenomatous polyps, or *in situ* lesions. Nevertheless, it is generally believed that colorectal carcinoma proceeds through stages of hyperplasia, adenomatous polyp, cellular atypia, and carcinoma.[33-37] In order to understand the evolution of the carcinoma and its precursor tissues, it is important to illustrate important features of normal colon mucosa. It is composed of test tube-shaped glands called crypts of Leiberkuhn; the surface of the mucosa is relatively flat and covered with mucus. The mucosa maintains a homeostasis of cell growth and differentiation. Cell division is limited to the lower one third of the crypts; daughter cells migrate upward in the crypts, losing their proliferative potential, and differentiate into mucous goblet cells. The luminal surface exhibits well-developed microvilli of similar length and equal spacing. Along the crypt lengths, neighboring cells are connected with intercellular interdigitations and desmosomes. The oval-shaped nuclei usually occupy the basal position, and supranuclear cytoplasm contains mucous vesicles of variable sizes. The cytoplasmic organelles such as mitochondria, endoplasmic reticulum, and Golgi are well developed. Muscularis mucosa adjoining the basement membrane forms a crucial boundary between the mucosal and interstitial tissue; a neoplasm that breaks through the muscularis is considered invasive.

Two types of hyperplastic lesions in colon mucosa have been described.[36,38] Hyperplasia, expressed as hyperplastic polyps and accounting for approximately 90% of all polyps, exhibits a slight imbalance in cell renewal involving increased cell proliferation and a slight expansion of the proliferation zone.[39] Cell division, however, remains restricted to the lower segments of the crypts, and cell differentiation into mucous goblet and absorptive cells is maintained. Diagnostically, such lesions are not considered of any consequence. Adenomatous polyps (dysplasia), which occur as pedunculated or sessile, show serious imbalance in cell renewal and differentiation.[36,38,39] Based upon the degree of cellular atypia, dysplasias are classified as mild, moderate, and severe.[40-42] Morphological alterations associated with these lesions have been described.[42] Adenomas with mild atypia contain crypts that vary slightly in size and shape, with a reduction in goblet cell production. The cells are somewhat elongated and contain rod-shaped nuclei. Luminal cell microvilli and intercellular interdigitations become relatively heterogeneous. Desmosomes and cytoplasmic organelles remain similar to those in cells of normal mucosa. In moderate atypia, hypertrophied crypts become irregularly arranged and consist of elongated cells that mostly lack mucus production. Their nuclei are pleomorphic, vary in shape from oval to elongated, and appear somewhat stratified. The most notable features of moderate atypia is an almost complete lack of intercellular interdigitations and a reduction of desmosomes between cells that appear closely apposed to one another. The number of lysosomes is greatly increased in these cells.[42] Adenomas with severe atypia exhibit a pronounced increase in cell proliferation and stratification. The nuclei show a high degree of pleomorphism and loss of polarity. Mitotic cells are frequent throughout the mucosa. Other cellular alterations in severe atypia are similar to those of moderate atypia.

Since polyps occur naturally at a relatively high frequency, the nature of adenomatous polyps that may harbor neoplasia remains controversial. Association between the polyp size and occurrence of carcinoma has been reported.[36-38] Focal carcinomas are frequently observed in larger polyps (greater than 1.5 cm) at a frequency of 10%, which increases to 30% in polyps greater than 5 cm. On the other hand, the occurrence of carcinomas in all polyps is about 0.1%. This is of particular interest, as increased expression of the *ras* oncogene has been reported in adenomas greater than 1 cm.

Adenoma development apparently proceeds through nonpolypoid phases, during which the mucosa must harbor abnormalities associated with the pathogenesis. There are, however, studies reporting abrupt transitions between normal mucosa and carcinoma.[43,44] Other studies have described specific mucosal abnormalities in nonpolypoid mucosa adjacent to carcinoma. These alterations include goblet cell hyperplasia, crypt dilation, basal cell hyperplasia and metaplasia.[37,45,46] Lee[37] examined the entire colonic mucosa of 51 cases of colorectal carcinoma; goblet cell hyperplasia was observed in 80.4% of the cases, crypt dilation in 57%, and basal cell hyperplasia and metaplasia in 14% of the cases. The occurrence of basal cell hyperplasia and metaplasia suggests that these are active lesions in the process of development. Similar changes were described by Oohara et al.,[46] who reported adenomatous changes in basal cells and hyperplastic glands as far as 10 mm away from the carcinomas. Histochemical studies also showed pertinent alterations in the mucosa adjacent to colorectal carcinoma. For instance, immature and intermediate cells were observed at higher levels in the crypts.[47] Further, an increase in sialomucin occurred in the metaplastic as compared to the normal mucosa in which sulfomucins predominated.[47-49] These observations support the view that the genesis of colorectal carcinoma is a multistep process which involves identifiable preneoplastic alterations.

As mentioned earlier, histogenesis of colorectal cancer remains a matter of controversy. One widely held view is that, except in ulcerative colitis, the neoplasm arises from polyps through the adenoma-carcinoma sequence.[36,38,40,41] It is, however, important to make a distinction between the adenomatous polyps and papillary adenomas, and it is generally accepted that the two types of lesions may represent stages in the adenoma-carcinoma continuum.[16,25,41,48] In contrast, some others believe that colorectal cancers arise *de novo* from flat mucosa without involving the preneoplastic adenomatous polyps.[50-52] Unfortunately, the two views are mutually exclusive, although the differing opinions may be caused by variable definitions applied to precursor lesions and degrees of atypia. Major support for the adenoma-carcinoma theory comes from studies demonstrating the frequent association of adenomas with carcinomas and a lack of microcancers unassociated with adenomatous tissue. It is realized that all polyps do not harbor carcinomas; only adenomas are relevant to the development of cancer.[38,39] Further, there is some evidence that the size of polypoid lesions determines whether a lesion is likely to be malignant. Thus, it is estimated that approximately 1% of all hyperplastic lesions which are termed large adenomas (>1.5 cm) are most likely to contain carcinomas. Such larger adenomas reportedly contain focal invasive cancer at a frequency of about 10%.[38] This concept has recently received additional support from molecular biology studies which have demonstrated increased expression[53] and mutations in the *ras* oncogene in approximately 50% of the carcinomas and adenomas greater than 1 cm in size.[74]

Another contention is whether the neoplasia originates from the adenomatous tissue or *de novo*. The observations that neoplasias unassociated with adenoma are rare suggest that the latter possibility is unlikely.[36-38] Further, very small *in situ* lesions (measuring a few millimeters) may not be considered of neoplastic origin. Thus, larger cancers (1 to 2 cm) showing a lack of adenomatous tissue are not a correct representation of very small lesions. The observation that large cancers infrequently lack adenomatous tissue may be because preexisting adenomatous tissue may have been destroyed by the rapidly growing cancer.[36,38] Investigators who believe that colon carcinoma arises *de novo* agree that numerous cases of the neoplasms lacking evidence of adenomatous tissue have been described.[51,52,54,55] If the carcinomas were derived only from adenomas, it is hardly unlikely that all remnants of the tissue would be destroyed. The *de novo* concept of colorectal carcinoma is based mainly on the existence of small carcinomas. Indeed, there are many reports in the literature describing cases of small carcinomas without evidence of preexisting adenomatous polyps. Recently, Shamsuddin et al.[55] have proposed that since the carcinoma originates from the mucosal cells, it could arise from both the flat mucosa and adenomatous polyps after exposure to carcinogenic stimuli. This is consistent with reports which described development of the carcinomas in nonpolypoid mucosa,[54-58] adenomatous polyps,[38-41] and even diverticula.[59,60] Since cells in the polyps exhibit a higher proliferative potential than the mucosal cells, it is only logical that the polypoid cells have a greater susceptibility to carcinogenesis. In this respect, it is interesting that in the initiation-promotion model of skin carcinogenesis, neoplastic transformation of polyps has been clearly defined.[61,62] Similarly, it is likely that initiation-promotion mechanisms also occur during colorectal neoplasia. *In vitro* studies have reported pertinent carcinogenic alterations in epithelial cells after treatment of colon explants with certain carcinogens.[63]

III. GENETIC ALTERATIONS IN COLORECTAL NEOPLASIA

It is generally believed that tumorigenesis is a multistep process involving a series of genetic manifestations causing abnormalities in normal growth-control mechanisms and, eventually, neoplastic development. Evidence now exists to indicate that colorectal neoplasms develop by mutational activation of certain protooncogenes and/or inactivation of tumor suppressor genes.[23,24] Additionally, alterations in multiple genes are apparently required for complete expression of the malignancy, and activation of protooncogenes may involve amplification and/or rearrangements. Genetic alterations, however, may not occur in any specific sequence; accumulation of the total changes is believed to be responsible for neoplasms.[24] Genetic alterations in colorectal neoplasms have been investigated more extensively than any other type of cancer because, in this case, it has been possible to characterize lesions at various stages of their existence which can be obtained in sufficient quantities for the analysis. Most results of genetic events involved in colorectal cancer come from the laboratory of Vogelstein. They have described two major types of genetic alterations. The first involves mutations in protooncogenes, particularly *ras* oncogenes, which reportedly occur in approximately 50% of both adenomas and carcinomas.[24-26] The second type involves deletions of specific chromosome regions, particularly the short arm of chromosome 17(17p), the long arm of chromosome 18(18q), and the long arm of chromosome 5(5q).[24,29,30] It is, however, important to note that such genetic alterations are present in various types of colon lesions: it is not certain whether these alterations are responsible for the development of the lesions or are coincidental. These questions can only be answered by a pertinent experimental model of neoplastic development of colon epithelial cells, which at the present time is not available.

Somatic mutations convert protooncogenes into oncogenic forms, products of which participate in neoplastic transformation. This hypothesis was based on the earlier observations showing that transfection of NIH-3T3 fibroblasts by DNA derived from human tumors caused the cells to exhibit transformed phenotypes.[64-67] Such transforming activity was also demonstrated by DNAs derived from human colorectal tumors and tumor cell lines.[67-69] The transforming activity was attributed to the products of the *ras* gene family. *Ras* genes encode 21-kDA proteins that are membrane bound and are reportedly involved in signal transduction activity.[70-76] Abnormal expression of c-*ras* oncogenes associated with quantitative and qualitative changes in their protein products may contribute to neoplastic transformation. In this respect, overexpression of c-*ras* genes has been reported in human colon tumors and tumor cell lines.[53,77,78] Expression of p21^{ras} was significantly elevated in human primary colon tumors compared to the adjoining normal colon, although the expression was relatively heterogeneous.[77] Of the primary tumors, 52% had significantly elevated levels, 41% had similar, and about 7% had less p21^{ras} compared to the normal tissue. Elevated expression of p21^{ras} was predominant at early stages of the tumor development because well-developed and metastatic tumors showed no significant increases.[77] The expression may also be related to the malignant potential of the colorectal lesions. For instance, in adenomatous polyps that are considered premalignant and subsequently show a malignant potential of about 5% for tubular adenomas and 40% for villus adenomas,[79] p21^{ras} was elevated in tubular adenomas and to a greater degree in villus adenomas compared to the hyperplastic polyps and normal colon mucosa.[78] Similarly, *ras* gene transcripts were elevated in premalignant and malignant colon tumors compared to the normal mucosa.[53] Expression of oncogenes using whole-tissue preparations, however, should be viewed with caution, as the benign and/or neoplastic tissues are generally intermixed to variable degrees with normal tissues, and the proportion of each cannot be effectively determined in such preparations. Nevertheless, the evidence clearly shows the involvement of *ras* oncogenes in colorectal neoplasms.

Activation of *ras* protooncogenes in human tumors, including that of colon, is believed to involve specific point mutations. *Ras* gene mutations have been localized to amino acids 12, 13, or 61 of c-Ha-*ras, *c-Ki-*ras,* or N-*ras* which result in the replacement of a normal glycine codon with aspartic acid, serine, valine, or cysteine codons.[25,26,76] Earlier studies using the NIH-3T3 cell transformation assay, however, reported low frequencies or even a lack of *ras* mutations in bladder carcinoma,[80] mammary adenocarcinoma,[81] and gastric carcinoma.[82] These observations may have been complicated by the low sensitivity of the assay and/or the presence of large amounts of nontumorous cells in the solid tumors and the lability of the DNA.[25] Some of these problems may have been alleviated by the recently developed techniques involving selective dissection of the tumor area consisting primarily of neoplastic cells (for DNA preparations), selective amplification of short segments of genomic DNA containing the pertinent sequences, and the high sensitivity of oligomer hybridization assays to detect gene mutations.[25,83] Using these methods, *ras* gene mutations were examined in various types of human colorectal lesions.[24,25] Of the 27 colorectal cancers,[25] 11 exhibited mutations in the *ras* genes: 9 tumors had mutations at codon 12 of the c-Ki-*ras* gene (replacing glycine with aspartic acid, serine, valine, or cysteine), 1 had a mutation at codon 61 of the c-Ki-*ras* (replacing glutamine with histidine), and 1 tumor had a mutated N-*ras* gene at position 12 (replacing glycine with cysteine). DNA derived from the normal mucosa of patients with tumor-exhibiting mutations showed no mutations in the normal DNA. Since many cancers also contained areas of adenomas, the *ras* gene mutations were analyzed in the DNA derived from the microdissected populations of the adenoma cells.[25] In five of the six observations, DNA derived from the adenomas and the carcinoma regions showed similar mutations. In another study,[24] *ras* gene mutations were analyzed in carcinomas and adenomas of various sizes, including some containing invasive lesions. Forty-seven percent of the carcinomas exhibited the *ras* gene mutations, most of them (88%) in the c-Ki-*ras* gene. Further, although a great majority of the mutations were in codon 12, some were present in codons 13 and 61 of c-Ki-*ras* and N-*ras.* No correlation was observed between the occurrence of a *ras* gene mutation and the degree of differentiation and invasiveness of the tumor, or age or sex of the patient.[24] *Ras* gene mutations also occurred in approximately 58% of the adenomas greater than 1 cm in size, while only 9% of the smaller lesions had these mutations. Additionally, the mutations in the adenomas were similar to those in the carcinomas. These observations are intriguing, and they support the view that carcinomas arise from adenomas and that activation of the c-*ras* oncogenes may occur at an early stage in tumor development. It should, however, be noted that only about half of the adenomas exhibited alterations in the c-*ras* genes. Additionally, smaller adenomas which had no mutations nevertheless were hyperplastic. This suggests that alterations in addition to the *ras* gene mutations must be involved in the initiation and progression of colorectal cancers. It is also possible that adenomas with the mutations exhibit a greater potential to undergo malignant changes. Alternatively, the presence of the mutations may represent a malignant alteration. In this respect, at least one case has been reported in which a mutation was present in the carcinoma, whereas it was not present in the adenoma which coexisted with the carcinoma.[25]

Other genetic alterations observed in human tumors involve deletions of specific chromosome regions believed to contain tumor suppressor genes.[84-86] In normal cells, products of tumor suppressor genes would inhibit cell proliferation, presumably through a negative feedback mechanism, and prevent abnormal and neoplastic growth. Inactivation of such genes through deletion and/or mutation would release the cells from their normal growth regulation and cause stimulation of cellular proliferation and neoplasia. Evidence for tumor suppression was first demonstrated with hybrids, cells produced by fusion between tumor cells and normal cells, which exhibit properties of the normal phenotype.[87-89] The tumor suppression hypothesis gained further support by experiments involving microcell transfer of a specific normal chromosome into tumor cells.[90] Most experimental studies on tumor suppression have been focused on chromosome 11, because the microcell studies have identified this chromosome to be an effective suppressor of Hela cell tumorigenesis[91] and because deletion of chromosome 11p is involved in Wilm's tumors[92] and bladder cancer.[93] Most tumor suppressor genes function recessively; therefore, both copies of the genes must be inactivated to completely eliminate the suppression function. Deletion of chromosome 13q has been identified in retinoblastoma,[94,95] 3p in small lung carcinoma,[96-98] and renal cell carcinoma and cervical carcinoma.[99,100] The most frequent losses of genetic sequences in human colon tumors are encountered in the regions of chromosomes 5q, 17p, and 18q.[23,29-31]

Familial polyposis, an autosomal-dominant disorder in humans, is characterized by the occurrence of numerous adenomas in the colon, a predisposition to the development of colorectal tumors.[101] The gene linked to familial polyposis has been localized to chromosome 5q.[102,103] Allelic losses on chromosome 5 of colonic adenomas and tumors derived from patients with and without familial polyposis were recently analyzed.[24] None of the 34 adenomas of polyposis patients showed an allelic loss. In contrast, 29% of the adenomas and 36% of the carcinomas from patients without polyposis had allelic losses on chromosome 5. Further, approximately 37% of the adenomas smaller than 1 cm also exhibited such allelic losses. These results suggested that the allelic deletions involved in the adenoma development are not the same as those associated with familial polyposis.[24] In fact, mitotic recombination and deletion experiments demonstrated that the alleles lost in the tumors are different from those previously linked to polyposis.[24] These data are at variance with other inherited disorders such as retinoblastoma, where specific allelic deletions are linked to the development of the tumors.[104,105] If the familial polyposis locus is not involved in colorectal tumors, how is the gene involved in adenoma production? It has been suggested that this locus may normally be responsible for negative regulation of the epithelial cell proliferation.[24] Inactivation of the one allele probably results in the inefficient control of cell proliferation, creating a selective growth advantage for cells harboring the allelic deletions. The transition from the hyperplastic epithelium to adenoma may involve additional events such as the *ras* gene mutations. In this respect, approximately 25% of the adenomas from polyposis patients reportedly had *ras* gene mutations or allelic losses of chromosomes 17 or 18.[24]

Deletion of chromosome 17p is reportedly most common in colorectal tumors, and occurred in approximately 75% of the specimens examined.[24,31] While the *ras* gene mutations and 5q deletions occur in adenomas and early stages of colorectal neoplasm development, 17p deletions were observed mainly in the tumors. Existing evidence suggests that 17p deletion is associated with tumor progression and may be involved in the transition from adenoma to carcinoma. DNAs derived from normal human colon mucosa and neoplastic cells of carcinoma have been analyzed in detail.[31] The two parental alleles were distinguished in all normal mucosa, whereas 75% of the tumors exhibited allelic losses. Further analysis of the tumor DNA showed that the common region of deletion extended from 17p12 to 17p13.3.[31] Similar deletions in chromosome 17p have also been reported in other human tumors, including those of lung,[106,107] breast,[108,109] bladder,[110] and brain.[111] Allelic deletions are generally believed to indicate the presence of tumor suppressor genes within the deleted regions, whose function is to prevent uncontrolled growth and tumorigenesis.[87,106,112] In human tumors, chromosome 17p contains the tumor suppressor genes for p53 protein.[31,113] Inactivation and/or loss of production of the wild-type p53 would be expected to result in transformation of the cells and tumorigenesis. Numerous studies, however, showed that the expression of the p53 gene was highly elevated in transformed and tumor cells and tissues, including colon tumors.[114,115] This contradiction was recently resolved with the demonstration that p53 gene sequences employed in earlier gene transfer experiments were in fact mutant forms of p53.[116] Thus, the increased levels of p53 protein observed in transformed and tumor cells were also of a mutant form.[117] Substantial evidence is now present to show that wild-type p53 inhibits the transformation of cells and that p53 may become oncogenic by cooperating with other oncogenes.[32] For instance, plasmids encoding wild-type p53 inhibited transformation of primary rat embryo fibroblasts induced by a combination of *ras* gene and mutant p53, whereas plasmid coding the mutant p53 had no effect or slightly stimulated the transformation.[118] The cellular p53 was originally identified as a protein that formed a stable complex with the SV40 large T-antigen and adenovirus E1b protein.[119-121] Introduction of p53 into primary cells was reported to result in immortalization and, in cooperation with the *ras* oncogene, transformation of the cells, although these alterations are now attributed to mutant forms of p53.[116]

Studies on human colorectal tumors have provided strong evidence to link abnormalities in the p53 gene and tumorigenesis.[31,113] As mentioned above, almost 75% of the colorectal cancers exhibit a complete loss of one of the two p53 alleles.[31] The second allele in the two tumors studied in detail was mutated. The mutations involved a substitution of alanine for valine at codon 143 of one tumor and a histidine for arginine at codon 175 of the second tumor. Most interestingly, the mutations occurred in this highly con-served protein domain. Mutations in these locations have previously been reported to greatly alter the biological properties of murine p53.[116,122,123] These results strongly implicate p53 mutations in colon carcinogenesis, although the associated mechanisms are a matter of speculation. It is possible that the normal p53 gene products interact with specific DNA and/or proteins to cause suppression of colon epithelial cell proliferation and neoplastic growth. Mu-tations in p53 genes may result in products that may prevent interaction with specific macromolecules or compete with normal p53 proteins to act in a negative dominant manner. Further, the effects of the mutated gene products may be more pronounced when the normal allele is lost, as occurs in colorectal tumors (loss of 17p). It is interesting to note that wild-type p53 was shown to stimulate transcription, whereas the mutated forms were unable to act as transcriptional activators.[124,125] Therefore, it was suggested that the inability of p53 mutant proteins to induce transcription may cause the transformation. Another possibility to consider is the interaction between *ras* oncogene and p53. There are reports that mutant mouse p53 genes can cooperate with *ras* oncogenes to transform rodent cells.[126,127] *Ras* oncogenes are highly expressed in approximately 50% of colorectal adenomas. The simultaneous occurrence of mutations in *ras* oncogenes and p53, which are present in colorectal lesions, presents intriguing possibilities.

Another allelic loss in colorectal tumors occurs in chromosome 18q. Such deletions have been detected in approximately 70% of the carcinomas and 50% of late adenomas.[24,29] As noted above, chromosome deletions frequently indicate the presence of tumor suppressor genes. Recently, a candidate gene, termed DCC, has been identified as a possible tumor suppressor gene in chromosome 18q.[29] This gene apparently encodes a protein with amino acid sequences similar to neural cell adhesion molecules and is related to plasma membrane glycoproteins. The identification of the gene was based on several parameters:[29] (1) one allele of the DCC gene was found deleted in 71% of the colorectal neoplasms, (2) the DCC gene was expressed in all normal mucosal tissues, but its expression was greatly reduced or absent in 88% of colon tumor cell lines, and (3) somatic mutations of the DCC gene occurred in almost 13% of the carcinomas. The mechanisms by which the DCC gene is involved in colorectal neoplasms is not understood; reduction in its expression could reduce the growth-inhibiting properties and other cell surface properties. Much evidence exists to suggest that neoplastic transformation involves cell surface alterations. For instance, loss of contact inhibition of growth *in vitro* is considered an important phenotype of preneoplastic and/or neoplastic transformation; malignant transformation is believed to involve cell-cell and cell-basement membrane interactions. Further studies, however, are necessary to define the role of the DCC gene in colorectal carcinoma.

Besides genetic modifications (deletions/mutations) in protooncogenes and tumor suppressor genes, abnormal expression of oncogenes caused by gene amplification, translocation, or rearrangements may contribute to tumorigenesis. Altered expression of oncogenes, other than those discussed above, have been reported in colorectal carcinomas. Among these, enhanced expression of c-*myc* oncogene in the adenomas and colon carcinomas relative to the normal mucosa have been observed.[27,128,131] No structural alterations of c-*myc* oncogene, however, have been reported. The c-*myc* encodes a protein with a 62-kDa molecular weight which is predominantly located in the nucleus and is believed to be involved in cellular proliferation and differentiation. Although the precise function of c-*myc* protein is not known, the cellular homologs of the gene are highly preserved throughout the species, indicating its considerable importance. Most cells have some expression of c-*myc*, but it is generally elevated when the cells are stimulated to divide and decreases in terminally differentiated cells. In the normal colon mucosa, c-*myc* expression was positive in the middle crypt zone and surface cells, while the basal crypts were essentially negative.[129-132] The staining was mainly cytoplasmic,[131] but another study reported it to be predominantly nuclear.[128] In adenomatous polyps, the expression was elevated, the staining being predominant in areas of dysplasia.[131] All colorectal tumors also exhibited higher levels of c-*myc* expression, which appeared unrelated to the clinical behavior or degree of differentiation of the tumors.[128,130] Well-differentiated tumors, however, showed more cytoplasmic staining than the poorly differentiated.[131] In another study, however, a direct correlation was observed between the expression of c-*myc* protein and mRNA and degree of differentiation of colorectal tumors.[129] Well-differentiated tumors exhibited higher expression relative to the poorly differentiated. Nevertheless, taken collectively, these studies clearly indicate that c-*myc* oncogenes have an important role in the production of colorectal carcinoma, although the mechanisms involved remain unclear. Other oncogenes with enhanced expression in colon tumors include c-*erb*B-2 and c-*myb*.[27,133,134] Hypomethylation of DNA has also been reported in colon tumors.[135,136] The loss of the methyl groups might cause alterations in the chromosomes and subsequent genetic instability.[137]

Studies on human colon tumorigenesis have identified important genetic events which may be associated with the specific stages of the multistep tumorigenesis. Four alterations, i.e., mutations in the c-*ras* genes and allelic deletions in chromosomes 5q, 17p, and 18q, are predominant. *Ras* gene mutations and allelic deletions in 5q occur predominantly during the early stage of the carcinogenesis, while the deletion in chromosome 17p and 18q occur mainly in late adenomas and carcinomas. Further, approximately 90% of the carcinomas contained two or more of the alterations, whereas only 7% of the early adenomas had more than one of the alterations. This number increased as the lesions progressed to intermediate- and late-stage adenomas.[23] Late-stage adenomas exhibited all four alterations, while all carcinomas also had additional allelic deletions. These observations suggest that the number of genetic alterations may be associated with the number of stages in the carcinogenesis,[23] although there is overwhelming evidence to indicate over-lapping genetic alterations.

IV. MARKERS IN COLORECTAL NEOPLASIA

Several markers have been associated with the development and/or pro-gression of colorectal tumors. Ornithine decarboxylase (ODC), a rate-limiting enzyme involved in polyamine biosynthesis, was reportedly increased in be-nign polyps, the degree of increase being directly related to the severity of dysplasia.[138] An eightfold increase in ODC activity occurred in colonic car-cinoma relative to the adjacent noninvolved mucosa.[139] In benign polyps, the increase was intermediate between the uninvolved mucosa and the carcinoma. Recently, ODC mRNA was shown to be elevated in colorectal neoplasms compared to the normal mucosa; no amplification of the ODC gene was observed, suggesting that the regulation of ODC activity probably occurs at the posttranscriptional level.[140] Similarly, increases in the levels of spermine and spermidine were observed in colorectal tumors, although no correlation was found between the polyamine levels and the degree of differentiation of the tumors.[141] Malignant alterations generally accompany changes in the pro-duction and secretion of plasminogen activator (PA). Several solid tumors, including those of lung,[142] prostate,[143] breast,[144] and colon,[144-146] were reported to secrete higher amounts of PA. Recently, it was also demonstrated that PA activity was increased severalfold in colonic adenocarcinoma compared to the normal mucosa, the activity being intermediate in adenomatous polyps.[147] It was suggested that the sequence of normal mucosa-adenomatous polyps-adenocarcinoma is associated with a parallel increase in the PA activity. Carcinoembryonic antigen (CEA), first described by Gold and Freedman,[148] has been widely used as a clinical marker for colorectal tumors.[149,150] The relationship between the cell-associated CEA and degree of tumor cell dif-ferentiation is controversial. For instance, CEA expression was found to be independent of histological differentiation of the tumors.[151] On the other hand, several studies have demonstrated a positive relationship between the degree of cellular differentiation in the tumor and CEA titers.[152] Several agents that enhanced cellular differentiation also increased the level of CEA.[153,154]

Histochemical and biochemical studies have demonstrated that carcino-genesis in colon mucosa involves important alterations in the glycoconjugate composition of the mucin. In particular, production of sialomucin reportedly increased in neoplastic tissue and transitional mucosa compared to the mucosa of normal individuals where the predominant glycoconjugates are sulfomucin. Furthermore, normal-appearing mucosa distant from the tumors also expressed higher levels of sialomucin relative to the mucosa of normal individuals.[47-49,155] These alterations were consistent with abnormalities in the epithelial cell differentiation and various proportions of different cell types composing the mucosal crypts. The number of absorptive cells was significantly reduced, while the immature and intermediate cells, which were observed at higher levels along the crypts, increased.[47] These studies suggested that alterations in mucin composition may reflect a transformation to a fetal type of epithelium and may indicate a preneoplastic alteration, although the application of these parameters to identify individuals at high risk of colon cancer remains unclear. There is, however, some disagreement on whether such changes in the transitional mucosa are secondary rather than primary phenomena.[156]

Cell proliferation has been of central importance in elucidation of the mechanisms involved in the initiation and progression of colorectal cancer and in correlating levels of cell proliferation with susceptibility of tumor development. Autoradiographic techniques have demonstrated significant differences in epithelial cell proliferation in the colon of normal individuals and patients at high risk of colorectal cancer. The latter population included members of families with polyposis[157,158] and patients with adenomatous polyps[159-161] or cancer.[162,163] Extensive evidence exists to show that the proliferative zone in normal colorectal mucosa is limited to the lower two thirds of the crypts. In the colons of high-risk patients, proliferative cells were observed throughout the lengths of the crypts. This alteration apparently occurs in two stages.[160] In the first stage, the proliferative activity is expanded to occur throughout the crypts, but the higher proliferation rate remains limited to the lower two thirds of the crypts. In the second stage, the major zone of DNA syntheses shifts to the upper portion of the glands and the proliferation occurs throughout the glands. Several investigators have also examined cell proliferative activity in normal-appearing colon at various distances from the carcinomas and adenomatous polyps.[164-166] Hyperplasia and an upward shift in the proliferative zone was observed throughout the colon of patients with cancer and adenomas. Further, in patients with adenomas <1 cm, the level of the proliferation was intermediate between values for normal individuals and patients with larger adenomas.[166] These are important observations and provide the basis for risk assessment for colon cancer.

V. *IN VITRO* MODELS OF HUMAN COLON EPITHELIA

Despite major advances in the elucidation of molecular events involved in colorectal cancer, mechanisms associated with the initiation and progression of the tumors remains unclear. The oncogene abnormalities discussed above may reflect consequences rather than causes of transformation processes. Alternatively, such abnormalities may merely provide a growth advantage in the transformed cells. Such gaps in information are mainly due to the lack of well-defined *in vitro* neoplastic transformation models, providing cell populations at various stages of the process. Extensive effort has been made by a number of laboratories to establish epithelial cell culture models of human and rodent colon epithelium[20-22,167-169] and to induce transformation with chemical carcinogens and tumor promoters and oncogenic DNAs. Epithelial cell lines from the normal human colon and patients with familial polyposis (nonpolypoid tissue) have been established.[20] The cell lines apparently did not exhibit any evidence of senescence and have been passaged more than 25 times. Interestingly, the cultures derived from the polyposis patients exhibited higher saturation density, CEA titers, and increased tetraploidy relative to the cells derived from normal individuals. Whether these parameters can be employed to identify individuals at high risk of colon cancer remains unclear. We have established long-term epithelial cultures from human fetal colon mucosa (13 to 14 weeks of gestation).[21,22] The cells were extensively characterized with respect to their epithelial nature and colonic origin. The cultures were initiated and maintained in medium consisting of 50% Ham's F-12 and 50% Dulbecco's minimum essential medium supplemented with 40 μg/ml ascorbic acid, 50 μg/ml isoleucine, 20 ng/ml EGF, 5 μg/ml insulin, 5 ng/ml cholera toxin, 1 μg/ml transferrin, 25 mm HEPES, 10% fetal bovine serum, and antibiotics. The cultures were incubated at 37°C in an atmosphere of 95% air and 5% CO_2. The cells exhibited microvilli on cell surfaces and showed junctional complexes and interdigitations between cells. Indented nuclei with dense chromatin and marginated heterochromatin, numerous mitochondria, rough endoplasmic reticulum, and Golgi complexes were conspicuous. The cells exhibited extensive production of mucopolysaccharides, stained intensely with periodic acid Schiff's reagent, and also showed CEA-positive titers. These long-term cultures were highly dependent on insulin, EGF, transferrin, and cholera toxin for their growth and differentiation phenotype.[21] Cultures were employed in a broad range of studies of DNA repair and transformation induced by bile acids and chemical carcinogens (see below). Another laboratory had also propagated epithelial cell cultures from various human gastrointestinal tissues, including colon.[167] The cells reportedly exhibited epithelial characteristics, but their growth patterns were atypical because they apparently grew in suspension. Nevertheless, the cultures have been employed in neoplastic transformation studies.[170,171]

Fetal colon epithelial cells were used to examine unscheduled DNA synthesis (UDS) after their exposure to carcinogens, *N*-methyl-*N*'-nitro-*N*-nitrosoguanidine (MNNG), and 4-nitroquinoline 1-oxide (4NQO) or bile acids (cholic, deoxycholic, chenodeoxycholic, or lithocholic acid).[172] The colon cells were grown to near confluency and then cultured in the serumless medium for 5 d in order to prevent the cells from entering the S-phase of the cell cycle. The chemicals, each tested at two different concentrations, were dissolved in DMSO and added to the cultures along with [³H]-thymidine (2 μCi/ml, 25 Ci/mmol) for a 4-h period. Appropriate controls were similarly treated, except they did not receive any test factors. The incorporation of radioactivity was determined by autoradiography. Silver grains indicating UDS were detected on the nuclei, and approximately 500 nuclei were counted from each group. Both 4NQO and MNNG induced UDS; the effect was highest with 0.5 μg/ml 4NQO, which increased the number of cells undergoing UDS by almost 100-fold. 4NQO also reduced the number of S-phase cells, but MNNG at 5 μg/ml was slightly stimulatory. All the bile acids tested also induced UDS. Also, bile acids stimulated labeling indices of the S-phase cells. These results indicated that the bile acids not only induced DNA damage in the colon epithelial cells, but also stimulated their proliferative activity. Numerous protocols to induce neoplastic transformation in the fetal colon epithelial cells by carcinogens alone or in combination with bile acids were attempted. No transformation of the cells, however, was observed, although the cells apparently have escaped the senescence phase. Morphological transformations of adult human colon epithelial cells by azoxymethane treatment, SV40 virus infection, or transfection with vectors containing T-antigen, have been reported.[170,171] The transformed cells exhibited increased longevity, decreased growth factor requirement, anchorage-independent growth, and altered cell surface properties. The significance of this work, however, is unclear because the so-called transformed cells were not immortalized, as they could not be continuously passaged.

In another approach to developing models of colorectal tumor progression, cell cultures were obtained from various-size adenomas from patients with familial polyposis, a genetic predisposition for colorectal cancer.[172,173] These experiments were carried out to determine if the adenoma-derived cells would yield immortal cell lines which may subsequently be neoplastically transformed. The view was based on the premise that certain adenomas may harbor premalignant cells with continuous growth potential. Results showed that epithelial cells derived from adenomas less than 1 cm could only be cultured for relatively short periods. On the other hand, cultures derived from larger adenomas showed a greater potential to be established *in vitro*. A cell line derived from a 3- to 4-cm adenoma escaped senescence and has been passaged for more than 4 years.[173,174] The cells exhibited typical characteristics of epithelial cells, including microvilli on cell surfaces and desmosomes between cells, and produced mucin. Further, the cells remained anchorage dependent and nontumorigenic. A proportion of the cells expressing CEA titers, however, showed a progressive increase with successive passaging. Karyotype analysis showed that at early passage, the cells were mostly diploid.[173] Late-passage cells, however, were aneuploid, with a modal number of 48 chromosomes. One of the most distinctive features of the late-passage cells was the presence of an isochromosome 1q, with all cells having both normal copies of chromosome 1. These results are consistent with the adenoma-carcinoma sequence and show that acquisition of immortality may be associated with a premalignant stage in tumorigenesis, but only when the adenomas are large and/or display dysplasia. Abnormalities in oncogene expression such as c-*ras* and c-*myc*, which are known to occur in the adenomas, were not examined in the immortalized cells. Abnormalities in chromosome 1 have been reported in other types of tumors, including those of the breast,[175] ovary,[176] and intestine,[177] and tumor cell lines.[178]

Recent studies have reported that a clonal variant of the immortalized colorectal cell line was neoplastically transformed after treatment with sodium butyrate and MNNG.[179] For instance, anchorage-independent growth which occurred after treatment of the cells with MNNG increased significantly after successive passaging of the cultures. At passage 65, the colony-forming efficiency was about 0.16%, which increased to 17.3% at passage 82. Tumorigenesis in nude mice was observed only after passage 70; size and number of tumors per animal increased with the later-passage cultures.[179] The tumors were heterogeneous, with areas consisting of moderately to well-differentiated and poorly differentiated cells. The transformed cells contained one to two copies of the abnormal chromosome 1 with rearrangements involving breakpoints at p32 and q23; other abnormalities noted were up to six copies of chromosomes 7, 9, and 13 and monosomy of chromosome 18. These studies thus demonstrate that neoplastic transformation of the adenoma-derived cells occurred in successive stages. Carcinogen treatment rendered the anchorage-dependent cells anchorage independent, and subsequent passaging resulted in their acquisition of tumorigenic potential. The data also supported the view that adenomas can serve as precursor lesions in colorectal tumorigenesis.

VI. *IN VITRO* NEOPLASTIC TRANSFORMATION OF RAT COLON EPITHELIAL CELLS

Human epithelial cells exhibit great resistance to neoplastic transformation *in vitro*. Recently, the transformation of certain epithelial cells by transfection with oncogenic DNA, alone or in combination with chemical carcinogens, has been demonstrated. Such models, however, are of limited potential value for elucidating mechanisms of neoplastic transformation, as the cells contain extraneous DNA. Rodent epithelial cells are apparently more sensitive to neoplastic transformation by chemical carcinogens. We have established epithelial cell lines from the colon mucosa of newborn rats[168,169] which are neoplastically transformed by MNNG.

The rat colon epithelial cell lines have been maintained in Eagle's minimum essential medium buffered with HEPES (25 mm) and supplemented with fetal bovine serum (2.5%), L-glutamine (2 mM), insulin (0.5 μg/ml), transferrin (5 μg/ml), fungizone (1.25 μg/ml), and antibiotics (Figure 1). Cultures were exposed to MNNG at a concentration of 1.25 μg/ml for a period of 28 d, and subsequently once a week for 24 h each time. Culture medium was changed three times per week. Control cultures were similarly treated, except they did not receive the carcinogen. Treatment of cultures (ninth passage) with MNNG continuously for 28 d, followed by four weekly treatments, caused formation of foci of overlapping cells (Figure 2). The foci appeared to release cells that floated into the culture medium. At this stage, the cells did not show anchorage-independent growth. Further culturing of the cells produced anchorage-independent growth in semisolid medium (fourteenth passage, Figure 3). Inoculation of seventeenth passage cultures into athymic nude mice resulted in the production of tumors in 20% of the animals (Figure 4). Histological examination of the tumors revealed two morphologically distinct patterns of growth. The first area consisted of irregular foci of tall columnar-type epithelial cells forming crypts and granular structures (Figures 5 and 6). The crypts were dilated and the lumens contained mucus and necrotic cellular debris. Frequently, the glandular lesions were surrounded by zones of immature mesenchymal cells (Figure 7). The lumens of the crypts and glandular structures were filled with secretory materials that stained positive with periodic acid Schiff's (PAS) reagent, compatible with the intestinal origins of the tumors. A focal area also contained vacuolated cells with eccentrically located nuclei (Figure 8). These lesions were suggestive of signet ring cells observed in adenocarcinoma,[63] but were not PAS positive. The second area was moderately dense and fibrotic with central necrosis and suppurative exudate. These results, for the first time, show that rat colon epithelial cells can be neoplastically transformed by a direct-acting carcinogen, MNNG. Most importantly, the transformed cells produced adenocarcinomas in nude mice, the structure of which was similar to those described for human colon adenocarcinomas and carcinogen-induced tumors in the colon of rat.[63]

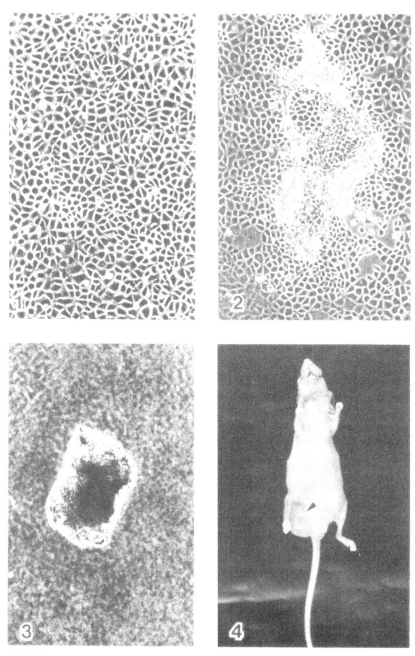

FIGURE 1.   An epithelial cell culture (control) at fourteenth passage. (Original magnification × 360.)

FIGURE 2.   Characteristic morphology of a focus formed in an MNNG-treated culture. (Original magnification × 360.)

FIGURE 3.   Morphology of a colony formed in semisolid medium by MNNG-treated cells.

FIGURE 4.   A large tumor mass (arrow) produced in a nude mouse by MNNG-treated cells.

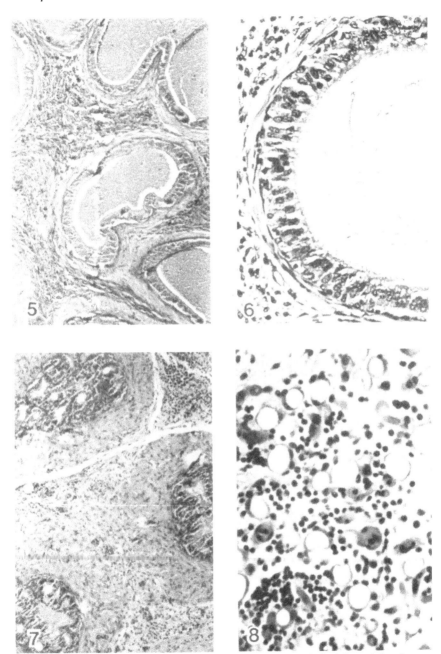

FIGURE 5. A section of the tumor shown in Figure 4. It showed glandular and ductal-type adenocarcinoma (arrows). (Original magnification × 400.)

FIGURE 6. A higher magnification picture of glandular epithelium of the tumor. (Original magnification × 1400.)

FIGURE 7. Tissue section showing an area of glandular lesions surrounded by mesenchymal tissue. (Original magnification × 400.)

FIGURE 8. Another area of tumor shown in Figure 4. The section shows lesions similar to signet ring carcinoma. (Original magnification × 360.)

REFERENCES

1. **Weisburger, J. H., Reddy, B. S., and Wynder, E. L.,** Colon cancer: its epidemiology and experimental production, *Cancer,* 40, 241, 1977.
2. **Carrol, K. K., Braden, L. M., Bell, J. A., and Kalamegham, R.,** Fat and cancer, *Cancer,* 58, 1818, 1986.
3. **Willett, W. C. and MacMahon, B.,** Diet and cancer: an overview, *N. Engl. J. Med.,* 310, 697, 1984.
4. **McKeown-Eyssen, G. E. and Bright-See, E.,** Dietary factors in colon cancer: international relationships, *Nutr. Cancer,* 6, 160, 1984.
5. **Kritchevsky, D. and Klurfeld, D. M.,** Fat and cancer, in *Nutrition and Cancer: Etiology and Treatment,* Newell, G. R. and Ellison, N. M., Eds., Raven Press, New York, 1981, 173.
6. **Doll, R.,** The geographical distribution of cancer, *Br. J. Cancer,* 23, 1, 1969.
7. **Atxell, L. M. and Chiazze, L., Jr.,** Changing relative frequency of cancers of the colon and rectum in the United States, *Cancer,* 19, 750, 1966.
8. **Armstrong, B. and Doll, R.,** Environmental factors and cancer incidence and mortality in different countries with special reference to dietary patterns, *Int. J. Cancer,* 15, 617, 1975.
9. **Wynder, E. L., Kajitani, T., Ishikawa, S., Dodo, H., and Takano, A.,** Environmental factors of cancer of the colon and rectum. II. Japanese epidemiological data, *Cancer,* 23, 1210, 1969.
10. **Gregor, O., Toman, R., and Prusova, F.,** Gastrointestinal cancer and nutrition, *Gut,* 10, 1031, 1969.
11. **Peuchant, E., Salles, C., and Jensen, R.,** Relationship between fecal neutral steroid concentrations and malignancy in colon cells, *Cancer,* 60, 994, 1987.
12. **Reddy, B. S. and Wynder, E. L.,** Large-bowel carcinogenesis: fecal constituents of populations with diverse incidence rates of colon cancer, *J. Natl. Cancer Inst.,* 50, 1437, 1973.
13. **Salyers, A. A., Sperry, J. F., Wilkins, T. D., Walker, A. P. R., and Richardson, N. J.,** Neutral steroid concentrations in the feces of North American white and South African black populations at different risks for cancer of the colon, *S. Afr. Med. J.,* 51, 823, 1976.
14. **Thompson, M.,** Aetiological factors in gastrointestinal carcinogenesis, *Scand. J. Gastroenterol.,* 104, 77, 1984.
15. **Weisburger, J. H., Reddy, B. S., Barnes, W. S., and Wynder, E. L.,** Bile acids, but not neutral sterols, are tumor promoters in the colon in man and rodents, *Environ. Health Perspect.,* 50, 101, 1983.
16. **Cruse, J. P., Lewin, M. R., Ferulano, G. P., and Clark, C. G.,** Co-carcinogenic effects of dietary cholesterol in experimental colon cancer, *Nature,* 276, 822, 1978.
17. **Narisawa, T., Magadia, N. E., Weisburger, J. H., and Wynder, E. L.,** Promoting effect of bile acids on colon carcinogenesis after intrarectal instillation of *N*-methyl-*N'*-nitro-*N*-nitrosoguanidine in rats, *J. Natl. Cancer Inst.,* 53, 1093, 1974.
18. **Weisburger, J. H.,** Colon carcinogens: their metabolism and mode of action, *Cancer,* 28, 60, 1971.
19. **Fiala, E. S.,** Investigations into the metabolism and mode of action of the colon carcinogens 1,2-dimethylhydrazine and azoxymethane, *Cancer,* 40, 2436, 1977.
20. **Danes, B. S.,** Long-term-cultured colon epithelial cell lines from individuals with and without colon cancer genotypes, *J. Natl. Cancer Inst.,* 75, 261, 1985.
21. **Siddiqui, K. M. and Chopra, D. P.,** Primary and long-term epithelial cell cultures from human fetal normal colonic mucosa, *In Vitro,* 20, 859, 1985.
22. **Chopra, D. P., Siddiqui, K. A., and Cooney, R. A.,** Effects of insulin, transferrin, cholera toxin and epidermal growth factor on growth and morphology of human fetal normal colon epithelial cells, *Gastroenterology,* 92, 891, 1987.

23. **Fearon, E. R. and Vogelstein, B.,** A genetic model for colorectal tumorigenesis, *Cell,* 61, 759, 1990.
24. **Vogelstein, B., Fearson, E. R., Hamilton, S. R., Kern, S. E., Preisinger, A. C., Leppert, M., Nakumura, Y., White, R., Smits, A. M. M., and Bos, J. L.,** Genetic alterations during colorectal-tumor development, *N. Engl. J. Med.,* 319, 525, 1988.
25. **Bos, J. L., Fearon, E. R., Hamilton, S. R., Verlaan-de Vries, M., van Boom, J. H., van der Eb, A. J., and Vogelstein, B.,** Prevalence of ras gene mutations in human colorectal cancers, *Nature,* 327, 293, 1987.
26. **Forrester, K., Almoguera, C., Han, K., Grizzle, W. E., and Perucho, M.,** Detection of high incidence of k-ras oncogenes during human colon tumorigenesis, *Nature,* 327, 298, 1987.
27. **Meltzer, S. J., Ahnen, D. J., Battifora, H., Yokota, J., and Cline, M. J.,** Protooncogene abnormalities in colon cancers and adenomatous polyps, *Gastroenterology,* 92, 1174, 1987.
28. **Solomon, E., Voss, R., Hall, V., Bodmer, W. F., Jass, J. R., Jeffries, A. J., Lucibello, F. C., Patel, I., and Rider, S. H.,** Chromosome 5 allele loss in human colorectal carcinomas, *Nature,* 328, 616, 1987.
29. **Fearon, E. R., Cho, K. R., Nigro, J. M., Kern, S. E., Simons, J. W., Ruppert, J. M., Hamilton, S. R., Preisinger, A. C., Thomas, G., Kinzler, K. W., and Vogelstein, B.,** Identification of a chromosome 18q gene that is altered in colorectal cancer, *Science,* 247, 49, 1990.
30. **Kern, S. E., Fearon, E. R., Tersmette, K. W. F., Enterline, J. P., Leppert, M., Nakamura, Y., White, R., Vogelstein, B., and Hamilton, S. R.,** Clinical and pathological associations with allelic loss in colorectal carcinoma, *J. Am. Med. Assoc.,* 261, 3099, 1989.
31. **Baker, S. J., Fearon, E. R., Nigro, J. M., Hamilton, S. R., Preisinger, A. C., Jessup, J. M., vanTuinen, P., Ledbetter, D. H., Baker, D. F., Nakamura, Y., White, R., and Vogelstein, B.,** Chromosome 17 deletions and p53 gene mutations in colorectal carcinomas, *Science,* 244, 217, 1989.
32. **Finlay, C. A., Hinds, P. W., and Levine, A. J.,** Protooncogene can act as a suppressor of transformation, *Cell,* 57, 1083, 1989.
33. **Helwig, E. B.,** Adenomas and the pathogenesis of cancer of the colon and rectum, *Dis. Colon Rectum,* 2, 5, 1959.
34. **Morson, B. C.,** Precancerous and early malignant lesions of the large intestine, *Br. J. Surg.,* 55, 725, 1968.
35. **Morson, B. C.,** The polyp-cancer sequence in the large bowel, *Proc. R. Soc. Med.,* 67, 451, 1974.
36. **Fenoglio, C. M. and Lane, N.,** The anatomical precursor of colorectal carcinoma, *Cancer,* 34, 819, 1974.
37. **Lee, Y.-S.,** Background mucosal changes in colorectal carcinomas, *Cancer,* 61, 1563, 1988.
38. **Lane, N.,** The precursor tissue of ordinary large bowel cancer, *Cancer Res.,* 36, 2669, 1976.
39. **Lane, N., Kaplan, H., and Pascal, R. R.,** Minute adenomatous and hyperplastic polyps of the colon: divergent patterns of epithelial growth with specific associated mesenchymal changes, *Gastroenterology,* 60, 537, 1971.
40. **Morson, B. C.,** Evolution of cancer of the colon and rectum, *Cancer,* 34, 845, 1974.
41. **Muto, T., Bussy, H. J., and Morson, B. C.,** The evolution of cancer of the colon and rectum, *Cancer,* 36, 2251, 1975.
42. **Matsuda, M., Misumi, A., Shimada, S., Nelson, R. L., and Akagi, M.,** Quantitative electron microscopic study on grades of atypia in adenomas of the human large bowel, *Dis. Colon Rectum,* 32, 57, 1989.
43. **Ackerman, L. V. and Spratt, J. S.,** Do adenomatous polyps become cancer?, *Gastroenterology,* 44, 905, 1963.

44. Ackerman, L. V., Spjut, H. J., and Spratt, J. S., The biological characteristics of colonic and rectal neoplasms with refutation of the concept that adenomatous polyps are highly premalignant, *Acta Unio Int. Cancer*, 20, 716, 1964.
45. Saffos, R. O. and Rhatigan, R. M., Benign (nonpolypoid) mucosal changes adjacent to carcinomas of the colon, *Hum. Pathol.*, 8, 441, 1977.
46. Oohara, T., Ihara, O., and Tohma, H., Background mucosal changes of primary advanced large intestinal cancer in patients without familial polyposis coli, *Dis. Colon Rectum*, 26, 91, 1983.
47. Dawson, P. A. and Filipe, M. I., An ultrastructural and histochemical study of the mucous membrane adjacent to and remote from carcinoma of the colon, *Cancer*, 37, 2388, 1976.
48. Filipe, M. I. and Cooke, K. B., Changes in composition of mucin in the mucosa adjacent to carcinoma of the colon as compared with the normal: a biochemical investigation, *J. Clin. Pathol.*, 27, 315, 1974.
49. Filipe, M. I., The value of a study of the mucosubstances in rectal biopsies from patients with carcinoma of the rectum and lower sigmoid in the diagnosis of premalignant mucosa, *J. Clin. Pathol.*, 25, 123, 1972.
50. Castleman, B. and Krickstein, H. L., Do adenomatous polyps of the colon become malignant?, *N. Engl. J. Med.*, 267, 469, 1962.
51. Castleman, B. and Krickstein, H. L., Current approach to the polyp-cancer controversy, *Gastroenterology*, 51, 108, 1966.
52. Spjut, H. J., Frankel, N. B., and Apple, M. F., The small carcinoma of the large bowel, *Am. J. Surg. Pathol.*, 3, 39, 1979.
53. Spandidos, D. A. and Kerr, I. B., Elevated expression of the human ras oncogene family in premalignant and malignant tumors of the colorectum, *Br. J. Cancer*, 49, 681, 1984.
54. Shamsuddin, A. K. M., Microscopic intraepithelial neoplasia in large bowel mucosa, *Hum. Pathol.*, 13, 510, 1982.
55. Shamsuddin, A. K. M., Kato, Y., Kunishima, N., Sugano, H., and Trump, B. F., Carcinoma in situ in nonpolypoid mucosa of the large intestine. Report of a case with significance in strategies for early detection, *Cancer*, 56, 2849, 1985.
56. Shamsuddin, A. K. M. and Elias, E. G., Rectal mucosa: malignant and premalignant changes after radiation therapy, *Arch. Pathol. Lab. Med.*, 105, 150, 1981.
57. Woda, B. A., Forde, K., and Lane, N., A unicryptal colonic adenoma, the smallest colonic neoplasm yet observed in a non-polyposis individual, *Am. J. Clin. Pathol.*, 68, 631, 1977.
58. Crawford, B. E. and Stromeyer, F. W., Small non-polypoid carcinomas of the large intestine, *Cancer*, 51, 1760, 1983.
59. Burkitt, D. P., Epidemiology of cancer of the colon and rectum, *Cancer*, 28, 3, 1971.
60. Hernandez, F. J. and Fernandez, B. B., Mucus-secreting carcinoid tumor in colonic diverticulum: report of a case, *Dis. Colon Rectum*, 19, 63, 1976.
61. Berenblum, I., Sequential aspects of chemical carcinogenesis: skin, in *Cancer, A Comprehensive Treatise*, Vol. 1, Becker, F. F., Ed., Plenum Press, New York, 1975, 323.
62. Balmain, A., Ramsen, M., Bowden, G. T. and Smith, J., Activation of the mouse cellular Harvey-ras gene in chemically-induced benign skin papillomas, *Nature*, 307, 658, 1984.
63. Shamsuddin, A. K. M. and Trump, B. F., Colon epithelium. III. In vitro studies of colon carcinogenesis in Fischer 344 rats: *N*-methyl-*N'*-nitro-nitrosoguanidine induced changes in rat colon epithelium in explant culture, *J. Natl. Cancer Inst.*, 66, 403, 1981.
64. Cooper, G. M., Cellular transforming genes, *Science*, 217, 801, 1982.
65. Der, C. J., Krontiris, T. G., and Cooper, G. M., Transforming genes of human bladder and lung carcinoma cell lines are homologous to the ras genes of Harvey and Kirsten sarcoma viruses, *Proc. Natl. Acad. Sci. U.S.A.*, 79, 3637, 1982.
66. Der, C. J. and Cooper, G. M., Altered gene products are associated with activation of cellular ras-K genes in human lung and colon carcinoma, *Cell*, 32, 201, 1983.

67. **Capon, D. J., Seeburg, P. H., McGrath, J. P., Hayflock, J. S., Edman, U., Levinson, A. D., and Goeddel, D. V.,** Activation of Ki-ras2 gene in human colon and lung carcinoma by two different point mutations, *Nature,* 304, 507, 1983.

68. **Pulciani, S., Snatos, E., Lauver, A. V., Long, L. K., Aaronson, S. A., and Barbacid, M.,** Oncogenes in solid human tumors, *Nature,* 300, 539, 1982.

69. **McCoy, M. S., Toole, J. J., Cunningham, J. M., Chang, E. H., Lowy, D. R., and Weinberg, R. A.,** Characterization of a human colon/lung carcinoma oncogene, *Nature,* 302, 79, 1983.

70. **Bishop, J. M.,** Cellular oncogenes and retroviruses, *Annu. Rev. Biochem.,* 52, 301, 1983.

71. **Willingham, M. C., Pastan, I., Shih, T. Y., and Scolnick, E. M.,** Localization of the src gene product of the Harvey strain of MSV to plasma membrane of transformed cells by electron microscopic immunocytochemistry, *Cell,* 19, 1005, 1980.

72. **Willumsen, B. M., Christensen, A., Hubbert, N. L., Papageorge, A. G., and Lowy, D. L.,** The p21 ras C-terminus is required for transformation and membrane association, *Nature,* 310, 583, 1984.

73. **McGrath, J. P., Capon, D. J., Goeddel, D. V., and Levinson, A. D.,** Comparative biochemical properties of normal and activated ras p21 protein, *Nature,* 310, 644, 1984.

74. **Gibbs, J. B., Sigal, I. S., Poe, M., and Scolnick, E. M.,** Intrinsic GTPase activity distinguishes normal and oncogenic ras p21 molecules, *Proc. Natl. Acad. Sci. U.S.A.,* 81, 5704, 1984.

75. **Gilman, A. G.,** G proteins and dual control of adenylate cyclase, *Cell,* 36, 577, 1984.

76. **Barbacid, M.,** Ras genes, *Annu. Rev. Biochem.,* 56, 779, 1987.

77. **Gallick, G. E., Kurzrock, R., Kloetzer, W. S., Arlinghaus, R. B., and Gutterman, J. U.,** Expression of p21 ras in fresh primary and metastatic human colorectal tumors, *Proc. Natl. Acad. Sci. U.S.A.,* 82, 1795, 1985.

78. **Michelassi, F., Leuthner, S., Lubienski, M., Bostwick, D., Rodgers, J., Handcock, M., and Block, G. E.,** Ras oncogene p21 levels parallel malignant potential of different human colonic benign conditions, *Arch Surg.,* 122, 1414, 1987.

79. **Naso, R. B., Arcement, L. J., and Arlinghaus, R. B.,** Biosynthesis of Rauscher leukemia viral proteins, *Cell,* 4, 3, 1975.

80. **Fujita, J., Srivastava, S. K., Kraus, M. H., Rhim, J. S., Tronick, S. R., and Aaronson, S. A.,** Frequency of molecular alterations affecting ras protooncogene in human urinary tract tumors, *Proc. Natl. Acad. Sci. U.S.A.,* 82, 3849, 1985.

81. **Kraus, M. H., Yuasa, Y., and Aaronson, S. A.,** A position 21-activated H-ras oncogene in all HS578T mammary carcinosarcoma cells but not normal mammary cells of the same patient, *Proc. Natl. Acad. Sci. U.S.A.,* 81, 5384, 1984.

82. **Sakamoto, H., Mori, M., Taira, M., Yoshida, T., Matsukawa, S., Shimizu, K., Sekiguchi, M., Terada, M., and Sugimura, T.,** Transforming gene from human stomach cancers and noncancerous portion of stomach mucosa, *Proc. Natl. Acad. Sci. U.S.A.,* 83, 3997, 1986.

83. **Saiki, R. K., Scharf, S., Faloona, F., Mullis, K. B., Horn, G. T., Erlich, H. A., and Arnheim, N.,** Enzymatic amplification of beta-globin genomic sequences and restriction site analysis for diagnosis of sickle cell anemia, *Science,* 230, 1350, 1985.

84. **Klein, G.,** The approaching era of the tumor suppressor genes, *Science,* 238, 1539, 1987.

85. **Stanbridge, E. J.,** A case for human tumor-suppressor genes, *BioEssays,* 3, 252, 1985.

86. **Sager, R.,** Tumor suppressor genes: the puzzle and the promise, *Science,* 246, 1406, 1989.

87. **Knudson, A. G., Jr.,** Hereditary cancer, oncogenes and antioncogenes, *Cancer Res.,* 45, 1437, 1985.

88. **Sager, R.,** Genetic suppression of tumor formation: a new frontier in cancer research, *Cancer Res.,* 46, 1573, 1986.

89. **Harris, H.,** The analysis of malignancy by cell fusion: the position in 1988, *Cancer Res.,* 48, 3302, 1988.

90. **Fournier, R. E. K. and Ruddle, F. H.,** Microcell-mediated transfer of murine chromosomes into mouse, Chinese hamster and human somatic cells, *Proc. Natl. Acad. Sci. U.S.A.,* 74, 319, 1977.

91. **Stanbridge, E. J., Der, C. J., Doersen, C. J., Nishimi, R. Y., Peehl, D. M., Weissman, B. E., and Wilkinson, J. E.,** Human cell hybrids: analysis of transformation and tumorigenicity, *Science,* 215, 252, 1982.

92. **Francke, U.,** Specific chromosome changes in the human heritable tumors retinoblastoma and nephroblastoma, in *Chromosomes and Cancer,* Rowley, J. D. and Ultman, J. E., Eds., Academic Press, New York, 1983, 99.

93. **Fearon, E. R., Feinberg, A. P., Hamilton, S. H., and Vogelstein, B.,** Loss of genes on the short arm of chromosome 11 in bladder cancer, *Nature,* 318, 377, 1985.

94. **Friend, S. H., Bernards, R., Rogelj, S., Weinberg, R. A., Rapaport, J. M., Albert, D. M., and Dryja, T. P.,** A human DNA segment with properties of the gene that predisposes to retinoblastoma and osteosarcoma, *Nature,* 323, 643, 1986.

95. **Lee, W. H., Bookstein, R., Hong, F., Young, L. J., Shew, J. Y., and Leee, E. Y.,** Human retinoblastoma susceptibility gene: cloning, identification and sequence, *Science,* 235, 1394, 1987.

96. **Harbour, J. W., Lai, S. L., Whang-Peng, J., Gazdar, A. F., Minna, J. D., and Kaye, F. J.,** Abnormalities in the structure and expression of the human retinoblastoma gene in SCLC, *Science,* 241, 353, 1988.

97. **Naylor, S. L., Johnson, B. E., Minna, J., and Sakaguchi, A.,** Loss of heterozygosity of chromosome 3 p markers in small-cell lung cancer, *Nature,* 329, 451, 1987.

98. **Takahashi, T., Nau, M. M., Chiba, I., Birrer, M. J., Rosenberg, R. K., Vincour, M., Levitt, M., Pass, H., Gazdar, A. F., and Minna, J. D.,** p53: a frequent target for genetic abnormalities in lung cancer, *Science,* 246, 491, 1989.

99. **Kovacs, G., Erlandsson, R., Boldog, F., Ingvarsson, S., Muller-Brechlin, R., Klein, G., and Sumegi, J.,** Consistent chromosome 3p deletion and loss of heterozygosity in renal cell carcinoma, *Proc. Natl. Acad. Sci. U.S.A.,* 85, 1571, 1988.

100. **Yokata, J., Tsukada, Y., Nakajima, T., Gotoh, M., Shimosato, Y., Mori, N., Tsunokawa, Y., Sigimura, T., and Tereda, M.,** Loss of heterozygosity on the short arm of chromosome 3 in carcinoma of the uterine cervix, *Cancer Res.,* 49, 3598, 1989.

101. **Haggitt, R. C. and Reid, B. J.,** Hereditary gastrointestinal polyposis syndrome, *Am. J. Surg. Pathol.,* 10, 871, 1986.

102. **Bodmer, W. F., Bailey, C. J., Bodmer, J., Bussey, H. J., Ellis, A., Gorman, P., Lucibello, F. C., Murday, V. A., Rider, S. H., Scambler, P., Sheer, D., Solomon, E., and Spurr, N. K.,** Localization of the gene for familial adenomatous polyposis on chromosome 5, *Nature,* 328, 614, 1987.

103. **Leppert, M., Dobbs, M., Scrambler, P., O'Connell, P., Nakamura, Y., Stauffer, D., Woodward, S., Burt, R., Hughes, J., Gardner, E., Lathrop, M., Wasmuth, J., Lalouel, J.-M., and White, R.,** The gene for familial polyposis coli maps to the long arm of chromosome 5, *Science,* 238, 1411, 1987.

104. **Cavenee, W. K., Dryja, T. P., Phillips, R. A., Benedict, W. F., Godbout, R., Gallie, B. L., Murphee, A. L., Strong, L. C., and White, R. L.,** Expression of recessive alleles by chromosomal mechanisms in retinoblastoma, *Nature,* 305, 779, 1983.

105. **Hansen, M. F. and Cavenee, W. K.,** Genetics of cancer predisposition, *Cance r Res.,* 47, 5518, 1987.

106. **Yokota, J., Wada, M., Shimosato, Y., Tereda, M., and Sugimura, T.,** Loss of heterozygosity on chromosomes 3, 13 and 17 in small-cell carcinoma and on chromosome 3 in adenocarcinoma of the lung, *Proc. Natl. Acad. Sci. U.S.A.,* 84, 9252, 1987.

107. **Weston, A., Wiley, J. C., Modali, R., Sugimura, H., McDowell, E. M., Resau, J., Light, B., Haugen, A., Mann, D. L., Trump, B. F., and Harris, C. C.,** Differential DNA sequence deletions from chromosomes 3, 11, 13 and 17, in squamous-cell carcinoma, large-cell carcinoma, and adenocarcinoma of the human lung, *Proc. Natl. Acad. Sci. U.S.A.,* 86, 5099, 1989.

108. **Mackey, J., Steel, C. M., Elder, P. A., Forrest, A. P. M., and Evans, H. J.,** Allele loss on short arm of chromosome 17 in breast cancer, *Lancet*, 2, 1384, 1988.

109. **Devillee, P., van der Brock, M., Kuipers-Dijkshoorn, N., Kolluri, R., Khan, P. M., Pearson, P. L., and Cornelisse, C. J.,** At least four different chromosomal regions are involved in loss of heterozygosity in human breast carcinoma, *Genomics*, 5, 554, 1989.

110. **Tsai, Y.-C., Nichols, P. W., Hiti, A. L., Williams, Z., Skinner, D. G., and Jones, P. A.,** Allelic losses of chromosomes 9, 11 and 17 in human bladder cancer, *Cancer Res.*, 50, 44, 1990.

111. **James, C. D., Carlbom, E., Nordenskjold, M., Collins, V. P., and Cavenee, W. K.,** Mitotic recombinations of chromosome 17 in astrocytomas, *Proc. Natl. Acad. Sci. U.S.A.*, 86, 2858, 1989.

112. **Murphree, A. and Benedict, W.,** Retinoblastoma: clues to human oncogenesis (review), *Science*, 223, 1028, 1984.

113. **Nigro, J. M., Baker, S. J., Preisinger, A. C., Jessup, J. M., Hostetter, R., Cleary, K., Bigner, S. H., Davidson, N., Baylin, S., Devillee, P., Glover, T., Collins, F. S., Weston, A., Modali, R., Harris, C. C., and Vogelstein, B.,** Mutations in the p53 gene occur in diverse human tumor types, *Nature*, 342, 705, 1989.

114. **Crawford, L. V.,** The 53,000-dalton cellular protein and its role in transformation (rev.,) *Int. Rev. Exp. Pathol.*, 25, 1, 1983.

115. **van den Berg, F. M., Tigges, A. J., Schipper, M. E. I., den Hartog-Jager, F. C. A., Kroes, W. G. M., and Walboomers, J. M. M.,** Expression of the nuclear oncogene p53 in colon tumors, *J. Pathol.*, 157, 193, 1989.

116. **Finley, C. A., Hinds, P. W., Tan, T. H., Eliyahu, D., Oren, M., and Levine, A. J.,** Activating mutations for transformation by p53 produce a gene product that forms an hsc70-p53 complex with an altered half-life, *Mol. Cell. Biol.*, 8, 531, 1988.

117. **Iggo, R., Gatter, K., Bartek, J., Lane, D., and Harris, A. L.,** Increased expression of mutant forms of p53 oncogene in primary lung cancer, *Lancet*, 335, 675, 1990.

118. **Eliyahu, D., Michalovitz, D., Eliyahu, S., Pinhasi-Kimhi, O., and Oren, M.,** Wild-type p53 can inhibit oncogene-mediated focus formation, *Proc. Natl. Acad. Sci. U.S.A.*, 86, 8763, 1989.

119. **Lane, D. P. and Crawford, L. V.,** T antigen is bound to a host protein in SV40-transformed cells, *Nature*, 278, 261, 1979.

120. **Linzer, D. I. H. and Levine, A. J.,** Characterization of a 54K dalton cellular SV40 tumor antigen present in SV40-transformed cells and uninfected embryonal carcinoma cells, *Cell*, 17, 43, 1979.

121. **Sarnow, P., Ho, Y. S., William, J., and Levine, A. J.,** Adenovirus E1b-58 kd tumor antigen and SV40 large tumor antigen are physically associated with the same 54 kd cellular protein in transformed cells, *Cell*, 28, 387, 1982.

122. **Eliyahu, D., Goldfinger, N., Pinhasi-Kinhi, O., Shaulski, G., Skurnik, Y., Arai, N., Rotter, V., and Oren, M.,** Meth A fibrosarcoma cells express two transforming mutant p53 species, *Oncogene*, 3, 313, 1988.

123. **Jenkins, J. R., Chumakov, P., Addison, C., Sturzbecher, H. W., and Wade-Evans, A.,** Two distinct regions of the murine p53 primary amino acid sequence are implicated in stable complex formation with simian virus 40 T antigen, *J. Virol.*, 62, 3903, 1988.

124. **Fields, S. and Jang, S. K.,** Presence of a potent transcription activating sequence in the p53 protein, *Science*, 249, 1046, 1990.

125. **Raycroft, L., Wu, H. Y., and Lozano, G.,** Transcriptional activation by wild-type but not transforming mutants of the p53 anti-oncogene, *Science*, 249, 1049, 1990.

126. **Eliyahu, D., Raz, A., Gruss, P., Givol, D., and Oren, M.,** Participation of p53 cellular tumor antigen in traansformation of normal embryonic cells, *Nature*, 312, 646, 1984.

127. **Parada, L. F., Land, H., Weinberg, R. A., Wolf, D., and Rotter, V.,** Cooperation between gene encoding p53 tumor antigen and ras in cellular transformation, *Nature*, 312, 649, 1984.

128. **Williams, A. R. W., Piris, J., and Willie, A. H.,** Immunocytochemical demonstration of altered intracellular localizaton of the c-myc oncogene product in human colorectal neoplasms, *J. Pathol.,* 160, 287, 1990.

129. **Sikora, K., Chan, S., Evan, G., Gabra, H., Markham, N., Stewart, J., and Watson, J.,** C-myc oncogene expression in colorectal cancer, *Cancer,* 59, 1289, 1987.

130. **Jones, D. J., Ghosh, A. K., Moore, M., and Schofield, P. F.,** A critical appraisal of the immunohistochemical detection of the c-myc oncogene product in colorectal cancer, *Br. J. Cancer,* 56,779, 1988.

131. **Stewart, J., Evan, G., Watson, J., and Sikora, K.,** Detection of c-myc oncogene in colonic polyps and carcinomas, *Br. J. Cancer,* 53, 1, 1986.

132. **Mariani-Costantini, R., Theillet, C., Hutzell, P., Merlo, G., Schlom, J., and Callahan, R.,** In situ detection of c-myc mRNA in adenocarcinomas, adenomas and mucosa of human colon, *J. Histochem. Cytochem.,* 37, 293, 1989.

133. **D'Emilia, J., Bulovas, K., D'Ercole, K., Wolf, B., Steele, G., Jr., and Summerhayes, I. C.,** Expression of the c-erbB-2 gene product (p185) at different stages of neoplastic progression in colon, *Oncogene,* 4, 1233, 1989.

134. **Alitala, K., Wingqvist, R., Lin, C. C., de al Chapelle, A., Schwab, M., and Bishop, J. M.,** Aberrant expression of an amplified c-myb oncogene in two cell lines from a colon carcinoma, *Proc. Natl. Acad. Sci. U.S.A.,* 81, 4534, 1984.

135. **Goelz, S. E., Vogelstein, B., Hamilton, S. R., and Feinberg, A. P.,** Hypomethylation of DNA from benign and malignant human colon neoplasms, *Science,* 228, 187, 1985.

136. **Feinberg, A. P., Gehrke, C. W., Kuo, K. C., and Erlich, M.,** Reduced genomic 5-methylcytosine content in human colonic neoplasia, *Cancer Res.,* 48, 1159, 1988.

137. **Schmid, M., Haaf, T., and Grunert, D.,** 5-azacytidine-induced undercondensations in human chromosomes, *Hum. Genet.,* 67, 257, 1984.

138. **Luk, G. D. and Baylin, S. B.,** Ornithine decarboxylase as a biological marker in familial colon polyposis, *N. Engl. J. Med.,* 311, 80, 1984.

139. **Porter, C. W., Herrera-Ornelas, L., Pera, P., Petrelli, N. F., and Mittelman, A.,** Polyamine biosynthetic activity in normal and neoplastic human colorectal tissues, *Cancer,* 60, 1275, 1987.

140. **Radford, D. M., Nakai, H., Eddy, R. L., Haley, L. L., Byers, M. G., Henry, W. M., Lawrence, D. D., Porter, C. W., and Shows, T. B.,** Two chromosomal locations for human ornithine decarboxylase gene sequences and elevated expression in colorectal neoplasia, *Cancer Res.,* 50, 6146, 1990.

141. **Upp, J. R., Jr., Saydjari, R., Townsend, C. M., Singh, P., Barranco, S. C., and Thomson, J. C.,** Polyamine levels and gastrin receptors in colon cancers, *Ann. Surg.,* 207, 662, 1988.

142. **Markus, G., Takita, H., Camiolo, S. M., Corasanti, J. G., Evers, J. L., and Hobika, G. H.,** Content and characterization of plasminogen activators in human lung tumors and normal lung tissue, *Cancer Res.,* 40, 841, 1980.

143. **Kirchheimer, J. C., Koller, A., and Binder, B. R.,** Isolation and characterization of plasminogen activators from hyperplastic and malignant prostate tissue, *Biochem. Biophys. Acta,* 797, 256, 1984.

144. **Tissot, J. D., Hauert, J., and Bachmann, F.,** Characterization of plasminogen activators from normal human breast and colon and from breast and colon carcinomas, *Int. J. Cancer,* 34, 295, 1984.

145. **Corasanti, J. G., Celik, C., Camiolo, S. M., Mittelman, A., Evers, J. L., Barbasch, A., Hobika, G. H., and Markus, G.,** Plasminogen activator content of human colon tumors and normal mucosae: separation of enzymes and partial purification, *J. Natl. Cancer Inst.,* 65, 345, 1980.

146. **de Bruin, P. A. F., Verspaget, H. W., Griffioen, C., Nap, M., Verheijen, J. H., and Lamers, C. B. H. W.,** Plasminogen activator activity and composition in human colorectal carcinomas, *Fibrinolysis,* 1, 57, 1987.

147. **de Bruin, P. A. F., Griffioen, G., Verspaget, H. W., Verheijen, J. H., and Lamers, C. B. H. W.,** Plasminogen activator profiles development in the human colon: activity levels in normal mucosa, adenomatous polyps, and adenocarcinomas, *Cancer Res.*, 47, 4654, 1987.

148. **Gold, P. and Freedman, S. O.,** Demonstration of tumor specific antigens in human colonic carcinomata by immunologic tolerance and absorption techniques, *J. Exp. Med.*, 121, 439, 1965.

149. **Zamcheck, N.,** The present status of CEA in diagnosis, prognosis and evaluation of therapy, *Cancer*, 36, 2460, 1975.

150. **Sikorska, H., Shuster, J., and Gold, P.,** Clinical application of carcinoembryonic antigen, *Cancer Detect. Prev.*, 12, 321, 1988.

151. **Itzkowitz, S. H., Shi, Z. R., and Kim, Y. S.,** Heterogeneous expression of two oncodevelopmental antigens, CEA and SSEA-1 in colorectal cancer, *Histochem. J.*, 18, 155, 1986.

152. **Goldenberg, D. M., Sharkey, R. M., and Primus, F. J.,** Carcinoembryonic antigen in histopathology: immunoperoxidase staining of conventional tissue sections, *J. Natl. Cancer Inst.*, 57, 11, 1976.

153. **Tsao, D., Shi, Z.-R., Wong, A., and Kim, Y. S.,** Effect of sodium butyrate on carcinoembryonic antigen produced by human colonic adenocarcinoma cells in culture, *Cancer Res.*, 43, 1217, 1983.

154. **Niles, R. M., Wilhelm, S. A., Thomas, P., and Zamcheck, N.,** The effects of sodium butyrate and retinoic acid on growth and CEA production in a series of human colorectal tumor cell lines representing different states of differentiation, *Cancer Invest.*, 6, 39, 1988.

155. **Robey-Cafferty, S. S., Ro, J. Y., Ordonez, N. G., and Cleary, K.,** Transitional mucosa of colon: a morphological, histochemical and immunohistochemical study, *Arch. Pathol. Lab. Med.*, 114, 72, 1990.

156. **Isaacson, P. and Attwood, P. R. A.,** Failure to demonstrate specificity of the morphological and histochemical changes in mucosa adjacent to colonic carcinoma (transitional mucosa), *J. Clin. Pathol.*, 32, 214, 1979.

157. **Deschner, E. E. and Lipkin, M.,** Proliferative patterns in colonic mucosa in familial polyposis, *Cancer*, 35, 413, 1975.

158. **Lipkin, M., Blattner, W. E., Fraumeni, J. F., Jr., Lynch, H. T., Deschner, E. E., and Winawer, S.,** Tritiated thymidine (phi p, phi h) labeling distribution as a marker for hereditary predisposition to colon cancer, *Cancer Res.*, 43, 1899, 1983.

159. **Deschner, E. E. and Raicht, R. F.,** Kinetics and morphologic alterations in the colon of a patient with multiple polyposis, *Cancer*, 47, 2440, 1981.

160. **Deschner, E. E.,** Early proliferative changes in gastrointestinal neoplasia, *Am. J. Gastroenterol.*, 77, 207, 1982.

161. **Kanomitsu, T., Koike, A., and Yamamoto, S.,** Study of the cell proliferation kinetics in ulcerative colitis, adenomatous polyps, and cancer, *Cancer*, 56, 1094, 1985.

162. **Deschner, E. E.,** Cell proliferation as a biological marker in human colorectal neoplasia, in *Colorectal Cancer: Prevention, Epidemiology and Screening*, Winawere, S. J., Schottenfeld, D., and Sher, P., Eds., Raven Press, New York, 1980, 133.

163. **Bleiberg, H., Buyse, M., and Galand, P.,** Cell kinetic indicators of premalignant stages of colorectal cancer, *Cancer*, 56, 124, 1985.

164. **Ponz de Leon, M., Roncucci, L., Di Donato, P., Tassi, L., Smerieri, O., Amorico, M. G., Malagoli, G., De Maria, D., Antonioli, A., Chahin, N. J., Perini, M., Rigo, G., Barberini, G., Manenti, A., Biasdco, G., and Barbara, L.,** Pattern of epithelial cell proliferation in colorectal mucosa of normal subjects and of patients with adenomatous polyps or cancer of the large bowel, *Cancer Res.*, 48, 4121, 1988.

165. **Bourry, J., Gioanni, J., Ettore, F., Giacomini, M. A., Simon, J. M., and Courdi, A.,** Labeling index and labeling distribution in the colonic crypts: a contribution to definition of patients at high risk for colorectal cancer, *Biomed. Pharmacother.*, 41, 151, 1987.

166. **Terpstra, O. T., van Blankenstein, M., Dees, J., and Eilers, G. A.**, Abnormal pattern of cell proliferation in the entire colonic mucosa of patients with colon adenoma or cancer, *Gastroenterology*, 92, 704, 1987.

167. **Moyer, M. P.**, Culture of human gastrointestinal epithelial cells, *Proc. Soc. Exp. Biol. Med.*, 174, 12, 1983.

168. **Yeh, K.-Y. and Chopra, D. P.**, Epithelial cell cultures from the colon of the suckling rat, *In Vitro*, 16, 976, 1980.

169. **Chopra, D. P. and Yeh, K.-Y.**, Long-term culture of epithelial cells from the normal rat colon, *In Vitro*, 17, 441, 1981.

170. **Moyer, M. P. and Aust, J. B.**, Human colon cells: culture and in vitro transformation, *Science*, 224, 1445, 1984.

171. **Moyer, M. P. and Aust. J. B.**, Phenotypic changes and gene expression in human colon mucosal epithelial cells upon transfection of a SV40 DNA-gpt recombinant, *In Vitro Cell. Dev. Biol.*, 23, 141, 1987.

172. **Siddiqui, K. M. and Chopra, D. P.**, Unscheduled DNA synthesis in colon epithelial cells exposed in vitro to carcinogens or bile acids, *Proc. Am. Assoc. Cancer Res.*, 27, 102, 1986.

173. **Paraskeva, C., Buckle, B. G., Sheer, D., and Wigley, C. B.**, The isolation and characterization of colorectal epithelial cell lines at different stages in malignant transformation from familial polyposis coli patients, *Int. J. Cancer*, 34, 49, 1984.

174. **Paraskeva, C., Finerty, S., and Powell, S.**, Immortalization of human colorectal adenoma cell line by continuous in vitro passage: possible involvement of chromosome 1 in tumor progression, *Int. J. Cancer*, 41, 908, 1988.

175. **Kovacs, G.**, Preferential involvement of chromosome 1q in a primary breast carcinoma, *Cancer Genet. Cytogenet.*, 3, 125, 1981.

176. **Atkin, N. B. and Pickthall, V. J.**, Chromosome 1 in 14 ovarian cancers: heterochromatin variants and structural changes, *Hum. Genet.*, 38, 25, 1977.

177. **Kovacs, G.**, Abnormalities of chromosome no. 1 in human solid malignant tumors, *Int. J. Cancer*, 21, 688, 1978.

178. **Willson, J. K. V., Bittner, G. N., Oberley, T. D., Meisner, L. F., and Weese, J. L.**, Cell culture of human colon adenomas and carcinomas, *Cancer Res.*, 47, 2704, 1987.

179. **Williams, A. C., Harper, S. J., and Paraskeva, C.**, Neoplastic transformation of a human colonic epithelial cell line: in vitro evidence for the adenoma to carcinoma sequence, *Cancer Res.*, 50, 4724, 1990.

Chapter 8

MULTISTEP CARCINOGENESIS AND HUMAN EPITHELIAL CELLS

Johng S. Rhim

TABLE OF CONTENTS

I. INTRODUCTION

It is now widely accepted that cancer arises in a multistep fashion and that environmental exposure, particularly to physical, chemical, and biological agents, is a major etiological factor.[1,2] Besides chemicals, irradiation, and viruses, other influences (e.g., genetic, hormonal, nutritional, and multifactor interactions) are also involved. While the majority of studies of carcinogenesis have relied on the use of rodent cells in culture, experimental models to define the role of carcinogenic agents in the development of human cancer must be established using human cells. Thus, the study of human cell transformation *in vitro* by carcinogenic agents is of particular importance for understanding the cellular and molecular mechanisms underlying human carcinogenesis. In keeping with the multistep development of human cancer *in vivo*, a stepwise approach to neoplastic transformation *in vitro* presents a reasonable strategy.

For many years, our interest has been in studying the mechanism of carcinogenesis *in vitro* using rodent cells and, more recently, human cells. Our interests have been (1) to develop suitable *in vitro* model systems, (2) to define the factors (enhancers or suppressors) that modulate cellular transformation, (3) to examine the usefulness of the defined *in vitro* model system for assaying carcinogenic agents, and (4) to study the molecular, cellular, and genetic mechanisms of the neoplastic process.

Unlike rodent cells, normal human cells in culture do not, or rarely, undergo spontaneous transformation and have generally proven resistant to neoplastic transformation by carcinogens.[3-7] Previous transformations of human cells have mostly been with fibroblastic cells, which are relatively easy to culture. While the use of DNA tumor viruses,[8,9] X-ray,[10] and chemical carcinogens[11,12] has led to the development of established, biologically abnormal lines of fibroblasts, neoplastic transformation has proven very difficult to achieve. Recently, neoplastic conversion of immortalized, nontumorigenic human fibroblasts expressing the SV40 tumor antigen[13] or induced by irradiation[14] was achieved by infection with murine sarcoma viruses. Possibly, transformation of human fibroblasts is complicated by the requirement, similar to that observed in primary rodent fibroblasts, of two separate genetic events, one for rescue from senescence and another for conversion to the tumorigenic phenotype.[15,16]

Since epithelial cells are the cells of origin for most human cancer, an epithelial cell culture system is a crucial tool for studying the cellular changes that take place in cells during malignant transformation and the molecular mechanisms whereby carcinogens and oncogenes induce the process. However, because of the inability until recently to grow human epithelial cells and subsequently transform them *in vitro*, it has been difficult to define the process of neoplastic transformation of human epithelial cells.[17,18]

We began to study human epithelial cell carcinogenesis by asking several simple questions: (1) Do highly oncogenic RNA or DNA tumor viruses induce morphological alterations or alter the growth properties of primary human epithelial cells? (2) Can virus-transformed human epithelial cells be maintained as stably established cell lines? (3) Do virus-transformed cell lines induce carcinomas when transplanted into nude mice? We used primary human foreskin epidermal keratinocytes to ascertain whether prototype RNA (Kirsten murine sarcoma virus, Ki-MSV) or DNA (Ad12-SV40 hybrid virus) tumor viruses could confer the malignant phenotype to normal primary human epithelial cells. In doing so, we were able to develop for the first time an

TABLE 1
Biologic Properties of Human Epidermal Keratinocytes Exposed to Ki-MSV or Ad12-SV40 Virus

Cells	Passage in culture (number)	Agar colony formation (%)	Nude mice with tumors[a]	
			10⁷	10⁶
Primary human keratinocytes	<3	<0.01		
+ Ki-MSV	<3	<0.01		
+ Ad12-SV40	>50	<0.01	0/20[b]	0/4
+ Ad12-SV40 + Ki-MSV	>50	0.1—0.50	16/18*	3/4*

Note: Asterisk (*) indicates tumors were reestablished in tissue culture and confirmed as human; their resemblance to the cells of origin was determined by karyologic analysis. Ki-MSV (BaEV) was produced in human nonproducer cells by superinfection with baboon endogenous virus. Ad12-SV40 virus was grown in Vero cells.

[a]　Nude mice were inoculated with 10⁶ or 10⁷ cells as indicated.
[b]　Number of tumors/number of mice.

in vitro multistep model suitable for the study of human epithelial cell carcinogenesis.[19,20]

In this chapter, we describe the derivation of our *in vitro* multistep human epidermal keratinocyte model, the factors involved in modulating this cellular transformation system, the usefulness of this model system for viral, chemical, and radiation carcinogenesis, and the multistep nature of human epithelial cell carcinogenesis.

II. DERIVATION OF NONTUMORIGENIC HUMAN EPIDERMAL KERATINOCYTE LINE (RHEK-1) BY INFECTION WITH Ad12-SV40 VIRUS

In an attempt to alter the growth properties of primary human epidermal keratinocytes, we used Ki-MSV, a prototype retrovirus whose K-*ras* oncogene has been detected in many human epithelial malignancies,[21,22] and Ad12-SV40 hybrid virus, which induces malignant transformation of fibroblasts in culture. Neither control nor Ki-MSV-infected human epithelial cultures could be propagated serially beyond two or three subcultures (Table 1). In contrast, primary cultures of human epithelial cells infected with Ad12-SV40 infection led to the appearance of actively growing colonies by weeks 3 to 4. By week 6, SV40 tumor (T)-antigen was revealed in the nuclei of a large fraction of the infected cultures by complement fixation and indirect immunofluorescence staining. A number of cell lines were obtained from colonies that proliferated at a growth-limiting dilution of cells. All lines but one released Ad12-SV40 virus, as indicated by the induction of a cytopathic effect in Vero cells. We selected the nonproducer line, designated Ad12-SV40 line 1 (RHEK-1), for

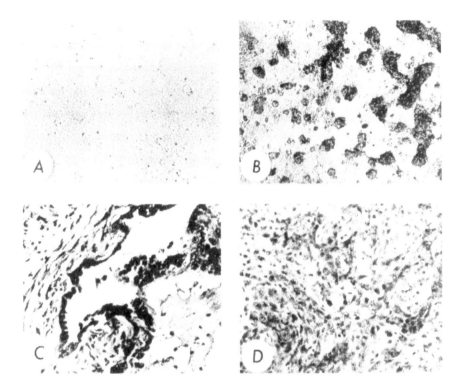

FIGURE 1. (A) Human epidermal keratinocyte line (RHEK-1) and (B) Ki-MSV-transformed RHEK-1 line. (C) Regressing cystic nodules containing epidermal cells induced by RHEK-1 cells. (D) *In vivo* tumor induced by Ki-MSV-transformed RHEK-1 cells. Invasive squamous cell carcinoma with central necrosis.

further characterization. The RHEK-1 line had a flat epithelial morphology (Figure 1A), a number of epithelial cell markers, and was not tumorigenic (see Table 1). In some cases, regressing small cystic nodules containing epidermoid cells appeared at the site of inoculation (Figure 1C).

In experiments to determine which, if any, of the transforming genes in the Ad12-SV40 hybrid virus genomes were actively transcribed in the altered human epithelial cells, molecular characterization of the RHEK-1 line was carried out (Figure 2). The Ad12-SV40-transformed human epithelial cell line had no detectable transcripts from the early region of Ad12, but had substantial amounts of messenger RNA (mRNA) from the transforming region of SV40. Further analysis of the resulting immunoprecipitates (Figure 2B) on a sodium dodecyl sulfate-polyacrylamide gel (SDS-PAGE) revealed both large T- and small t-antigens of SV40, showing that the transforming proteins of SV40 were expressed in this human epithelial cell line. Thus, only the SV40 T-antigens, the transforming proteins coded for by Ad12-SV40 virus, could be responsible for inducing and maintaining the growth properties of this

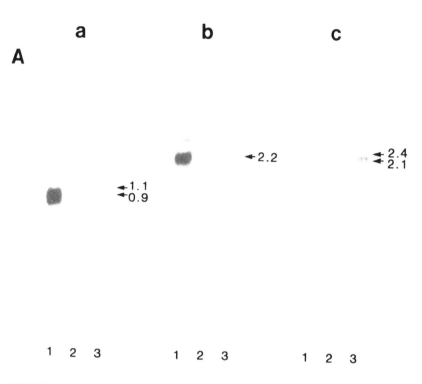

FIGURE 2.   Characterization of Ad12-SV40 early gene expression in Ad12-SV40-altered human epidermal cell line. (A) Detection of virus-specific mRNAs. Poly(A)$^+$ RNA from the Ad12-transformed cell line (lane 1 in each panel), the Ad12-SV40 human epithelial cell line (lane 2 in each panel), or the SV40-transformed cell line (lane 3 in each panel) was fractionated by electrophoresis on a 0.9% agarose gel in the presence of formaldehyde. The RNA was then transferred to a nitrocelluloase membrane and hybridized to one of three ^{32}P-labeled DNA probes: Ad12 EIA (panel a), Ad12 EIB (panel b), or SV40 (panel c). The positions of the viral transcripts are indicated (kilobases). (B) Immunoprecipitation of SV40 T-antigen. Extracts from cells that had been labeled with [^{35}S] methionine for 4 h were subjected to indirect immunoprecipitation with either control hamster serum (lane 1) or hamster SV40 tumor antiserum (lanes 2 and 3). The cell extracts used were from either the Ad12-SV40 human epithelial cell lines (lanes 1 and 2) or SV40-infected African green monkey kidney cells (lane 3). The precipitates were analyzed on a 10% SDS-polyacrylamide gel. The molecular weight markers (in thousands) were phosphorylase B, bovine serum albumin, ovalbumin, carbonic anhydrase, and cytochrome *c*.

established human keratinocyte cell line. This "flat" nonproducer cell line (RHEK-1) (see Figure 1A) has proven useful in our laboratory for studying multistage carcinogenesis.

III. NEOPLASTIC TRANSFORMATION OF HUMAN EPIDERMAL KERATINOCYTES BY Ad12-SV40 VIRUS AND Ki-MSV

The flat epithelial morphology and lack of tumorigenicity of the Ad12-SV40 nonproducer cell line (RHEK-1) led us to inquire whether its growth

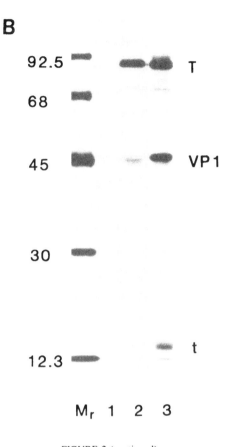

FIGURE 2 (continued).

properties might be further altered by addition of a virus containing an activated *ras* oncogene. Infection of the RHEK-1 line at passage 10 with Ki-MSV (BaEV) resulted in a striking alteration in cell morphology. As early as 5 to 6 d after infection, the cells began to pile up in focal areas, forming small projections and releasing round cells from the foci (see Figure 1B). The absence of any detectable alterations induced by helper virus (BaEV) alone implied that Ki-MSV was responsible for the rapid induction of the transformed morphology.

The Ki-MSV-altered RHEK-1 cells released focus-forming viruses and expressed the K-*ras* p21 protein. They not only produced colonies in soft agar, but were tumorigenic in nude mice. When athymic nude mice were inoculated with as few as 10^6 Ki-MSV-transformed RHEK-1 cells (Tables 1 and 2), the animals developed invasive, rapidly progressive tumors within 3 weeks. Such tumors were diagnosed as squamous cell carcinomas (Figure 1D) with characteristic keratin pearls. Cell lines established from the tumors were readily transplantable, and were confirmed as being of human origin

TABLE 2
Ki-MSV-Induced Growth Alterations of Epidermal RHEK-1 Cells

Properties	Ki-MSV (BaEV)	Uninfected
Morphology	Transformed	Flat
Saturation density	9.1×10^6	3.1×10^6
Reverse transcriptase	Positive	Negative
Type C virus particles	Positive	Negative
Induction of foci after inoculation of culture supernate in human cells	Positive	Negative
p21 *ras* protein	High	Low
Tumorigenicity in nude mice	Positive	Negative

and resembling the cells of origin by karyological analysis. These findings demonstrate the malignant transformation of human primary epithelial cells in culture by the combined action of SV40 T-antigen and Ki-MSV, and support a multistep process for neoplastic conversion.

Several investigators have reported that primary rodent fibroblasts can undergo neoplastic conversion in response to the combined action of two viral or cellular oncogenes.[15,16] Our study is, to our knowledge, the first to show neoplastic conversion of human epithelial cells in culture and to define the minimum number of transforming genes that appear to be required.

IV. HYDROCORTISONE ENHANCES Ki-MSV-INDUCED FOCUS FORMATION IN RHEK-1 CELLS

In an attempt to achieve maximum transformation efficiency, the effect of hydrocortisone on focus formation by Ki-MSV in human epidermal keratinocytes was examined. Hydrocortisone has previously been shown to significantly enhance Ki-MSV-induced transformation in human skin fibroblasts.[23] The results showed that hydrocortisone significantly enhances focus formation in RHEK-1 cells (Table 3). The maximum effect, a 20-fold increase in focus formation, was seen at a hydrocortisone concentration of 5µg/ml. A concentration as low as 1 µg/ml also had a significant effect. In the hydrocortisone-treated human epidermal cells, Ki-MSV produced larger and well-defined foci which could be counted 7 d after infection. In contrast, in untreated human epidermal cells, foci were small and barely visible, and could not be counted until 14 d after infection. Therefore, the medium containing a hydrocortisone concentration of 5 µg/ml was used throughout our transformation experiments.

V. NEOPLASTIC CONVERSION OF HUMAN EPIDERMAL KERATINOCYTES BY Ad12-SV40 VIRUS AND CHEMICAL CARCINOGENS

Since certain carcinogenic polycyclic hydrocarbons have been identified

TABLE 3
Effect of Hydrocortisone on the Incidence of Transformed Foci in Ki-MSV (BaEV)-Infected RHEK-1 Human Epidermal Line[a]

Hydrocortisone (μg)	Virus titer (FFU[b]/ml)	Enhancement (fold)
0	1.0×10^3	—
1.0	1.2×10^4	12
2.5	1.5×10^4	15
5.0	2.0×10^4	10

Note: Ki-MSV (BaEV) was produced in human nonproducer cells by superinfection with baboon endogenous virus (BaEV).

[a] One day after the 15th subculture, the RHEK-1 line was plated at 5×10^5 cells per 100 mm in a Falcon plastic dish with DMEM + 10 % FBS with and without various concentrations of hydrocortisone. The infected cultures were incubated at 37° in 5% CO_2, refed with the same medium, and foci counted 20 d after infection.

[b] Focus-forming units.

in our environment and some chemicals are known definitely to cause cancers in humans, it is important to study the response of human cells to such compounds. However, there was no reproducible system for carcinogen-induced neoplastic transformation of human epithelial or fibroblastic cells in culture.[6,7] The availability of a human epithelial cell line (RHEK-1) that could undergo neoplastic conversion in response to a *ras* oncogene led us next to inquire whether this system might be useful in detecting chemical carcinogens for human epithelial cells.[24]

After exposure of primary human epidermal keratinocytes to various doses of the chemical carcinogen N-methyl-N-nitro-N-nitrosoguanidine (MNNG) or 4-nitroquinoline-1-oxide (4NQO), no morphological differences between treated and untreated control cultures could be seen. Neither control nor treated cultures were able to grow serially beyond two to three subcultures. The cells underwent progressive deterioration and were lost. In the RHEK-1 line exposed to MNNG at either 0.1 or 0.01 μg/ml, morphological alterations of cells and an abnormal pattern of growth were noted by the sixth subculture, 52 to 62 d after treatment (Table 4). Similar changes were not observed in the control RHEK-1 cells treated with dimethyl sulfoxide (DMSO) only. When RHEK-1 cells were treated with 4NQO at 0.1 μg/ml, morphological alteration was observed in the seventh subculture, 62 d after treatment (see Table 4). A dose of 4NQO at 1.0 μg/ml was lethal. The morphological changes observed in these cultures were similar to those observed with Ki-MSV,[19,20] namely, the transformed cells piled up in focal areas, formed small projections, and

TABLE 4
Morphological Alteration of Human Epithelial Cells (RHEK-1) Treated
with Chemical Carcinogens

Passage	Cumulative number of days in tissue culture after chemical treatment	Morphological changes				
		MNNG (1.0 mg/ml)	MNNG (0.1 mg/ml)	MNNG (0.01 mg/ml)	4NQO (0.1 mg/ml)	DMSO (0.5%)
1	14	—	—	—	—	—
4	35	—	—	—	—	—
6	49	—	+	—	—	—
7	55	—	+	+	+	—
9	72	—	+	+	+	—

Note: One day after plating 1 × 10⁶ cells per ml from the 13th subculture of the RHEK line in Falcon plastic dishes, the medium (DMEM + 15% FGS + HC) was removed and replaced with medium containing MNNG at various concentrations (1.0 to 0.01 mg/ml) or 4NQO (1.0 and 0.1 mg/ml) in 0.5% dimethyl sulfoxide (DMSO). The control medium contained 0.5% DMSO. After 1 d of treatment with carcinogens, the cultures were washed, fed again with carcinogen-free growth medium, and subsequently passaged by trypsin treatment every 7 to 10 d. When changes in morphology and growth patterns appeared, some cultures were fixed in alcohol and stained with Giemsa for further microscopy. Both transformed and untransformed cultures were established as continuous cell lines.

released round cells from the foci (Figure 3A, B, and C). These foci grew in chains or as islets that stained heavily with Giemsa. In contrast, the cellular morphology remained unchanged in the untreated human RHEK-1 epithelial cell line, which continued to grow as nonoverlapping, round to polygonal adherent cells that were flat and cobblestone-like in appearance (Figure 3D). The saturation densities of the chemical transformants were approximately three or four times higher than those of the untreated RHEK-1 cells (Table 5). Moreover, the chemical transformants grew in soft agar with colony-forming efficiencies of 0.6 to 0.8%, whereas the untreated cells did not grow in soft agar (see Table 5). When nude mice were inoculated subcutaneously with 10⁷ chemically transformed cells, the animals developed tumors within 3 to 4 weeks. Such tumors were diagnosed as squamous cell carcinomas (Figure 3E). The cells were arranged in rows or columns, and individual cells often contained keratohyalin granules or prekeratin. Cultures established from the tumors resembled the carcinogen-treated cells (Figure 3F), were confirmed as human, and resembled the cells of origin by karyological analysis. In contrast, subcutaneous inoculation of 10⁷ untreated RHEK-1 cells into nude mice produced regressing cystic nodules that contained epidermal cells (see Table 5). Evidence of the human origin of all the cell lines was obtained by isoenzyme analysis and species-specific cell membrane immunofluorescence. The relatedness of the transformed cells to the parent RHEK-1 cells was

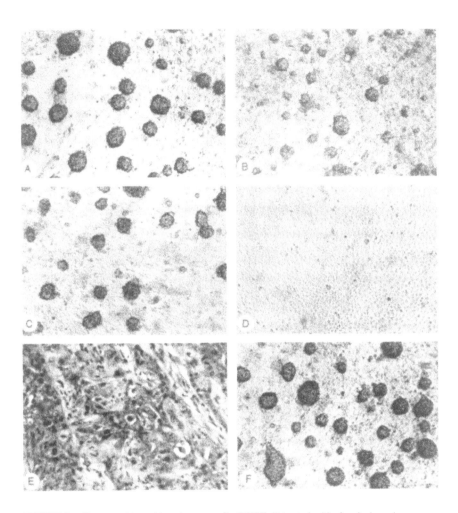

FIGURE 3. Human epidermal keratinocyte cells (RHEK-1) treated with chemical carcinogens for 1 d followed by seven subcultures in nutrient media. Note the morphological alterations in human epithelial cells treated with (A) 0.01 μg/ml MNNG, (B) 0 1 μg/ml MNNG, and (C) 0.1 μg/ml 4NQO. (D) Untreated human cells; (E) *in vivo* tumor induced by RHEK-1 cells treated with 4NQO (0.1 μg/ml), with moderately well-differentiated squamous cell carcinoma; (F) typical field of a culture originated from a primary tumor induced by RHEK-1 cells treated with 4NQO (0.1 μg/ml).

further established by chromosome analysis. Untreated RHEK-1 cells and transformed cells had similar marker chromosomes, as detected by conventional staining.[20] Moreover, there were no major changes in chromosome number. The cells for the most part were near diploid, like the RHEK-1 parent. A small fraction of polyploid cells was observed in both treated and untreated cultures.

These results appear to represent the first induction of human epithelial

TABLE 5
Biological Properties of the RHEK-1 Human Epidermal Line
Transformed by Chemical Carcinogens

Cell line	Saturation density[a] ($\times 10^5/cm^2$)	Soft-agar colony formation[b] (%)	Number with tumors/number inoculated
DMSO (0.5%)	1.9	<0.01	0/19
MNNG (0.1 µg/ml)	5.8	0.6	16/16[c]
MNNG (0.01 µg/ml)	6.4	0.8	18/18[c]
4NQO (0.1 µg/ml)	8.2	0.7	16/17[c]

[a]   Maximum number of cells obtained after initial planting with 5×10^3 cells per square centimeter.

[b] 1×10^5 cells per milliliter plated in 0.33% soft agar.

[c]   Tumors were reestablished in culture and confirmed as human; their resemblance to the cells of origin was determined by karyological analysis.

cancer cells in culture by the concerted action of a DNA tumor virus and chemical carcinogens. At least two and possibly more alterations in cell growth properties seem to be required. The significance of the combined action of Ad12-SV40 virus and chemical carcinogens in the induction of neoplastic transformation of human epithelial cells is emphasized by the inability of chemical carcinogens alone to induce continued proliferation of primary epithelial cells under our assay conditions. Thus, chemical carcinogens are similar to Ki-MSV in their ability to complement Ad12-SV40 virus in fully transforming human epidermal cells.

Unlike the rapid transformation of RHEK-1 cells observed after Ki-MSV infection,[19,20] growth alterations associated with chemical carcinogen treatment were delayed in their appearance and required several subcultures for visualization. These findings suggest that multiple cell divisions are required for fixation and expression of the transformed phenotype in response to the carcinogen. It is possible that more than one genetic lesion may be required as well. Cooperating cellular or viral oncogenes have also been shown to induce malignant transformation of embryonic rodent fibroblasts.[15,16] In addition, the combined action of tumor viruses and chemical carcinogens has been shown to produce neoplastic transformation of rodent fibroblasts.[25-27] Our ability to obtain malignant transformants as a result of chemical carcinogen treatment of Ad12-SV40-altered human epidermal cells provides additional support for a multistep process of neoplastic conversion. This system may be useful in evaluating the carcinogenic potential of environmental chemicals, and for studying genes which are activated and suppressed in the multistep process leading to malignancy.

VI. *Ras* ONCOGENES WERE NOT ACTIVATED IN THE CHEMICALLY TRANSFORMED HUMAN EPIDERMAL (RHEK-1) LINE

The detection and identification of cellular transforming genes from chemical carcinogen-induced animal tumors[28-33] and chemically transformed cells *in vitro*[34-36] by DNA-mediated gene transfer studies with NIH-3T3 cells have made it possible to study the molecular and genetic basis of chemical carcinogenesis. Most transforming genes so far detected by these studies are related to three highly conserved members of the *ras* gene family — H-, Ki-, and N-*ras* — all of which encode closely related proteins generically designated p21.[37,38] Most *ras* oncogenes analyzed have been activated by point mutations in the codons for amino acids 12 or 61.[21,22] These carcinogen-activated *ras* oncogenes have the same type of activating mutation as those present in human tumors.[29,39]

Earlier studies have shown that a continuous line derived from a human osteosarcoma (HOS TE85 clone F-5) treated with MNNG[40] acquired altered growth properties, including tumorigenicity in nude mice. This altered phenotype was later shown to be associated with the activation of a previously uncharacterized cellular transforming gene, designated *met*.[41,42] The *met* gene is activated by gene rearrangement. The 5' end of the activated *met* gene is derived from chromosome 1, while the 3' end is derived from chromosome 7.[43] Additional studies have demonstrated that the region of activated *met* derived from chromosome 7 is homologous to a family of genes that encode protein kinases[44] and is linked closely to the genetic market for cystic fibrosis.[45] Activation of the H-*ras* oncogene has also been demonstrated in a 3-methylcholanthrene (3MC)-transformed human 312H-HOS cell line. Analysis of the p21 *ras* oncogene product in this transformant by immunoprecipitation and gel electrophoresis suggested that this gene was activated by a point mutation at codon 61.[46]

Since RHEK-1 cells can be transformed by Ki-MSV infection and become tumorigenic,[20] we analyzed the *ras* oncogene p21 product in the chemically transformed as well as in the Ki-MSV-transformed RHEK-1 cells by using antibody to p21 and sodium dodecyl sulfate-polyacrylamide gel electrophoresis (SDS-PAGE).[47] In contrast to the findings in the Ki-MSV-transformed cells, neither altered mobility nor increased expression of p21 was observed in the chemically transformed RHEK-1 cells (Figure 4). Moreover, the DNA from these chemically altered cells failed to induce detectable transformed foci upon transfection of NIH-3T3 cells. These results indicate that *ras* oncogenes, which have been implicated in chemical carcinogen-induced animal tumors, spontaneous human tumors, and 3MC-induced human transformed cell lines, were not activated in the chemically transformed human epithelial cell lines so far analyzed. Thus, this system may be useful in efforts to detect and characterize other cellular genes that can contribute to the neoplastic phenotype of human epithelial cells.

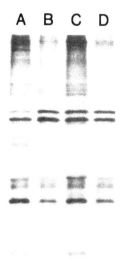

]p21

FIGURE 4.   Analysis of *ras* oncogene p21 product in RHEK-1 cells exposed to chemical carcinogens. [³⁵S]methionine-labeled cell extracts from (A) untreated RHEK-1 cells, (B) Ki-MSV-transformed RHEK-1 cells, (C) MNNG (0.01 μg/ml)-transformed RHEK-1 cells and (D) 4NQO (0.1 μg/ml)-transformed RHEK-1 cells were immunoprecipitated with anti-p21 monoclonal antibody Y13-259 and analyzed by SDS-PAGE as described previously.[47]

VII. MALIGNANT CONVERSION OF HUMAN EPIDERMAL KERATINOCYTES BY Ad12-SV40 VIRUS AND RETROVIRAL ONCOGENES

The availability of a human epithelial cell line (RHEK-1) that could undergo neoplastic conversion in response to a *ras* oncogene[20] has led to investigations to determine whether this system might be useful in assaying other retroviral oncogenes in human epithelial cells.[48]

Analysis of human cell susceptibility to transformation by various retroviral oncogenes has been impaired by the difficulty in introducing these viruses

TABLE 6
Biological Properties of the RHEK-1 Human Epidermal Line Transformed by Retroviral Oncogenes

Transforming gene	Cell line	Morphological alteration	Soft-agar colony formation[a]	Tumorigenicity in nude mice[b]
	Uninfected	—	—	—
v-K-*ras*	Ki-MSV (BaEV)	+	+	+ (10 d)[c]
v-H-*ras*	H-MSV (AP)	+	+	+ (10 d)
v-*bas-ras*	B-MSV (AP)	+	+	+ (10 d)
v-*fes*	ST-FeSV (AP)	+	+	+ (10 d)
v-*fms*	McD FeSV (AP)	+	+	+ (30 d)
v-*erb*B	AEV (AP)	+	+	+ (30 d)
v-*src*	RSV (AP)	+	+	+ (30 d)
v-*fgr*	GR FeSV (AP)	—	—	—
v-*abl*	Abelson MuLV (AP)	—	—	—
v-*sis*	SSV (AP)	—	—	—

[a] Cell suspensions (1 × 10⁵ cells per milliliter) were plated in 0.33% soft-agar medium containing 10% fetal bovine serum. Visible colonies were scored at 21 d.

[b] Nude mice were inoculated with 10⁷ cells.

[c] Tumors were reestablished in tissue culture and confirmed as human. The resemblance to the cells of origin was determined by karyological analysis.

into human cells. Therefore, we used various viral oncogenes, pseudotyped with amphotropic murine leukemia virus (MuLV),[49,50] to facilitate entry into human cells. MuLV pseudotypes of replication-defective viruses containing various oncogenes were obtained by superinfection of appropriate nonproducer cells with a clonal strain of amphotropic MuLV. They include retroviruses containing genes for growth factors (*sis*), growth factor receptors (*erb*B, *fms*), tyrosine kinase (*fes*, *fgr*, *abl*, and *src*) and GTP-binding proteins (Ki-*ras*, *bas*, and H-*ras*) (Table 6).

Approximately 5 to 7 d after infection with viruses containing K-*ras*, H-*ras*, *bas*, *erb*B, *fes*, *fms*, and *src* oncogenes, foci consisting of round cells began to appear in the infected cells. These foci increased in size during the following week and became distinct, forming small projections and releasing round cells; but no foci were observed in uninfected cells or in cells infected with viruses containing *fgr*, *sis*, and *abl* oncogenes (see Table 6). The transformed foci induced by H-*ras* and *bas* v-oncogenes were similar to those obtained with Ki-MSV (BaEV) in RHEK-1 cells.[20] The round foci induced by the v-*fes* oncogene were more diffuse (Figure 5F). The most striking characteristic observed in the v-*erb*B-infected cells was the presence of many ridge formations (Figure 5G). Similar morphological alterations were also observed with various retroviruses in another human epidermal keratinocyte line (designated 11367) established from primary human keratinocytes by pSV₃*neo* transfection.[51]

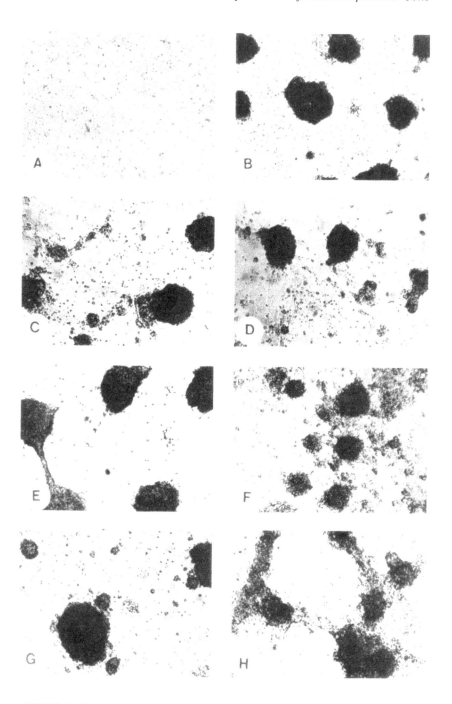

FIGURE 5.   Human epidermal keratinocytes (RHEK-1) infected with retroviruses containing various oncogenes. Foci in RHEK-1 cells 14 d after infection: (A) Uninfected; (B) v-K-*ras*; (C) v-H-*ras*; (D) v-*bas-ras*; (E) v-*fms*; (F) v-*fes*; (G) v-*erb*B; (H) v-*src*.

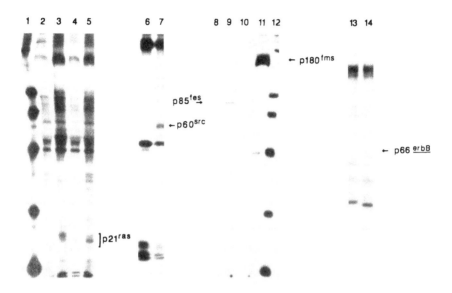

FIGURE 6. Analysis of retroviral-specific protein products in RHEK-1 cells infected with various retroviral oncogenes. [³⁵S]methionine-labeled cell extracts from uninfected RHEK-1 cells (lanes 2, 6, 8, 10, and 13), v-Ki-*ras*-transformed RHEK-1 cells (lane 3), v-H-*ras*-transformed RHEK-1 cells (lane 4), v-*bas-ras*-transformed RHEK-1 cells (lane 5), v-*src*-transformed RHEK-1 cells (lane 7), v-*fes*-transformed RHEK-1 cells (lane 9), v-*fms*-transformed RHEK-1 cells (lane 11), and v-*erb*B-transformed RHEK-1 cells lane (14) were immunoprecipitated with respective anti-retrovirus antibodies and analyzed by SDS-PAGE. Protein size markers (lanes 1 and 12) were myosin heavy-chain (200 kDa), phosphorylase B (97.4 kDa), bovine serum albumin (68 kDa) or albumin (43 kDa), α-chymotrypsinogen (25.7 kDa), and β-lactoglobin (18.4 kDa).

The retroviral oncogene-transformed cells were further characterized. The viral transformants grew in soft agar and were found to release virus continuously. Cell-free preparations of supernatant fluid from the *in vitro* transformed cells produced foci in NIH-3T3 cells. The altered cells contained reverse transcriptase activity, and analysis of immunoprecipitates of cell extracts by SDS-PAGE revealed virus-specific proteins (p21^{ras}, p60^{src}, p85^{fes}, p180^{fms}, and p66^{erbB}), confirming the presence and expression of the respective retroviral oncogene (Figure 6).[52-56] All the transformants induced carcinomas when transplanted into nude mice. Cultures established from the tumors resembled the original transformed cells and were confirmed to be of human origin by karyological analysis. In contrast, subcutaneous inoculation of 10⁷ uninfected RHEK-1 cells into nude mice produced occasional regressing cystic nodules that contained epidermal cells.[20] These findings demonstrate the malignant transformation of human primary epithelial cells in culture by the combined action of Ad12-SV40 virus and retroviral oncogenes, and support a multistep process for neoplastic conversion. This *in vitro* system may be useful in studying the interaction of a variety of retroviral oncogenes and human epithelial cells.

VIII. NEOPLASTIC TRANSFORMATION OF HUMAN KERATINOCYTE LINE (RHEK-1) BY POLYBRENE-INDUCED, DNA-MEDIATED TRANSFER OF AN ACTIVATED HUMAN ONCOGENE

Calcium phosphate transfection has been widely used for introducing genes into cells in culture.[57] When we employed this procedure, as modified by Sutherland and Bennet,[58] on RHEK-1 cells to study the uptake and expression of a plasmid CNA carrying a dominant selectable marker (pSV$_2$*neo*),[59] the frequency of transfectants per 1×10^5 cells ranged from none to a few. Because of this low frequency, we had to develop an alternative method. Polybrene, in conjunction with DMSO shock, has been shown to increase the frequency of DNA transfection of mammalian cells compared to the frequency obtained with calcium phosphate transfection.[60,61] We have successfully adapted this procedure for use with the RHEK-1 cells.[62] To identify a parameter that would yield the maximum number of geneticin-resistant colonies per dish, RHEK-1 cells were exposed to a polybrene concentration ranging from 0 to 30 μg/ml and a DMSO concentration of 10, 20, or 30%. The transfection frequency was dependent on the concentration of both polybrene and DMSO. The maximum transfection frequency (defined as the highest average number of geneticin-resistant colonies observed per dish) was obtained when the cells were treated with 10 mg/ml polybrene and shocked with 30% DMSO. Under these conditions, a frequency of approximately 100 resistant colonies per 1×10^5 cells transfected was obtained (Figure 7). This was about 100 times higher than the frequency we observed using the same RHEK-1 cells with the calcium phosphase method. Therefore, for our oncogene transfection study, the RHEK-1 cells were exposed for 24 h in medium containing polybrene at a concentration of 10 μg/ml and followed by a shock for 4 min inmedium containing 30% DMSO.

The RHEK-1 cells were neoplastically transfected, using polybrene at a concentration of 10 μg/ml followed by a 4-min shock with 30% DMSO, with a plasmid carrying the activated H-*ras* gene from the EJ bladder carcinoma cell line.[63] The transfected cells showed morphological alterations (Figure 8) and induced carcinomas (Figure 9) when transplanted into nude mice. They contained integrated copies of the transfected H-*ras* gene (Figure 10A) and expressed high levels of the p21 protein (Figure 10B). Polybrene-induced DNA transfection, therefore, offers the opportunity to transfer genes effectively into human epidermal keratinocytes and should accelerate the study of the interaction between oncogenes and human epithelial cells.

IX. NEOPLASTIC CONVERSION OF Ad12-SV40-IMMORTALIZED HUMAN EPIDERMAL LINE (RHEK-1) BY IONIZING RADIATION

The carcinogenic action of ionizing radiation in humans has been well

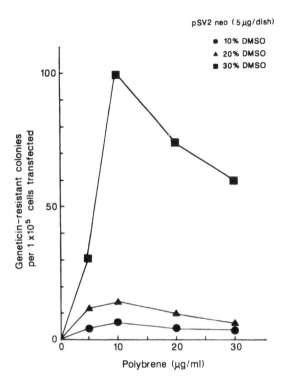

FIGURE 7. Conditions for transfection of RHEK-1 cells with pSV$_2$*neo* DNA. RHEK-1 cells were treated with 2.5 ml of Dulbecco's modified minimum essential medium (DMEM) containing 10% FBS, 5.0 μg pSV$_2$*neo* DNA, and the indicated dose of polybrene. After a 6-h incubation at 37°C, the cells were shocked for 4 min at room temperature with DMEM containing 10% FBS and the indicated dose of DMSO. The cells were washed twice and refed. Geneticin was added 18 to 24 h later.

recognized from epidemiological data. Despite this fact, there has been no model to study the radiation-induced neoplastic transformation of human cells, particularly those of epithelial origin. We have therefore examined the susceptibility of the RHEK-1 cell line to X-ray radiation.[64] The RHEK-1 line irradiated twice at either 2 or 4 Gy showed morphological alteration by the third subculture 6 to 7 weeks later (Table 7). Similar changes were not observed in either the unirradiated or twice (6 and 8 Gy)-irradiated RHEK-1 cells. The morphological changes observed in these cultures were similar to those observed with chemical carcinogens (Figure 11, 1a). Such transformants formed colonies in soft agar (Figure 11, 2a) and induced carcinomas when transplanted into nude mice (Table 8). The transformed lines derived from soft-agar colonies were highly tumorigenic. All the mice inoculated with

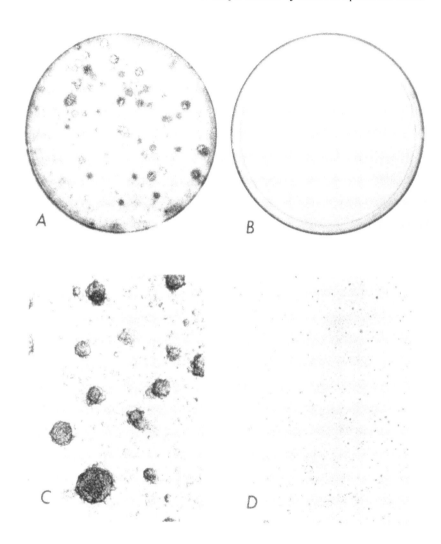

FIGURE 8.   Transfection of human epidermal keratinocyte cell line (RHEK-1) with pSV₂*ras*. (A,C) Foci in RHEK-1 cells 21 d after transfection; (B,D) untransfected RHEK-1 cells.

as few as 10^6 2 × 2 Gy-irradiated transformed cells developed progressively growing tumors within 4 weeks (see Table 8). Colonies established from those tumors resembled the radiation-treated cells, were confirmed as human, and resembled the cells of origin by karyological analysis. In contrast, subcutaneous inoculation of 10^7 unirradiated RHEK-1 cells into nude mice produced usually regressing cystic nodules containing epidermal cells. These findings demonstrate the malignant transformation of human primary epithelial cells in culture by the combined action of a DNA tumor virus and radiation, indicating a multistep process for radiation-induced neoplastic conversion.

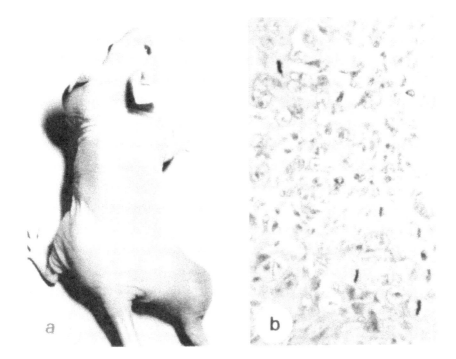

FIGURE 9. Tumorigenicity of pSV₂*ras*-transformed RHEK-1 cells in athymic nude mice. (a) Gross appearance of subcutaneous (left neck) and intramuscular (right thigh) tumor masses at 14 d after inoculation of cells. (b) Histological examination of a paraffin-fixed, hematoxylin and eosin-stained section of a tumor mass.

This *in vitro* system may be useful as a tool for dissecting the process of radiation-induced neoplastic transformation of human epithelial cells.

While the activation of cellular *ras* oncogenes has been demonstrated in rodent tumors induced by ionizing radiation,[33,39,65] the activation of unique non-*ras* oncogenes has been shown in malignant, radiogenic-transformed rodent cells.[66] The neoplastic transformation of the RHEK-1 human epithelial line by X-ray irradiation suggests that cellular oncogenes may be activated as part of the process. Therefore, we analyzed the *ras* oncogene p21 product in the radiation-transformed as well as the Ki-MSV-transformed RHEK-1 cells by using antibody to p21 and SDS-PAGE. In contrast to the findings in the Ki-MSV-transformed cells, neither altered mobility nor increased expression of p21 was observed in the radiation-transformed RHEK-1 cells. Moreover, the DNA from these radiation-altered cells has failed so far to induce detectable transformed foci upon transfection of NIH-3T3 cells.[64] These findings indicate that the activation of *ras* oncogenes is not involved in the radiation-induced human epithelial cell lines analyzed. Thus, this system may be useful in efforts to detect and characterize other cellular genes that can contribute to the neoplastic phenotype of human epithelial cells.

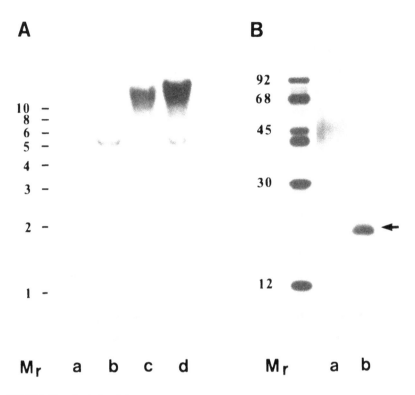

FIGURE 10. Analysis of the presence of the transfected H-*ras* gene and its expression in the transformed RHEK-1 cells. (A) Southern blot hybridization analysis of DNA from nontransfected RHEK-1 cells (lane a), transfected RHEK-1 cells (lane b), uninvolved tissue from a nude mouse (lane c), and tumored tissue from a nude mouse inoculated with transformed RHEK-1 cells (lane d). The hybridization probe used was derived from the v-H-*ras* gene. The arrow indicates a component of 5.2 kilobase in size. (B) Western immunoblot analysis of total protein obtained from either nontransfected (lane a) or transformed (lane b) RHEK-1 cells, using a pan-reactive rabbit antibody against p21. The immunoreactive component was visualized by binding of ^{125}I-labeled protein A. The arrow indicates the position of the p21 protein.

X. TRANSFORMING GENES FROM RADIATION-TRANSFORMED HUMAN EPIDERMAL KERATINOCYTES DETECTED BY A TUMORIGENICITY ASSAY

DNA-mediated gene transfer studies using rodent cells as recipients have demonstrated the presence of transforming genes in radiation-induced tumors and rodent cells transformed by radiation.[33,39,65,66] As described above,[64] there were no detectable transformed foci upon transfection of NIH-3T3 cells with the DNAs from the radiation-altered human epidermal cells. Therefore, we have now tested the DNAs from these transformants by a tumorigenicity assay, since the tumorigenicity assay has been shown to detect weak transforming genes.[67,68] The DNAs from two highly tumorigenic radiation-altered

TABLE 7
Morphological Alteration of RHEK-1 Human Epidermal Cells by Exposure to X-Ray Irradiation

Passage	Cumulative number of days in tissue culture after X-ray irradiation	Morphological changes				
		4 Gy (2 Gy × 2)	8 Gy (4 Gy × 2)	12 Gy (6 Gy × 2)	16 Gy (8 Gy × 2)	None
1	10	—	—	—	—	—
2	25	—	—	—	—	—
3	42	—	+	—	—	—
4	52	+	+	—	—	—
8	72	+	+	—	—	—

Note: One-day old cultures of the RHEK-1 cells (plated at 5×10^5 cells per 80-cm² flask) were irradiated with graded doses of X-rays (0.2, 4, 6, and 8 Gy). Following irradiation, the cultures were allowed to grow to confluence with a change of medium every 3 d, subsequently passaged by trypsin treatment, and irradiated again with same doses. Cultures were subcultured every 7 to 10 d and observed biweekly for changes in morphology or growth pattern.

soft-agar clones (4 and 8 Gy) induced *Alu*-positive tumors in nude mice. Both positive primary nude mouse tumor DNAs were retransmitted to a second round of analysis in the tumorigenicity assay with high frequency and short latency, and were found to be *Alu* positive. The DNAs from the *Alu*-positive secondary nude mouse tumors were screened for homology with probes for the *ras* gene family. None of the *Alu*-positive bands were found to be N-, K-, or H-*ras*. Subsequent analysis has also eliminated the c-*raf* gene. Further characterization of these transforming genes is in progress. The results so far indicate that members of the *ras* oncogene family are not activated in the radiation-transformed human epidermal lines.

XI. MORPHOLOGICAL TRANSFORMATION OF HUMAN KERATINOCYTES (RHEK-1) EXPRESSING THE LMP GENE OF EPSTEIN-BARR VIRUS (EBV)

The cross-association of EBV with nasopharyngeal carcinoma (NPC) has been known for some time,[69] but the precise role of EBV in this cancer is poorly understood, due partly to the lack of an *in vitro* system for studying NPC cells and the effect of EBV on epithelial cells. Biopsies of NPC tumors have revealed expressions of the EBV latent membrane protein (LMP) in 65% of cases,[70] suggesting that in at least some NPC tumors, LMP may cause cell transformation. Here, we addressed the questions of the effect of LMP expression on human epithelial cells.[71]

We have transfected an immortalized, nontumorigenic human keratinocyte line (RHEK-1) with the EBV-encoded LMP and the EBNA-2 gene,

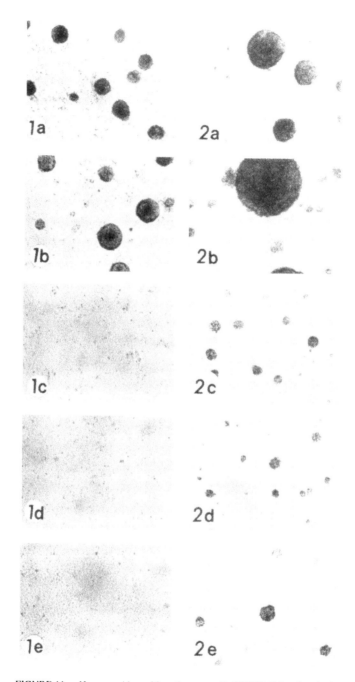

FIGURE 11.   Human epidermal keratinocyte cells (RHEK-1) irradiated with
X-ray twice, followed by a third subculture in nutrient medium (1a) 4 Gy
(2 Gy × 2), (1b) 8 Gy (4 Gy × 2), (1c) 12 Gy (6 Gy × 2), (1d) 16 Gy (8
Gy × 2), and (1e) unirradiated. The colonies produced in soft agar by these
cells: (2a) 4 Gy (2 Gy × 2), (2b) 8 Gy (4 Gy × 2), (2c) 12 Gy (6 Gy ×
2), (2d) 16 Gy (8 Gy × 2), and (2e) unirradiated RHEK-1 cells.

TABLE 8
Biological Properties of RHEK-1 Human Epidermal Line Transformed by X-Ray Irradiation

X-ray total dose	Saturation density[a] (10⁵/cm²)	Soft agar colony formation (%)[b]	Nude mice with tumors (tumors)[c] 10⁷	10⁶
4 Gy (2 Gy × 2)	4.5	0.56	2/4[d]	ND
4 Gy (2 Gy × 2) SA[e]	5.2	0.63	4/4[f]	4/4
8 Gy (4 Gy × 2)	4.6	0.22	2/4[f]	ND
8 Gy (4 Gy × 2) SA[e]	4.8	0.30	3/4	
12 Gy (6 Gy × 2)	2.5	0.02	0/4	ND
16 Gy (8 Gy × 2)	2.7	<0.01	0/4	ND
None	2.1	<0.01	0/4	0/4

[a] Saturation density was measured as the maximum number of cells obtained after initial plating with 5×10^3 cells per cm² and then incubating at 36°C, with growth medium changed every 3 d.

[b] Cell suspension (1×10^5 cells per milliliter) were plated on 0.33% soft-agar medium containing 10% fetal bovine serum.

[c] Nude mice were inoculated with 10^6 or 10^7 cells as indicated.

[d] Number of tumors/number of mice; ND = test not done.

[e] Lines derived from soft-agar colonies

[f] Tumors were reestablished in tissue culture and confirmed as human; their resemblance to the cells of origin was determined by karyological analysis.

respectively. EBNA-2 is known to play an essential role in the activation of virally infected B-cells, but is not expressed in the two main EBV-carrying tumors, Burkitt's lymphoma and NPC. The majority of the EBNA-2-transfected cells expressed the nuclear antigen, but showed no detectable morphological change. The LMP-transfected cells expressed the full-size (63 kDa) membrane protein and a striking morphological change (Figure 12). Immunofluorescence and Western blot analysis with anti-cytokeratin monoclonal antibodies, PKK-1 and PKK-2, revealed a total down-regulation of most cytokeratins in the LMP-transfected cells. Only one band was retained at approximately 40 kDa, detected by the pooled AE1/AE3 mixture of anti-cytokeratin antibodies. Both the original wild-type and the EBNA-2-transfected subline expressed the cytokeratins detected by the PKK-1 and PKK-2 and AE1/AE3 antibodies at a high level. Further experimentation showed that the LMP (pSV$_2$*gpt* MTLM)-transfected cells induced tumors when transplanted into nude mice.[107] Our results suggest that LMP expression may be an important causal factor in the development of NPC.

XII. IMMORTALIZATION OF OTHER HUMAN EPITHELIAL CELLS BY Ad12-SV40 VIRUS TRANSFORMATION

Recent advances in the cultivation of human epithelial cells has made it

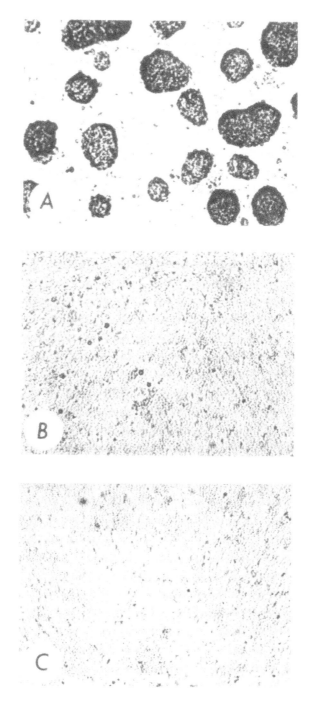

FIGURE 12. (A) RHEK-1 cells transfected with the LMP gene.
Note the change in morphology. (B) RHEK-1 cells transfected with
the EBNA-2 gene; (C) RHEK-1 cells.

possible to study problems related to carcinogenesis and differentiation in cell culture systems. Primary cultures of epithelial cells can now be established from various human tissue biopsies without difficulty even in the absence of serum supplement. However, the usefulness of such cell cultures is limited by factors such as cellular "senescence", slow growth rates, and small numbers of available cells. Many of these limitations can now be overcome by our ability to transform epithelial cells.

Besides the human foreskin epidermal cells described above, we were able to successfully establish lines from primary cultures by Ad12-SV40 virus infection of (1) human bronchial epithelial cells, (2) human salivary gland epithelial cells, (3) nasal polyp epithelial cells from cystic fibrosis (CF) patients, and (4) normal and CF bronchial epithelial cell lines.

A. ESTABLISHMENT OF HUMAN BRONCHIAL EPITHELIAL LINES BY INFECTION WITH Ad12-SV40 HYBRID VIRUS

Lung cancer is the most common cause of cancer deaths in many western countries, and most lung cancers arise in the epithelium of the bronchial type. The study of bronchial epithelial cells is, therefore, of considerable interest. Procedures have been devised for culturing normal human bronchial epithelial (NHBE) cells grown from explants of bronchial tissues obtained at autopsy.[72] This has allowed the study of many aspects of the cell biology of NHBE *in vitro*, including control of growth and squamous differentiation, the metabolic activation and effects of carcinogens and tumor promoters, and DNA repair. NHBE cells can be maintained in culture for relatively short times before cellular senescence occurs; however, in order to be able to study multistage bronchial carcinogenesis *in vitro*, an important objective was to develop cultures with indefinite life-span.

NHBE cells were infected with SV40 virus or an Ad12-SV40 hybrid virus, or transfected with strontium phosphate coprecipitated with plasmids containing the SV40 early-region genes (pRSV-T). Colonies of morphologically altered cells were isolated and cultured; these cells had extended culture life-spans compared to NHBE cells. All cultures eventually underwent senescence, with the exception of the Ad12-SV40-infected cells (BEAS-2B), which appear to have unlimited proliferative potential. Colonies arising after virus infection were screened for virus production by cocultivation with Vero cells; only viral nonproducer cultures were analyzed further. The cells retained electron microscopic features of epithelial cells, and keratin and SV40 T-antigen were detected by indirect immunofluorescence. All of the cultures were aneuploid, with karyotypic abnormalities characteristic of SV40-transformed cells. No tumors formed after subcutaneous injection of the cells in nude mice. These cells should be useful for studies of multistage bronchial epithelial carcinogenesis.[73]

B. ESTABLISHMENT OF SALIVARY GLAND EPITHELIAL CELL LINES FROM PATIENTS WITH SJOGREN'S SYNDROME AND NORMAL INDIVIDUALS

Sjogren's syndrome (SS) is an autoimmune disorder characterized by lymphocytic infiltration of salivary and lacrimal glands.[74] To determine whether EBV might play a role in the pathogenesis of this disorder, Fox et al.[75] used monoclonal antibodies and DNA probes to detect evidence of viral gene products and genomes in these patients' tissue biopsies and saliva. He demonstrated an elevated content of EBV in the salivary glands of SS patients and suggested that EBV may play a role in pathogenesis. In order to further study the relationship of EBV or human B-cell leukemia virus (HBLV) to SS, we have attempted to culture epithelial cells from the biopsies of patients with SS in a serum-free medium. We reported successful cultivation of primary epithelial cultures from salivary gland biopsies of patients with SS and of normal individuals in a serum-free medium, and further establishment of stable cell lines. Characterization of these cell lines has been described.[76]

C. NASAL POLYP EPITHELIAL CELL LINES FROM CYSTIC FIBROSIS PATIENTS

CF is a lethal inherited disease with a high incidence in the Caucasian population. The primary cause of the disease is thought to be a defective regulation of apical chloride channels in epithelial cells.[77] The molecular basis of the disease has not been explained; however, it is clear from studies performed so far that chloride channels are regulated in a complex way.[78] Which of the components involved is mutated in CF is unknown. The limited availability of suitable cell material presents a problem in research on CF. Epithelial cells appear to be the material of choice and, indeed, cultured airway and sweat gland cells are frequently used.[77,78] However, this material is available in small quantities only, has a limited proliferative capacity in culture, and is often heterogeneous. Therefore, we felt the need to develop a continuously growing epithelial cell line with CF genotype and phenotype. Because spontaneously transformed epithelial cell lines from CF patients are not available, we chose to use the protocol developed for immortalization of human epithelial cells using Ad12-SV40 virus.[20]

We developed epithelial CF cell lines by infecting nasal polyp cells with Ad12-SV40 virus. The cell lines obtained are epithelial in nature, as shown by cytokeratin production and morphology, although cytokeratins 4 and 13 typical of primary nasal polyp cells are produced at a much reduced level. Using chamber experiments, we showed that the precrisis CF cell line NCF3 was able to perform transcellular chloride transport when activated by agents which elevate intracellular calcium. cAMP agonists had no effect on chloride flux in NCF3 cells, as expected for CF cells. The apical chloride channels found with the patch clamp technique in NCF3 cells and in the postcrisis cell line NCF3A have a conductance similar to that of chloride channels found earlier in normal and CF epithelial cells. The channels show a delay in the onset of activity in off-cell patches and are not activated by increased cAMP levels in the cell. This indicates that CF epithelial cell lines will provide a useful model for the study of CF.[79]

D. NORMAL AND CF BRONCHIAL EPITHELIAL CELL LINES ESTABLISHED BY Ad12-SV40 VIRUS TRANSFORMATION

Primary bronchial epithelial cultures were established from tissues obtained at heart-lung transplantation of a 7-year-old male with CF and from a surgical resection of a chronically atelectatic right lower lobe from a 7-year-old male with a normal sweat chloride level. The cells were isolated by proteolytic digestion and propagated in LHC-8 medium.[72] The cultures were passaged twice and when they reached 80% confluence, two T25 flasks from each specimen were exposed to Ad12-SV40 virus.[20] The cells were then rinsed twice and passaged when confluent. Within 4 to 6 weeks, foci of clonal growth were visible. The CF cells were now at passage 17 and 8 months since infection. Uninfected cells did not survive beyond five passages. The normal cells are now at passage six and 3 months postinfection. Uninfected normal cells did not survive.

Patch clamp analysis of CF cells revealed Cl^- channels that did not open during exposure to either protein kinase A or C, but could be activated with a positive applied voltage of $+80$ mV. Normal cells had Cl^- channels that opened with both enzymes. Thus, each cell line obtained the electrophysiological characteristics of the native tissue. Transformed CF cells were serially cloned by plating at a concentration of 5000 per 100-mm dish. Indirect immunofluorescence with antibody to the epithelial cell marker, keratin, was positive. Chromosomal analysis of the CF cells revealed a modal number of 45. Normal and CF cells both originated from the bronchial epithelia of pediatric donors and were both infected with the same SV40 large T-containing virus. Thus, they should be ideal for characterizing the defect that distinguishes CF and normal airway epithelial cells.[80]

XIII. EVIDENCE FOR THE MULTISTEP NATURE OF *IN VITRO* HUMAN EPITHELIAL CELL CARCINOGENESIS

Besides the Ad12-SV40-immortalized human epidermal (RHEK-1) model already described, we would like to present other multistep models for human epithelial cell transformation.

A. NEOPLASTIC CONVERSION OF NORMAL HUMAN EPIDERMAL LINE (11367) ESTABLISHED BY psV₃*neo* TRANSFECTION WITH Ki-MSV INFECTION

Three plasmids, psV₃*gpt*, psV₃*neo* (American Type Culture Collection, Rockville, MD), and clone 4(E) pKi-MSV (supplied by Dr. S. Tronick;

obtained originally from Dr. N. Tsuchida) were used singly or in combination in 15 transfection experiments on secondary cultures. The plasmids pSV₃gpt and pSV₃neo contain the SV40 origin, early promoter region polyadenylation sequences, and sequences coding for large and small tumor-antigens,[81,82] whereas clone 4(E) pKi-MSV contains the *ras* oncogene from Ki-MSV.

In 15 experiments carried out over an 18-month period, 200 cultures were transfected, 115 with pSV₃gpt or -neo and 85 with both pSV₃ and pKi-MSV; 101 cultures served as controls. Transfections which included pSV₃ increased the life-span usually by three subcultures (1:2 splits); however, with one exception, epithelial cells ultimately ceased proliferation after approximately six subcultures (1:2 splits). Although an occasional colony developed in agarose, cells of these colonies failed to survive subculture. Transfection with the two plasmids (pSV₃ + pKi-MSV) did not increase the life-span beyond that obtained with pSV₃ alone. Only one of the 200 cell lines followed showed continuous, apparently infinte life-span. Cells of this line, designated NCTC 11367, maintained characteristics of epidermal keratinocytes, showing desmosomes and tonofilaments. The line was subcultured weekly at a 1:4 split ratio and cryopreserved after 52 passages, without evidence of declining proliferation rate. It originated in one of the last four experiments in which the cells were transfected with pSV₃neo following polyethylene glycol treatment. This transfection procedure, however, proved inhibitory to growth in most control and treated cultures.

Thus, alteration to a continuous cell line with apparently infinite life-span was a rare event under the transfection conditions used. Although we have no direct evidence for integration of the SV40 plasmid DNA, large and small SV40 tumor antigens were detected at passage 24 by indirect immunoprecipitation.

The successful neoplastic transformation of the continuous line RHEK-1 with Ki-MSV containing the *ras* oncogene prompted us to apply a similar approach to cells of line NCTC 11367. Infection of the line at passage 20 with Ki-MSV(BaEV) (K11367) readily produced foci of piled up, rounded cells similar to those observed in RHEK-1 infected with Ki-MSV.[19,20] The cultures released focus-forming viruses, proliferated rapidly, and showed increased expression of K-*ras* p21 protein, confirming the presence of the K-*ras* oncogene.

Cells of NCTC line 11367 were injected at passage 20 into four nude mice and gave rise 2 months later to nonprogressive epidermal cysts at all inoculation sites. Cells of K11367 were injected at passage 25 into five nude mice and gave rise 1 month later to tumors diagnosed as poorly differentiated epidermoid carcinomas. All tumors were cytogenetically identified as of human origin. Thus, the nontumorigenic human epidermal line (11367) established by pSV₃neo transfection could be transformed neoplastically by a *ras* oncogene.[51]

B. NEOPLASTIC TRANSFORMATION OF A HUMAN BRONCHIAL EPITHELIAL CELL LINE BY v-Ki-*ras*

An SV40 T-antigen-positive immortalized human bronchial epithelial cell line, BEAS-2B,[73] was infected with Ki-MSV or transfected with a plasmid containing the transforming region of Ki-MSV.[83] These cells formed poorly differentiated adenocarcinomas in athymic nude mice. Cell lines established from these tumors expressed v-Ki-*ras* p21 protein and were highly tumorigenic. Whereas serum or transforming growth factor induced the BEAS-2B cells at clonal density to undergo growth arrest and squamous differentiation, BEAS-2B cells containing activated *ras* genes were unaffected by TGF-β and were mitogenically stimulated by serum. It has been reported that full transformation of NHBE can be obtained with the v-Ha-*ras* oncogene, but long-term passage of the cells seemed to be necessary to obtain this neoplastic phenotype.[84] Neoplastic transformation of NHBE cells by v-Ha-*ras* oncogene thus appears to be a rare event, and the occurrence of a lengthy culture crisis and chromosomal aberrations suggests that one or more unidentified additional genetic events are required.

C. MALIGNANT CONVERSION OF HUMAN FORESKIN KERATIN-OCYTES BY HUMAN PAPILLOMA VIRUS TYPE 16 DNA AND v-K-*ras* ONCOGENE

Human papilloma viruses (HPVs) are know etiological agents of benign proliferations of skin and mucosa (papillomas or warts), and are implicated in the development of cervical dysplasia and anogenital carcinoma. The close association of HPV type 16 DNA with a majority of cervical carcinomas implies some role of the virus in this type of cancer.[85] To define the role of HPV in the development of human cancer, a model must be established to study the interaction between HPV and human epithelial cells. We recently developed an *in vitro* multistep model for human epithelial cell carcinogenesis.[86] Primary human epidermal keratinocytes acquired an indefinite lifespan in culture, but did not undergo malignant conversion in response to transfection with HPV type 16 DNA.[87] Addition of Ki-MSV, which contains a K-*ras* oncogene, to these cells induced morphological alterations associated with the acquisition of neoplastic properties (Table 9, Figure 13). The transformed Ki-MSV/HPK-1A line (Figure 14B) expressed high levels of the activated K-*ras* gene, whose product was distinguishable by its slower migration on SDS-polyacrylamide gel. A similar pattern of expression of the activated K-*ras* gene was also detected in the cell line 129 Nu 2409 (Figure 14C), which had been adapted from a nude mouse bearing a tumor derived by subcutaneous injection of Ki-MSV/HPK-1A cells. These findings demonstrate the malignant conversion of human primary epithelial cells in culture by the cooperation of a HPV DNA and a retroviral gene, and support a multistep process for neoplastic conversion. Thus, the availability of a human epithelial cell transformation model should facilitate studies of the interaction between HPV and human epithelial cells.

TABLE 9
Biological Properties of Human Epidermal Keratinocytes Transfected by HPV 16 DNA and Exposed to Ki-MSV

Cells	Passage in culture (passage number)	Saturation density[a] ($\times 10^5/cm^2$)	Number of animals bearing tumors[b]
Primary human keratinocytes	<3		
+ Ki-MSV	<3		
+ HPV-16 DNA (HPK-1A)	>50	1.9	0/5[c]
+ HPV-16 + Ki-MSV (Ki-MSV/HPK-1A)	>50	4.2	5/5

Note: Ki-MSV (BaEV) was produced in human nonproducer cells by superinfection with baboon endogenous virus.[13]

[a] Saturation density was measured as the maximum number of cells obtained after initial plating of 5×10^3 cells per square centimeter and then incubating at 37°C, with changes of growth medium every 3 d.

[b] Nude mice were inoculated with 10^7 cells. Tumors cells were reestablished in tissue culture and were found to resemble the cells of origin by karyotypic analysis.

[c] Number of tumors/number of mice.

XIV. DISCUSSION AND FUTURE PROSPECTS

The immortalization and transformation of cultured human epithelial cells has far-reaching implications for both cell and cancer biology. Human epithelial cell transformation studies will increase our understanding of the mechanisms underlying carcinogenesis and differentiation. The neoplastic process can now be studied in a model human epithelial cell culture system. The accompanying biochemical and genetic changes, once identified, will help define the relationship between malignancy and differentiation.

The present studies indeed demonstrate that the neoplastic process can now be studied in a human epithelial cell model system. Primary human epidermal keratinocytes infected with Ad12-SV40 virus became immortalized in culture, but were not tumorigenic. Additional exposure to either retroviruses, chemical carcinogens, or X-ray irradiation to these cells induced morphological alterations associated with the acquisition of neoplastic properties. These findings demonstrate the malignant transformation of human primary epithelial cells in culture by the combined action of either a DNA-transforming virus and a retrovirus or a DNA virus and a chemical or X-ray irradiation, and support a multistep process for neoplastic conversion.

The first demonstration that retroviruses containing growth factor receptor genes (*erb*B and *fms*), tyrosine kinase genes (*src* and *fes*), and GTP-binding protein genes (Ki-*ras, bas,* and H-*ras*) can infect and malignantly transform nontumorigenic human epidermal cells indicates that the human epithelial (RHEK-1) cells described here are very useful as a transforming assay of a

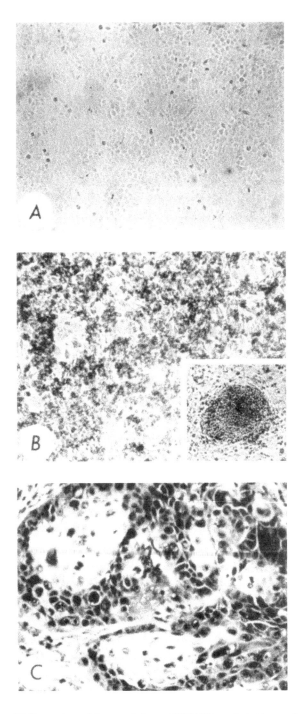

FIGURE 13. (A) Comparison of the morphology of HPV-16 transfected human keratinocytes (HPK-1A) at the 59th passage; (B) focus of Ki-MSV-infected HPK-1A cells 21 d after infection with Ki-MSV (BaEV); (C) invasive squamous cell carcinoma induced in nude mice inoculated with HPK-1A cells infected with Ki-MSV.

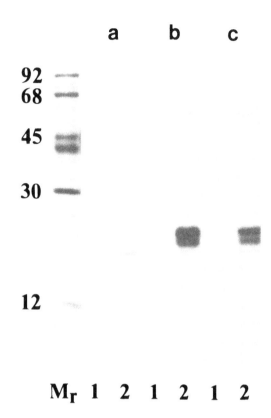

FIGURE 14.   Expression of p21 protein in Ki-MSV transformants. Cell lysates from HPK-1A (a), Ki-MSV/HPK-1A
passage 10 (b), and 129 Nu 2409 (c) were immunoprecipitated with either control mouse serum (lane 1) or a pan-
reactive mouse antibody against p21 (lane 2) and then subjected to Western immunoblot analysis using the same pan-
reactive antibody against p21 and ^{125}I-labeled protein A.
The molecular weight (M) markers used are shown on the
left.

variety of retroviral oncogenes. In addition, this line can be transfected neoplastically with an activated EJ H-*ras* human oncogene. Thus, the availability
of a rapid assay for transformation of human epithelial cells should facilitate
studies of the interaction of a variety of oncogenes and human epithelial cells.
Based on these findings, the RHEK-1 cell line may be extended to the study
of human viruses which have been implicated in the derivation of human
malignancies. Among these are EBV in nasopharyngeal carcinoma and HPV
in cervical carcinoma. In each of these cases, lack of a well-defined human
epithelial cell culture system has impeded our understanding of the roles of
these viruses in the development of malignancies. It has recently been shown
that the EBV-encoded LMP gene induced transformed phenotypes in the
RHEK-1 human keratinocyte line.[71]

Certain carcinogenic polycyclic aromatic hydrocarbons have been identified in our environment, and some chemicals or radiation are known definitely to cause cancer in human. Since it is estimated that 80% of human cancers are epithelial in origin, the human epithelial cell system described here may be a useful *in vitro* tool for screening potential carcinogenic agents. The molecular mechanisms of chemical or radiation-induced neoplastic transformation of human epithelial cells are poorly understood. Studies to identify the genes that are differentially expressed and to characterize activated transforming genes or suppressor genes during *in vitro* transformation by chemicals or irradiation are in progress. Analysis of the expression of known transforming genes in these transformed human epithelial cells are also in progress.

We have shown here the successful immortalization of various human epithelial cells by a simple infection with Ad12-SV40 virus. Thus, various methods are available for establishing human epithelial cells even in serum-free medium. An advantage of a transformed epithelial cell system can best be demonstrated by the example of CF cells. CF epithelial cells have abnormal regulation of chloride secretion.[77,78] Immortalized, transformed CF cells displaying the CF defective ion transport are essential to biochemical and genetic analysis of CF. Cultured CF and normal epithelial cells can be monitored for chemical modulation of defective chloride transport. Such a system is now available and will provide useful information for the development of treatment regimens. Immortalized cells will also facilitate the complementation studies employing potential candidate genes. Once a gene product is identified, it is possible that the immortalized cells can be used as a source to produce the desired product. This approach can be applied to other genetic disorders of the epithelium as well as characterizing components of differentiated functions.

Clinical observations have implied that cancer is a multistep process. In keeping with the multistep development of cancer *in vivo*, a stepwise approach to neoplastic transformation *in vitro* presents a reasonable strategy.[20,24] We have further shown the evidence for the nature of neoplastic transformation of human epithelial cells in culture. Apart from the Ad12-SV40 virus immortalized human epidermal (RHEK-1) multistep model described initially,[20] we have subsequently succeeded in developing other multistep model systems (Table 10). Malignant transformation of Ad12-SV40 virus immortalized human bronchial epithelial cells was achieved by a *ras* gene infection or transfection.[83,88] Malignant conversion of the HPV-16 DNA immortalized human epidermal line (HPK-1A) was succeeded by a *ras* gene infection.[86] It has been known that normal human cells in culture are remarkably resistant to experimentally induced tumorigenicity. However, as shown above, normal human cells could now be transformed into tumorigenic cells.

Since our initial report,[20,23] the list of successful reports on the neoplastic transformation of normal human cells, including fibroblasts, have been growing (see Table 10). These were achieved in a stepwise fashion. Human primary

TABLE 10
In Vitro Multistep Models for Human Cell Carcinogenesis

Cells transformed (epithelial)	Immortalization step	Transformation step	Ref.
	Stage of carcinogenesis		
Keratinocytes	Ad12-SV40	Ki-MSV	20
	Ad-12-SV40	MNNG or 4NQO	24
	Ad12-SV40	X-ray	64
	Ad12-SV40	c-H-*ras*	62
	pSV$_3$*neo*	Ki-MSV	51
	SV40	c-H-*ras*	90
	HPV-16	Ki-MSV	86
Bronchial	Ad12-SV40	Ki-MSV	83
	Ad12-SV40	v-H-*ras*	88
Mammary	BP	Retroviruses	91
Amniotic	SV40	Ki-MSV	92
Cervical	HPV-16	v-H-*ras*	93
Urinary tract	SV40	3MC	94
	SV40	c-H-*ras*	95
Liver	SV40	—	96
Kidney	Nichel II	v-H-*ras*	97
Thyroid	Adeno EIA	—	98
	SV40 *Ori-*	—	99
Colon	SV40 *Ori*	—	100
	MNNG and sodium butyrate	—	101
Tracheal gland	Ad12-SV40	—	102
Letinal pigment	SV40	—	103
Esophagus	SV40	—	104
Melanocyte	SV40	—	105
Fibroblasts	SV40	Ki-MSV	13
	γ-ray	H-MSV or c-H-*ras*	14
	γ-ray	c-H-*ras*	106

cells immortalized by a variety of means (viruses, chemicals, irradiation, or spontaneously without any treatment) could be transformed neoplastically by a carcinogenic agent (see Table 10). Thus, these studies demonstrate that neoplastic transformation of normal human cells in culture is indeed a multistep process. In all these cases, the initial event seemed to be immortalization of the cells followed by neoplastic conversion. As postulated for rodent fibroblasts,[89] the immortalization step is a critical initial step and rate limiting for *in vitro* neoplastic transformation of human epithelial cells.

ACKNOWLEDGMENTS

I would like to acknowledge my main collaborators involved in the different phases of this work: Gilbert Jay, Paul Arnstein, Katherine Sanford, Stuart A. Aaronson, Curt Harris, Anatoly Dritschilo, Matthias Durst, Ward Peterson, and George Klein.

REFERENCES

1. **Farber, E.,** The multistep nature of cancer, *Cancer Res.,* 44, 4217, 1984.
2. **Klein, G. and Klein, E.,** Evolution of tumors and the impact of molecular oncology, *Nature,* 315, 190, 1985.
3. **Sager, R., Tanaka, K., Lau, C. C., Ebina, Y., and Anisowicz, A.,** Resistance of human cells to tumorigenesis induced by cloned transforming genes, *Proc. Natl. Acad. Sci. U.S.A.,* 80, 7601, 1983.
4. **Rhim, J. S.,** Transformation of human cells in culture by chemical carcinogens, in *Prevention and Detection of Cancer,* Nieburgs, H. E., Ed., Marcel Dekker, New York, 1976, 1337.
5. **DiPaolo, J. A.,** Relative difficulties in transforming human and animal cells *in vitro, J. Natl. Cancer Inst.,* 70, 3, 1983.
6. **Harris, C. C.,** Human tissues and cells in carcinogenesis research, *Cancer Res.,* 47, 1, 1987.
7. **Chang, S. E.,** *In vitro* transformation of human epithelial cells, *Biochim. Biophys. Acta,* 823, 161, 1986.
8. **Girardi, A. J., Jensen, F. C., and Koprowski, H.,** SV40-induced transformation of human diploid cells: crisis and recovery, *J. Cell. Comp. Physiol.,* 65, 69, 1965.
9. **Graham, F. L., Smiley, J., Russell, W. C., and Nairn, R.,** Characteristics of a human cell line transformed by DNA from human adenovirus type 5, *J. Gen. Virol.,* 36, 59, 1977.
10. **Borek, C.,** X-ray induced *in vitro* neoplastic transformation of human diploid cells, *Nature,* 283, 776, 1980.
11. **Milo, G. and DiPaolo, J.,** Neoplastic transformation of human diploid cells *in vitro* after chemical carcinogen treatment, *Nature,* 275, 130, 1978.
12. **McCormick, J. J. and Mahr, V. M.,** Toward an understanding of malignant transformation of diploid human fibroblasts, *Mutat. Res.,* 199, 273, 1988.
13. **O'Brien, W., Stenman, G., and Sager, R.,** Suppression of tumor growth by senescence in virally transformed human fibroblasts, *Proc. Natl. Acad. Sci U.S.A.,* 83, 8659, 1986.
14. **Namba, M., Nishitani, K., Fukushima, F., Kimoto, T., and Nose, K.,** Multistep process of neoplastic transformation of normal human fibroblasts by ^{60}Co gamma rays and Harvey sarcoma viruses, *Int. J. Cancer,* 37, 419, 1986.
15. **Land, H., Parada, L. V., and Weinberg, R. A.,** Tumorigenic conversion of primary embryo fibroblasts requires at least two cooperating oncogenes, *Nature,* 304, 596, 1983.
16. **Ruley, H. E.,** Adenovirus early region 1A enables viral and cellular transforming genes to transform cells in culture, *Nature,* 304, 602, 1983.
17. **Steinberg, M. L. and Defendi, V.,** Altered pattern of growth and differentiation in human keratinocytes infected by simian virus 40, *Proc. Natl. Acad. Sci. U.S.A.,* 76, 801, 1979.
18. **Banks-Schlegel, S. P. and Howley, P. M.,** Differentiation of human epidermal cells transformed by SV40, *Cell. Biol.,* 96, 330, 1983.
19. **Rhim, J. S., Sanford, K. K., Arnstein, P., Fujita, J., Jay, G., and Aaronson, S. A.,** Human epithelial cell carcinogenesis: combined action of DNA and RNA tumor viruses produces malignant transformation of primary human epidermal keratinocytes, in *Carcinogenesis,* Vol. 9, Barrett, J. C. and Tennant, R. W., Eds., Raven Press, New York, 1985, 57.
20. **Rhim, J. S., Jay, G., Arnstein, P., Price, F. M., Sanford, K. K., and Aaronson, S. A.,** Neoplastic transformation of human epidermal keratinocytes by Ad12-SV40 and Kirsten sarcoma viruses, *Science,* 227, 1250, 1985.
21. **Cooper, G.,** Cellular transforming genes, *Science,* 217, 801, 1982.
22. **Weinberg, R. A.,** Oncogenes of spontaneous and chemically induced tumors, *Adv. Cancer Res.,* 36, 149, 1982.

23. **Rhim, J. S.,** Glucocorticoids enhance viral transformation of mammalian cells, *Proc. Soc. Exp. Biol. Med.,* 174, 217, 1983.

24. **Rhim, J. S., Fujita, J., Arnstein, P., and Aaronson, S. A.,** Neoplastic conversion of human epidermal keratinocytes by adenovirus 12-SV40 virus and chemical carcinogens, *Science,* 232, 385, 1986.

25. **Freeman, A. E., Price, P. J., Igel, H. J., Young, J. C., Maryak, J. M., and Huebner, R. J.,** Morphological transformation of rat embryo cells induced by diethylnitrosamine and murine leukemia viruses, *J. Natl. Cancer Inst.,* 44, 65, 1970.

26. **Rhim, J. S., Vass, W., Cho, H. Y., and Huebner, R.,** Malignant transformation induced by 7,12-dimethylbenz(a)anthracene in rat embryo cells infected with Rauscher leukemia virus, *Int. J. Cancer,* 7, 65, 1971.

27. **Price, P. J., Freeman, A. E., Lane, W. T., and Huebner, R. J.,** Morphological transformation of rat embryo cells by the combined action of 3-methylcholanthrene and Rauscher leukemia virus, *Nature New Biol.,* 230, 144, 1971.

28. **Sukumar, S., Notario, V., Martin-Zanca, D., and Barbacid, M.,** Induction of mammary carcinomas in rats by nitromethylurea involves malignant action of H-*ras*-1 locus by single point mutation, *Nature,* 306, 658, 1983.

29. **Zarbl, H., sukumar, S., Arthur, A. V., Martin-Zanca, D., and Barbacid, M.,** Direct mutagenesis of Ha-*ras*-1 oncogene by *N*-nitroso-*N*-methylurea during initiation of mammary carcinogenesis in rats, *Nature,* 315, 382, 1985.

30. **Balmain, A. and Pragnell, I. B.,** Mouse skin carcinomas induced *in vivo* by chemical carcinogens have a transforming Harvey-*ras* oncogene, *Nature,* 303, 72, 1983.

31. **Balmain, A., Ramsden, M., Bowden, G. T., and Smith, J.,** Activation of cellular Harvey-*ras* gene in chemically induced benign skin papillomas, *Nature,* 207, 658, 1984.

32. **Eva, A. and Aaronson, S. A.,** Frequent activation of c-*kis* as a transforming gene in fibrosarcomas induced by methyl-cholanthrene, *Science,* 220, 955, 1983.

33. **Guerrero, I., Calzada, P., Mayer, A., and Pellicer, A.,** A molecular approach to leukemogenesis:mouse lymphomas contain an activated c-*ras* oncogene, *Proc. Natl. Acad. Sci. U.S.A.,* 81, 202, 1984.

34. **Padhy, L. C., Shih, C., Cowing, D., Finkelstein, R., and Weinberg, R. A.,** Identification of a phosphoprotein specifically induced by the transforming DNA of rat neuroblastomas, *Cell,* 28, 865, 1982.

35. **Parada, L. and Weinberg, R. A.,** Presence of a Kirsten murine sarcoma virus *ras* oncogene in cells transformed by 3-methylcholanthrene, *Mol. Cell. Biol.,* 3, 2298, 1983.

36. **Sukumar, S., Pulciani, S., Doriger, J., DiPaolo, J. A., Evans, C. H., Zbar, B., and Barbacid, M.,** A transforming *ras* gene in tumorigenic guinea pig cell lines initiated by diverse chemical carcinogens, *Science,* 223, 1197, 1984.

37. **Varmus, H.,** The molecular genetics of cellular oncogenes, *Annu. Rev. Genet.,* 18, 553, 1984.

38. **Bishop, J. M.,** Cellular oncogenes and retroviruses, *Annu. Rev. Biochem.,* 52, 301, 1983.

39. **Guerrero, I., Villassante, A., Corces, V., and Pellicer, A.,** Activation of c-K-*ras* oncogene by somatic mutation in mouse lymphomas induced by gamma radiation, *Science,* 225, 1159, 1984.

40. **Rhim, J. S., Park, D. K., Arnstein, P., Huebner, R. J., Weisburger, E. K., and Nelson-Rees, W. A.,** Transformation of human cells in culture by *N*-methyl-*N'*-nitro-*N*-nitrosoguanidine, *Nature,* 256, 751, 1975.

41. **Cooper, C. S., Blair, d. G., Oskarsson, M. K., Tainsky, M. A., Eader, L. A., and Vande Woude, G. F.,** Characterization of human transforming genes from chemically transformed, teratocarcinoma, and pancreatic carcinoma cell lines, *Cancer Res.,* 44, 1, 1984.

42. **Cooper, C. S., Park, M., Blair, D. G., Tainsky, M. A., Huebner, K., Croco, C. M., and Vande Woude, G. F.,** Molecular cloning of a new transforming gene from a chemically transformed human cell line, *Nature,* 311, 29, 1984.

43. Park, M., Dean, M., Cooper, C. S., Schmidt, M., O'Brien, S. J., Blair, D. G., and Vande Woude, G. F., Mechanisms of *met* oncogene activation, *Cell,* 45, 895, 1986.

44. Dean, M., Park, M., LeBeau, M. M., Robins, T. S., Diaz, M. O., Rowley, J. D., Blair, D. G., and Vande Woude, G. F., The human *met* oncogene is related to the tyrosine kinase oncogenes, *Nature,* 318, 385, 1985.

45. White, R., Woodward, S., Leppert, M., O'Cornell, P., Hoff, M., Herbst, J., Lalonel, J. M., Dean, M., and Vande Woude, G. F., A closely linked genetic marker for cystic fibrosis, *Nature,* 318, 382, 1985.

46. Rhim, J. S., Fujita, J., and Park, J. B., Activation of H-*ras* oncogene in 3-methylcholanthrene-transformed human cell line, *Carcinogenesis,* 8, 1165, 1987.

47. Srivastava, S. K., Yuasa, Y., Reynold, S. H., and Aaronson, S. A., Effects of two major activity lesions on the structure and conformation of human *ras* oncogene products, *Proc. Natl. Acad. Sci. U.S.A.,* 82, 38, 1985.

48. Rhim, J. S., Kawakami, T., Pierce, J., Sanford, K., and Arnstein, P., Cooperation of v-oncogenes in human epithelial cell transformation, *Leukemia,* 2, 1515, 1988.

49. Hartley, J. W. and Rowe, W. P., Naturally occurring murine leukemia viruses in wild mice: characterization of a new "amphotropic" class, *J. Virol.,* 19, 19, 1976.

50. Rasheed, S., Gardner, M. B., and Chan, E., Amphotropic host range of naturally occurring wild mouse leukemia virus, *J. Virol.,* 19, 13, 1976.

51. Gantt, R., Sanford, K. K., Parshad, R., Price, F. M., Peterson, W. D., and Rhim, J. S., Enhanced G$_2$ chromatid radiosensitivity, an early stage in the neoplastic transformation of human epidermal keratinocytes in culture, *Cancer Res.,* 47, 1390, 1987.

52. Furth, M. E., Davis, L. J., Fleurdelys, B., and Scolnick, E. M., Monoclonal antibodies to the p21 products of the transforming gene of Harvey murine sarcoma virus and of the cellular *ras* gene family, *J. Virol.,* 43, 294, 1982.

53. Anderson, S. J., Furth, M., Wolff, L., Ruscetti, S. K., and Sherr, C. J., Monoclonal antibodies to the transformation-specific glycoprotein encoded by the feline retroviral oncogene v-*fms,* *J. Virol.,* 44, 692, 1982.

54. Veronese, F., Kelloff, G. J., Reynolds, T. H., Jr., Hill, R. W., and Stephenson, J. S., Monoclonal antibodies specific to transforming polyproteins encoded by independent isolates of feline sarcoma virus, *J. Virol.,* 43, 896, 1982.

55. Lipsich, L. A., Lewis, A. J., and Brugge, J. S., Isolation of monoclonal antibodies that recognize the transforming proteins of avian sarcoma viruses, *J. Virol.,* 48, 352, 1983.

56. Xu, Y. H., Rickert, N., Ito, S., Merchino, G. T., and Pastan, I., Characterization of epidermal growth factor receptor gene expression in malignant and normal human cell lines, *Proc. Natl. Acad. Sci. U.S.A.,* 81, 7308, 1984.

57. Graham, F. L. and van der Eb, A. J., A new technique for the assay of infectivity of human adenovirus 5 DNA, *Virology,* 52, 456, 1973.

58. Sutherland, B. M. and Bennett, P. V., Transformation of human cells by DNA transfection, *Cancer Res.,* 44, 2769, 1984.

59. Berg, P. and Southern, P. j., Transformation of mammalian cells to antibiotic resistance with a bacterial gene under control of the SV40 early region promoter, *Mol. Appl. Genet.,* 1, 327, 1982.

60. Kawai, S. and Nishizawa, M., New procedure for DNA transfection with polycation and dimethyl sulfoxide, *Mol. Cell. Biol.,* 4, 1172, 1984.

61. Morgan, T. L., Maher, V. M., and McCormick, J. J., A procedure of high efficiency DNA-mediated gene transfer in normal human fibroblasts, *In Vitro Cell. Dev. Biol.,* 22, 317, 1986.

62. Rhim, J. S., Park, J. B., and Jay, G., Neoplastic transformation of human keratinocytes by polybrene-induced DNA-mediated transfer of an activated oncogene, *Oncogene,* 4, 1403, 1989.

63. **Shih, C. and Weinberg, R. A.,** Isolation of a transforming sequence from a human bladder carcinoma cell line, *Cell,* 29, 161, 1982.

64. **Thraves, P., Salehi, Z., Dritschilo, A., and Rhim, J. S.,** Neoplastic transformation of immortalized human epidermal keratinocytes by ionizing radiation, *Proc. Natl. Acad. Sci. U.S.A.,* 87, 1174, 1990.

65. **Sawey, M. J., Hood, A. T., Burns, F. J., and Garte, S. J.,** Activation of c-*myc* and c-K-*ras* oncogenes in primary rat tumors induced by ionizing radiation, *Mol. Cell. Biol.,* 7, 932, 1987.

66. **Borek, C., Ong, A., and Mason, H.,** Distinctive transforming genes in X-ray transformed mammalian cells, *Proc. Natl. Acad. Sci. U.S.A.,* 84, 794, 1987.

67. **Fasano, O., Birnbaum, D., Edlund, L., Fogh, J., and Wigler, M.,** New human transforming genes detected by a tumorigenicity assay, *Mol. Cell. Biol.,* 4, 1695, 1984.

69. **Yuasa, Y., Kaniyama, T., Kato, M., Iwana, T., Ikenchi, T., and Tonomura, A.,** Transforming genes from familial adenomatous polyposis patient cells detected by a tumorigeneicity assay, *Oncogene,* 5, 589, 1990.

69. **Klein, G., Giovanella, B. C., Lindahl, T., Fialkow, P. J., Sing, S., and Stehlin, J. S.,** Direct evidence for the presence of Epstein-Barr virus DNA and nuclear antigen in malignant epithelial cells from patient with poorly differentiated carcinoma of the nasopharynx, *Proc. Natl. Acad. Sci U.S. A.,* 71, 4737, 1974.

70. **Fahraeus, R., Hu, L. F., Ernberg, I., Finke, J., Rowe, M., Klein, G., Faek, K., Yadav, M., Busson, P., Tursz, T., and Kallin, B.,** Expression of Epstein-Barr virus encoded proteins in nasopharyngeal carcinoma, *Int. J. Cancer,* 42, 329, 1988.

71. **Fahraeus, R., Rymo, L., Rhim, J. S., and Klein, G.,** Morphological transformation of human keratinocytes expressing the LMP gene of Epstein-Barr virus, *Nature,* 345, 447, 1990.

72. **Lechner, J. F. and LaVeck, M. A.,** A serum-free method for culturing normal human bronchial epithelial cells at clonal density, *J. Tissue Culture Methods,* 9, 43, 1985.

73. **Reddel, R. R., Ke, Y., Gerwin, B. I., McMenamin, M., Lechner, J. F., Su, R.-T., Brash, D. E., Park, J.-B., Rhim, J. S., and Harris, C. C.,** Transformation of human bronchial epithelial cells by infection with SV40 or adenovirus 12-SV40 hybrid virus, or transfection via strontium phosphate coprecipitation with a plasmid containing SV40 early region gene, *Cancer Res.,* 48, 1904, 1988.

74. **Fox, R.,** Epstein-Barr virus and human autoimmune diseases: possibilities and pitfalls, *J. Virol. Methods,* 21, 19, 1988.

75. **Fox, R. I., Pearson, G., and Vaughan, J. H.,** Detection of Epstein-Barr virus associated antigens and DNA in salivary gland biopsies from patients with Sjogren's syndrome. *J. Immunol.,* 137, 3162, 1986.

76. **Rhim, J. S., Fox, R. I., Ablashi, D. V., Salahuddin, S. Z., Buchbinder, A., and Josephs, S. F.,** Establishment of salivary gland epithelial cell lines from patients with Sjorgen's syndrome and normal individuals, in *Epstein-Barr Virus and Human Disease II,* Ablash, D. V., Faggioni, A., Krueger, G. R. F., Pagano, J. S., and Pearson, G. R., Eds., Humana Press, New Jersey, 1988, 155.

77. **Frizzell, R. A., Rechkemmer, G., and Shoemaker, R. L.,** Altered regulation of airway epithelial cell chloride channels in cystic fibrosis, *Science,* 233, 560, 1986.

78. **Welsch, M. J. and Liedke, C. M.,** Chloride and potassium channels in cystic fibrosis epithelia, *Nature,* 322, 467, 1986.

79. **Scholte, B. J., Bijman, J., Hoogeveen, A. T., Willemse, R., Rhim, J. S., and Van der Kamp, W. M.,** Immortalization of nasal polyp epithelial cells from cystic fibrosis patients, *Exp. Cell. Res.,* 182, 559, 1989.

80. **Zeitlin, P. L., Lu, L., Rhim, J., Cutting, G., Stetten, G., Kieffer, K. A., Craig, R. and Guggino, W. B.,** A cystic fibrosis bronchial epithelial cell line: immortalization by adeno-12-SV40 infection, *Am. J. Resp. Cell Mol. Biol.,* 4, 313, 1991.

81. **Mulligan, R. C. and Berg, P.**, Selection for animal cells that express the *Escherichia coli* gene coding for xanthine-guanine phosphoribosyltransferase, *Proc. Natl. Acad. Sci. U.S.A.*, 78, 2072, 1981.

82. **Southern, P. J. and Berg, P.**,Transformation of mammalian cells to antibiotic resistance with a bacterial gene under control of the SV40 early region promoter, *J. Mol. Appl. Genet.*, 1, 327, 1982.

83. **Reddel, R. R., Ki, Y., Kaighn, E., Malan-Shibley, L., Lechner, J. F., Rhim, J. S., and Harris, C. C.**, Human bronchial epithelial cells neoplastically transformed by v-Ki-*ras* altered response to inducers of terminal squamous differentiation, *Oncogene Res.*, 3, 401, 1988.

84. **Yoakum, G. H., Lechner, J. F., Gabrielson, E., Korba, B. E., Malan-Shibley, L., Willey, J. C., Valerio, M. G., Shamsuddin, A. K., Trump, B. F., and Harris, C. C.**, Transformation of human bronchial epithelial cells transfected by Harvey *ras* oncogene, *Science*, 227, 1174, 1985.

85. **zur Hausen, H. and Schneider, A.**, The role of papillomaviruses in human anogenital cancer, in *The Papovaviridae: The Papillomaviruses*, Howley, P. M. and Salzman, N., Eds., Plenum Press, New York, 245, 1987.

86. **Durst, M., Gallahan, D., Jay, G., and Rhim, J. S.**, Glucocorticoid-enhanced neoplastic transformation of human keratinocytes by human papillomavirus type 16 and activated *ras* oncogene, *Virology*, 73, 767, 1989.

87. **Durst, M., Dzarlievea-Petrusevska, R. T., Boukamp, P., Fusenig, N. E., and Gissmann, L.**, Molecular and cytogenetic analysis of immortalized human primary keratinocytes obtained after transfection with human papillomavirus type 16 DNA, *Oncogene*, 1, 251, 1987.

88. **Amstad, P., Reddle, R. R., Pfeifer, A., Malan-Shibley, L., Mark, G. E., and Harris, C. C.**, Neoplastic transformation of a human bronchial epithelial cell line by a recombinant retrovirus encoding viral Harvey *ras*, *Mol. Carcinogenesis*, 1, 151, 1988.

89. **Newbold, R. F. and Overell, R. W.**, Fibroblast immortality is a prerequisite for transformation by EJ c-Ha-*ras* oncogene, *Nature*, 304, 648, 1983.

90. **Fusenig, N.E., Boukamp, P., Breitkreutz, D., Karjetta, S., and Petrusevska, R. T.**, Oncogenes and malignant transformation of human keratinocytes, in *Anticarcinogenesis and Radiation Protection*, Cerutti, P., Ed., Plenum Press, New York, 1987, 227.

91. **Clark, R., Stampfer, M. R., Milley, R., O'Rourke, E., Walen, K. H., Kriegler, M., Kopplin, J., and McCormick, F.**, Transformation of human mammary epithelial cells by oncogenic retroviruses, *Cancer Res.*, 48, 4689, 1988.

92. **Walen, K. H. and Arnstein, P.**, Induction of tumorigenesis and chromosomal abnormalities in human amniocytes infected with simian virus 40 and Kirsten sarcoma virus, *In Vitro Cell. Dev. Biol.*, 22, 57, 1986.

93. **DiPaolo, J. A., Woodworth, C. D., Popescu, N. C., Notario, V., and Doniger, J**, Induction of human cervical squamous cell carcinoma by sequential transfection with human papilloma virus type 16 DNA and viral Harvey *ras*, *Oncogene*, 4, 395, 1989.

94. **Reznikoff, C. A., Loretz, L. J., Christian, B. J., Wu, S.-O., and Meisner, L. F.**, Neoplastic transformation of *SV40*-immortalized human urinary tract epithelial cells by in vitro exposure to 3-methyl-cholanthrene, *Carcinogenesis*, 9, 1427, 1988.

95. **Christian, B. C., Kao, C., Wu, W., Meisner, L. F., and Reznikoff, C. A.**, Transformation of SV40-immortalized human uroepithelial cells by transfection with *ras* oncogene, *Proc. Am. Assoc. Cancer Res.*, 29, 459, 1988.

96. **Cole, K. E. Pfeifer, A. M. A., Weston, A., Vignaud, J. M., Harris, C. C., and Lechner, J. F.**, Development of a differentiated human liver epithelial cell line, *Proc. Am. Assoc. Cancer Res.*, 31, 19, 1990.

97. **Hauger, A., Ryberg, D., Hansteen, I.-L. and Amstad, P.**, Neoplastic transformation of human kidney epithelial cell line transfected with v-Ha-*ras* oncogene, *Int. J. Cancer*, 45, 572, 1990.

98. **Cone, R. D., Platzer, M., Piccinini, L. A., Jaramillo, M., and Davies, T. F.**, HLA-DR gene expression in aproliferating human thyroid cell clone (12S), *Endocrinology*, 123, 2067, 1988.

99. **Lemoine, N. R., Mayall, E. S., Jones, T., Shear, D., McDermid, S., Kendall-Taylor, P., and Wynford-Thomas, D.**, Characterization of human thyroid epithelial cells immortalized *in vitro* by simian virus 40 DNA transfection, *Br. J. Cancer*, 60, 897, 1989.

100. **Berry, R. D., Powell, S. C., and Paraskeva, C.**, *In vitro* culture of human fetal colonic epithelial cells and their transformation with origin virus SV40 DNA, *Br. J. Cancer*, 57, 287, 1988.

101. **Williams, A. C., Harper, S. J., and Paraskeva, C.**, Neoplastic transformation of a human colonic epithelial cell line: *in vitro* evidence for the adenoma to carcinoma sequence, *Cancer Res.*, 50, 4724, 1990.

102. **Chopra, D. P., Taylor, G. W., Mathien, P., Hukka, B., and Rhim, J. S.**, Immortalization of human tracheal gland epithelial cells by Adenovirus 12-SV40 virus, *In Vitro Cellu. Develop. Biol.*, in press.

103. **Dutt, K., Scott, M., Del Monte, M., Agarwal, N., Sternberg, P., Srivastava, S. K., and Srinivasan, A.**, Establishment of human retinal pigment epithelial cell lines by oncogenes, *Oncogene*, 5, 195, 1990.

104. **Stoner, G. D., Kaighn, M. E., Reddel, R. R., Resan, J. H., Bowman, D., Naio, Z., Matsukura, M., You, M., Galati, A. J., and Harris, C. C.**, Establishment and characterization of SV40 T-antigen immortalized human esophageal epithelial cells, *Cancer Res.*, 51, 365, 1991.

105. **Melber, K., Zhu, G., and Diamond, L.**, SV40-transfected human melanocyte sensitivity to growth inhibition by the phorbol ester 12-0-tetradecanoylphorbol-13-acetate, *Cancer Res.*, 49, 3650, 1989.

106. **Namba, M., Nishitani, K., Fukushima, F., Kimoto, T., and Yuasa, Y.**, Multi-step neoplastic transformation of normal human fibroblasts by Co-60 gamma rays and Ha-*ras* oncogenes, *Mutat. Res.*, 199, 415, 1988.

Chapter 9

MORPHOLOGIC AND MOLECULAR CHARACTERIZATIONS OF PLASTIC TUMOR CELL PHENOTYPES

Charles F. Shuler and George E. Milo

TABLE OF CONTENTS

I. INTRODUCTION

The cellular composition of human squamous cell carcinomas includes several subpopulations of parenchymal cells.[1-3] These subpopulations exhibit a range of morphologic changes from the cytology of the normal epithelial cells, from which the tumors developed. Thus, most tumors contained some cells with morphologic changes consistent with epithelial cell terminal differentiation and other cells which were undifferentiated and had invaded the surrounding stromal tissue. This range in cell morphologies indicates that, even in a clonally derived tumor, multiple cell phenotypes are present which possess different biologic potentials.

The different tumor cell phenotypes probably represent a continuum with respect to their aggressive behavior. Examinations of populations of cells specific for the different stages in this progression would permit characterization of the molecular features associated with the different phenotypes. In addition, it would permit an analysis of the molecular mechanisms which are operative in the transition of cells between different phenotypes. These characterizations would allow a determination of the features which are either reversible or irreversible in each of the phenotypes and the capacity of tumor cells to be induced to adopt a less aggressive behavioral phenotype.

A model human carcinoma cell system developed in our laboratory permits us to examine both carcinogen-initiated normal human keratinocytes and tumor-derived cell populations at different stages of malignant progression. The model which we have developed is depicted in Figure 1. Similar cell populations have been isolated, from both tumors and chemically exposed keratinocytes, which were unique and displayed phenotype-specific patterns of growth. All of the cell populations exhibited anchorage-independent growth (AIG) by their capacity to form colonies in soft agar.[4-7] However, the different AIG-positive cell populations could be further divided by additional differences in growth potential (see Figure 1). Many of the isolated cells fell into a category called AIG-term (terminal differentiation). These cells were not immortal *in vitro*, but, rather, followed a pattern of epithelial cell terminal differentiation. These AIG-term cells could not be induced to adopt an alternate phenotype. A second phenotypic category, designated AIGT, contained cells that were capable of producing a progressively growing tumor in a xenogeneic host.[8,9] These cells were fixed in the tumorigenic phenotype and could not be induced to adopt an alternative phenotype. The most interesting phenotype which has been isolated was designated AIGNT and represented cells which were anchorage independent, immortal *in vitro*, but unable to produce a tumor in a xenogeneic host.[8-10,11] The AIGNT phenotype was not fixed; rather, these cells could be induced to adopt an alternate phenotype following exposure to appropriate stimuli.[8,10,11] Thus, we have succeeded in isolating specific subpopulations from both human carcinomas and clones of carcinogen-initiated cells which represented different stages in the tumorigenic progression.

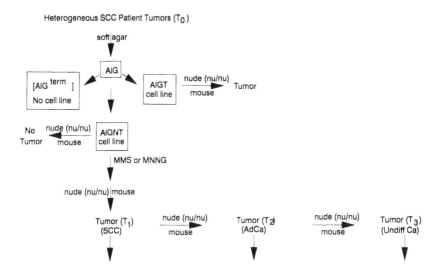

FIGURE 1. Schematic representation of the phenotypic characterization of tumor cell subpopulations. Cellular phenotypes: AIG, anchorage-independent growth; AIGNT, anchorage-independent growth not tumorigenic; AIGT, anchorage-independent growth tumorigenic; AIG^term, anchorage-independent terminal differentiation in monolayer culture.

The plasticity of the AIGNT phenotype cells provided the means to examine the molecular changes that occurred during the transition between different tumor cell phenotypes.[10,11] AIGNT cells treated with either MMS or MNNG became converted to an AIGT phenotype.[8] However, the phenotypic transition of these cells was not permanent; rather these cells reverted to the AIGNT phenotype after several population doublings (PDs) in monolayer culture. The onset of the tumorigenic phenotype in the chemically exposed AIGNT cells was thus not associated with a permanently heritable change. The MMS/MNNG-induced AIGNT-AIGT conversion could also be inhibited or reversed by treatment of the cells with benzamide. These types of treatments permitted the control of the phenotype expressed by these cells and thus a means for analyzing molecular events associated with the progression from one phenotype to another.

The AIGNT phenotype which we have characterized provides a model system to examine changes in cellular phenotype. The results which are presented further examine the cellular changes which accompany the progression of cells through different phenotypic stages and the plasticity of these stages. The studies which will be described should begin to determine the molecular changes associated with tumor cell phenotypic plasticity.

II. MATERIALS AND METHODS

A. ISOLATION OF ANCHORAGE-INDEPENDENT CELLS BY GROWTH IN SOFT AGAR

Tumor cell subpopulations capable of anchorage-independent growth were initially isolated by their ability to grow in soft agar.[5-7] Human squamous cell carcinomas were obtained directly from the operating room. The tumor tissues were minced and the minced tissue was filtered through a sieve. The individual cells were suspended in 2 × LoCal Dulbecco's Modified Eagle's Medium, supplemented with 40% fetal bovine serum at a density of 500,000 cells per milliliter. The cell suspension was mixed with an equal volume of 0.6% agar in water at 39°C. A portion (2 ml) of this cell/agar suspension was layered over 5 ml of a 2% agar base in a 60-mm tissue culture dish. The dishes were incubated at 37°C in a 3% CO_2-enriched air atmosphere with a high relative humidity. The agar cultures were evaluated by phase contrast microscopy for the development of colonies. Cell colonies which developed to greater than 60 μm in diameter were removed from the soft agar after 14 d and seeded *in vitro* to establish monolayer cultures. The AIG cells isolated by this procedure contained a mixed popoulation of cells with differing biologic potentials, but all had the capacity of AIG in common.

B. MONOLAYER CELL CULTURE

Tumor cells isolated from colonies which developed in soft agar were placed in monolayer culture.[5,12] Single cell suspensions of the AIG cells were seeded in plastic tissue culture dishes and maintained in growth medium (GM) (minimum essential medium [MEM], essential amino acids, 1.0 mM sodium pyruvate, 2.0 mM glutamine, 0.1 mM nonessential amino acids, 10% fetal bovine serum, and antibiotics). The cultures were maintained in a humidified 5% CO_2 atmosphere at 37°C. The growth of the cells was monitored by phase-contrast microscopy. Actively growing cells had the medium changed three times per week. Cells were passaged at a 1:5 dilution prior to confluency. The monolayer cultures contained a mixed population of tumor cells. In some cases, the epithelial cells stratified and underwent terminal differentiation. In other cases, clones of cells were isolated that demonstrated a prolonged longevity in culture and did not display cellular senescence. The cell lines were further isolated by cloning and the clones of cells examined for their capability to produce tumors in a xenogeneic host.

Tumor cells were also established in monolayer culture after recovery from progressively growing tumors in nude mice. The tumors (≥2.0 cm) were recovered from the mice as described below and minced into 1 × 1-mm fragments. The fragments were digested with 0.5% collagenase for 30 min. The isolated tumor cells were plated on plastic tissue culture dishes and maintained as previously described.

C. TUMORIGENICITY EVALUATION

The tumorigenicity of the squamous cell carcinoma cell lines which were isolated was evaluated in the nude mouse xenogeneic host model system.[7,9,13] Gnotobiotic male NCr/sed (nu/nu) nude mice (4 to 6 weeks old) were splenectomized and treated with 0.1 ml of mouse antilymphocyte serum (ALS) twice weekly prior to the use of these mice in the tumorigenic evaluations. The animals were used as recipients of the AIG cell lines 4 weeks after splenectomy. The nude mice were pretreated and the cells were administered identically regardless of whether the cells were derived from cloned squamous cell carcinoma (SCC) cell lines, carcinogen-exposed keratinocytes, or minced tumor tissue. The isolated cells were injected subcutaneously in the suprascapular region on the back of the nude mice. In each case, 10^7 cells suspended in MEM were injected. The animals were observed either daily or weekly for the development of a progressively growing tumor at the site of injection. A tumor was described as progressively growing and the cell line designated as AIGT if a $\geq$2.0-cm tumor developed within 4 to 8 weeks of the injection. The progressively growing tumors were recovered and the tissue was used for histological identification of the tumor type, establishment of secondary cell lines in monolayer culture, passage to additional nude mice, and the extraction of macromolecules for molecular analyses. Based on the results of the tumorigenicity evaluation, cloned tumor cell lines were designated with AIGNT or AIGT. These designations were subsequently used in other studies.

D. METHYLMETHANE SULFONATE (MMS) TREATMENT

AIGNT cell lines were treated with MMS to convert the cells to a different phenotype.[8,10,11,14] AIGNT cells in monolayer culture were exposed to the chemicals 24 h after splitting and passaging of the cultures. MMS (50 μg/ ml) was added to the GM and the cultures were maintained under the conditions previously described for 24 h. The treatment regimen and time of treatment with MMS has been previously described by Kerbel et al.[14] Following a 24-h exposure to MMS, the cultures were rinsed three times with GM. The cells were allowed to grow until the plates had reached 90% confluency (3 to 4 weeks). The cultures were then split 1:4 for three passages (1:4 split ratio = 2 PDs), after which 5×10^6 cells were injected subcutaneously into the nude mice as previously described.

E. BENZAMIDE TREATMENT

Benzamide treatment was used to inhibit the phenotypic conversion of AIGNT cells following exposure to MMS.[15] AIGNT cells treated with MMS were allowed to proliferate for 48 to 72 h, after exposure to the chemical. The proliferating cells were then exposed to benzamide (1, 2.5, 5, or 10.0 mM in GM) for 5 d. The benzamide treatment was repeated every 5 d for 2 weeks. Thereafter, the doubly treated (MMS and benzamide) cells were allowed to proliferate as described for the MMS-only treated cells. The tumorigenicity of the benzamide-treated cells was evaluated following subcutaneous injection into nude mice as previously described.

F. *IN SITU* HYBRIDIZATION

Frozen tumor sections (5 μm thick) were placed on poly-L-lysine-coated slides and fixed in 4% paraformaldehyde. The sections were preincubated in 0.3% H_2O_2 in methanol for 15 min to inactivate endogenous peroxidase. The sections were prehybridized at 37°C for 3 h. The cDNA probes H-*ras*, c-*myc* (American Type Culture Collection), Type I keratin (Dr. Elaine Fuchs, University of Chicago), and pBR 322 (Bethesda Research Labs) were labeled with biotinylated dUTP by nick translation.[16] Individual hybridization mixes for each probe were prepared at a probe concentration of 1 μg/ml in a standard hybridization buffer mix. The sections were incubated with the hybridization mix in a humidified chamber at a stringent temperature for 6 h. After hybridization, the slides were washed sequentially in 2 × SSC-50% formamide, 1 × SSC-50% formamide, and 1 × SSC.[17-20] For detection of the pattern of probe hybridization, the slides were incubated in Vectastain ABC at 37°C for 10 min, washed, and incubated with the chromagen, diaminobenzidine-HCl, H_2O_2. The sections were counterstained with eosin and examined by light microscopy.[11,20]

G. POLYMERASE CHAIN REACTION (PCR) EXPANSION OF DNA FROM CELL LINES AND DIRECT DNA SEQUENCING

Genomic DNA from anchorage-independent cells bearing the SCC-associated cell surface antigen was amplified by the polymerase chain reaction at *ras*-specific regions with the Onco-Lyzer core kit (Clontech Labs).[21,22] Each PCR mixture contained genomic DNA (0.5 μg), specific primers (0.6 μ*M* each), all four dNTPs (0.2 m*M* each), 1 × reaction buffer with 1.5 m*M* MgCl$_2$, and 1.25 U of AmpliTaq DNA polymerase (Perkin-Elmer Cetus). The genomic DNA was amplified in 30 cycles, each cycle including a 1-min denaturation step at 94°C, a 1-min primer annealing step at 65°C, and a 1-min primer extension step at 74°C. The amplified products were then purified by centrifugation through a Ventricon 100 microconcentrator (Amicon) and an aliquot was used in an asymmetric PCR assay to generate single-stranded DNA that was directly sequenced. The asymmetric PCR process was carried out exactly as above, except that one primer was limiting (0.6 μ*M* vs. 0.06 μ*M*).

The amplified products were purified by centrifugation through a Centricon 100 microconcentrator (Amicon) and then sequenced with the Sanger-dideoxy nucleotide method.[23,24] First, an equimolar amount of the limiting primer in the asymmetric PCR process was annealed to the amplified DNA in a 10-μl reaction volume by heating to 70°C for 3 min, then to 42°C for 10 min in the presence of a 5 × annealing buffer (35 m*M* MgCl$_2$ and 250 m*M* Tris, pH 8.8). To begin the synthesis of DNA chains, 0.5 μl of (a-[^{35}S]thio)dATP (>1000 Ci/mmol), 2 μl of labeling mix (2.5 μM each of dCTP, dGTP, and dTTP), 2 U of Sequenase, and 3 ml of dH$_2$O were added to the annealed DNAs and incubated at 42°C for 5 min, then cooled to room temperature. The 4-μl aliquots of this mixture were added to 4 μl of the A, C, G, or T termination mixes (20 μM of all four dNTPs and 60 to 800 μM of the particular ddNTP) and incubated at 70°C for 5 min. The sequencing products were then run out on an 8% urea-polyacrylamide gel which was exposed to Kodak X-omat AR film overnight at room temperature. The DNA sequences were read from the exposed X-ray film, entered into a computer, and analyzed by a computer-assisted sequence analysis program.

H. NORTHERN BLOT HYBRIDIZATION ANALYSIS

Total cellular RNA was isolated by the guanidium isothiocyanate/CsCl method and the RNA molecules electrophoretically separated on a 1.2% agarose denaturing gel.[25,26] Gels were stained with ethidium bromide and visualized for equivalent loading of RNA among lanes by comparison of the intensities of the 28S and 18S bands. The RNA was transferred to nylon membranes. The RNA immobilized on the membranes was hybridized with ^{32}P-labeled cDNA probes at high stringency. Following hybridization, the filters were washed and the pattern of hybridization detected on Kodak X-omat AR film.

III. RESULTS

A. ISOLATION OF A CELL LINE WITH A CAPACITY FOR PHENOTYPIC MODULATION

Several cell lines were isolated from human tumors that developed at different body sites.[8,9] All of the cell lines were capable of anchorage-independent growth and prolonged proliferation in monolayer culture. Tumorigenicity analysis of these cell lines showed that some of them were no longer capable of producing tumors in nude mice (Table 1).[8] These cell lines retained many characteristic features of transformed cells, but had lost the capacity for tumorigenesis. These AIG-positive, immortalized but nontumorigenic cell lines were designated AIGNT since they no longer produced tumors in the xenogeneic hosts. The AIGNT cell lines were examined for the effect of subsequent chemical exposure and the reacquisition of the tumorigenic phenotype.

B. EFFECT OF MMS EXPOSURE OF AIGNT CELL LINES

The AIGNT cell lines were exposed to MMS while in monolayer culture as previously described. The cells were treated with a nonmutagenic dose of MMS to determine whether the phenotype could be converted to AIGT. The responses of the different AIGNT cell lines to the MMS exposure were not identical (Table 2).[8] Only two of the cell lines were converted to the AIGNT

TABLE 1
Generation of Cell Lines with an AIGNT Phenotype

| | Frequency of tumors/experiment | | | |
Cell line	1	2	3	Total
SCC-83-01-82	0/2	0/3	0/3	0/8
OSU-83-5-45	0/3	0/3	0/3	0/9
HEII	0/4	0/4	ND	0/8
HET-1A	0/4	0/8	ND	0/8
Chondrosarcoma	0/3	0/3	ND	0/6

Note: All five cell lines exhibited anchorage-independent growth (AIG) by forming colonies in soft agar. The soft-agar colonies were isolated and the cells amplified by growth in monolayer culture. All five cell lines exhibited extended growth in monolayer culture and, following 70 population doublings, continued to proliferate. The tumorigenic potential of the cell lines was assessed by injecting 5×10^6 cells subcutaneously on the flank region of nude mice. The injection site was observed weekly. The failure to produce a tumor was determined when no visible or histologic growth was present 6 months after the injection.[8] SCC-83-01-82, tumor cell line developed from a human squamous cell carcinoma; OSU-83-5-45, tumor cell line developed from a human oat cell carcinoma of the lung; HEII, an immortalized cell line developed at Ohio State University by infection of human keratinocytes with SV40; HET-1A, a cell line developed by Dr. Gary Stoner at the Medical College of Ohio by transfection of human esophageal epithelial cells with a transforming DNA; chondrosarcoma, tumor cell line developed from a human chondrosarcoma of the heel; ND, experiment not done.

phenotype following the MMS treatment. In the MMS-converted AIGNT cell lines which became tumorigenic, the progressively growing tumors were >2.0 cm in size within 4 to 6 weeks following injection of the cells into the nude mice. The histopathology of the tumors which developed in the nude mice was consistent with the diagnosis of the human tumors from which the cell lines were derived (Figure 2).[8] Thus, the chemical treatment converted the phenotype of the AIGNT cells to one with a more aggressive biologic behavior. These MMS-converted cells resulted in the development of a tumor in nude mice that was histopathologically similar to the original origin of the cell line.

C. EFFECT OF BENZAMIDE TREATMENT FOLLOWING MMS EXPOSURE OF AIGNT CELLS

Conversion of the AIGNT phenotype to an AIGT phenotype following exposure to a nonmutagenic dose of MMS showed that these cells could be converted to tumorigenicity. These data did not show whether this change was either inhibitable or irreversible. We selected the SCC-83-01-82 cell line (AIGNT phenotype) to examine the effects of benzamide on the inhibition of the conversion to the AIGT phenotype. MMS-exposed cells were treated with benzamide as described in Section II. Treatment of the MMS-converted AIGNT

TABLE 2
Conversion of AIGNT Phenotype to AIGT Phenotype Following Exposure to MMS

| Cell line | Treatment | Frequency of tumors/experiment | | | Total |
		1	2	3	
SCC-83-01-82	MMS	1/2	2/3	2/3	5/8
OSU-83-5-45	MMS	1/2	2/3	2/3	5/8
HEII	MMS	0/2	0/4	ND	0/6
HET-1A	MMS	0/4	0/4	ND	0/8
Chondrosarcoma	MMS	0/4	0/3	ND	0/7

Note: All five cell lines exhibited anchorage-independent growth (AIG) by forming colonies in soft agar and extended growth in monolayer culture. Monolayer cultures were exposed to MMS (50 μg/ml) for 24 h, washed, and allowed to proliferate for eight population doublings. The tumorigenic potential of the cell lines was assessed by injecting 5 $\times$ 10^6 cells subcutaneously on the flank region of nude mice. The injection site was observed weekly. The development of the AIGT phenotype was determined by the presence of a progressively growing tumor ($\geq$2.0 cm in diameter) within 4 to 6 weeks after injection of the cells.[8] SCC-83-01-82, tumor cell line developed from a human squamous cell carcinoma; OSU-83-5-45, tumor cell line developed from a human oat cell carcinoma of the lung; HEII, an immortalized cell line developed at Ohio State University by the infection of human keratinocytes with SV40; HET-1A, a cell line developed by Dr. Gary Stoner at the Medical College of Ohio by transfection of human esophageal epithelial cells with a transforming DNA; chondrosarcoma, tumor cell line developed from a human chondrosarcoma of the heel; ND, experiment not done.

cells with benzamide (1, 2.5, 5, and 10.0 mM) delayed both the time of appearance of the progressively growing tumors and the time necessary to produce a $\geq$2.0-cm tumor (Table 3). The highest doses of benzamide, 5.0 and 10.0 mM, inhibited the development of tumors in the nude mice. Treatment of AIGNT cells with benzamide alone had no effect on the capacity of these cells for growth in soft agar, prolonged proliferation in monolayer culture, or nontumorigenicity in a xenogeneic host. Treatment of the AIGNT cells with benzamide prior to exposure to MMS did not alter the MMS-induced conversion to an AIGT phenotype. Thus, the effects of the MMS treatment could be reversed by a subsequent exposure of the cells to benzamide.

Modulation of the phenotype of AIGNT cell line 83-01-82 appeared to be the result of several separate events. Exposure to MMS was capable of altering the phenotype to AIGT, but this step could be inhibited by a subsequent treatment of the cells with benzamide. These results showed that the effects of MMS treatment were inhibitable, but they did not determine whether they were irreversible. Thus, the heritability of the tumorigenic conversion step required examination.

D. TRANSITION OF AIGT PHENOTYPE TO AIGNT PHENOTYPE

Perpetuation of the AIGT phenotype in MMS-converted AIGNT cells

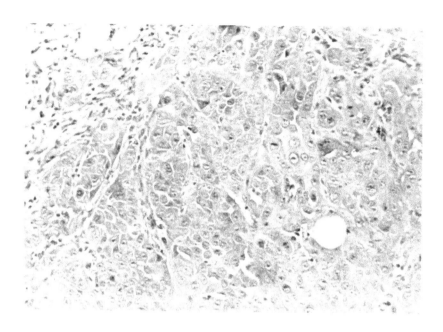

FIGURE 2. Histopathology of a nude mouse tumor which developed from MMS-converted AIGNT cell line SCC-83-01-82. The AIGNT cell line SCC-83-01-82 was exposed in monolayer culture to MMS. After eight population doublings, 5×10^6 cells were injected subcutaneously into the flank region of a nude mouse. A progressively growing tumor (≥ 2.0 cm) developed in 4 to 6 weeks following injection of the MMS-converted cells. The tumor was removed, fixed in formalin, embedded in paraffin, sectioned, and stained with hematoxylin and eosin. The histologic interpretation of the tumor is consistent with squamous cell carcinoma.

was examined in cells isolated from the tumors which developed in the nude mice.[10,11] Portions of the tumors were handled in different ways to examine the effects that environmental conditions had on the expression of the phenotype of these cells. Portions of the tumors (T_1) were subpassaged *in vivo*, immediately after removal, by injection of isolated tumor cells into another nude mouse. Additional portions of the tumors were used to obtain isolated cells for the generation of monolayer cultures. These cells were propagated *in vitro* and subsequently injected into nude mice to examine their tumorigenicity.

Serial passage of the tumor cells *in vivo* resulted in the continued development of tumors (Table 4).[10,11] The development of the tumors occurred at the same pace in the *in vivo* subpassaged cases, and with later subpassages, the histopathology of the tumors changed to a more anaplastic morphology (Figure 3). Maintenance of the cells in an *in vivo* environment resulted in the continued expression of the AIGT phenotype and apparently selected for a set of cells with a less differentiated morphology.

Establishment of cell lines from the tumors that developed resulted in a change of the phenotype in some of the cell lines. Initial culturing resulted

TABLE 3
Effect of Benzamide on the Development of the AIGT Phenotype in
MMS-Treated AIGNT Cells

Exp. #	mM BZ	Number of tumors/ number of mice injected	Time for growth of ≥ 2.0-cm tumor after BZ	Metastases
1	0.0	4/4	—	Yes
	1.0	4/4	6	No
	2.5	4/4	9	No
	5.0	1/4	14	No
	10.0	0/4	No tumors	NA
2	0.0	3/3	—	Yes
	1.0	3/3	10	No
	2.5	3/3	10	No
	5.0	1/5	14	No
	10.0	0/5	No tumors	NA

Note: SCC-83-01-82 cells were exposed to MMS and benzamide in monolayer culture as described in Section II. The cells were allowed to proliferate for eight population doublings after treatment, and then 5×10^6 viable cells were injected into each nude mouse. The development of a tumor at the site of injection was followed by weekly observation. NA = not applicable.

TABLE 4
Characterization of the SCC Tumors Developed by Serial Passage *In Vivo*

Tumor origin	Mice receiving implants/mice with tumors	mRNA presence by *in situ* hybridization H-*ras*	c-*myc*	Tumor histology
MMS-SCC (T_1)	4/4	+	+	SCC
SCC-T_1 (T_2)	4/4	+	+	AdCA
SCC-T_2 (T_3)	4/4	+	+	SCC-pd
SCC-T_3 (T_4)	4/4	−	+	SCC-pd
SCC-T_4 (T_5)	4/4	−	+	SCC-pd

Note: SCC-83-01-82 cells were treated in monolayer culture with MMS. The cells (5×10^6) were injected subcutaneously into nude mice and the tumor MMS-SCC (T_1) developed. Portions of the MMS-SCC (T_1) tumor were subpassaged *in vivo* by implantation of tumor cells subcutaneously in another nude mouse. The tumors which developed were similarly serially passaged.[10,11] T_2, tumor derived from SCC-T_1 tumor fragments; T_3, tumor derived from SCC-T_2 tumor fragments; T_4, tumor derived from SCC-T_3 tumor fragments; T_5, tumor derived from SCC-T_4 tumor fragments; SCC, moderately differentiated squamous cell carcinoma; AdCA, undifferentiated adenocarcinoma; SCC-pd, poorly differentiated SCC.

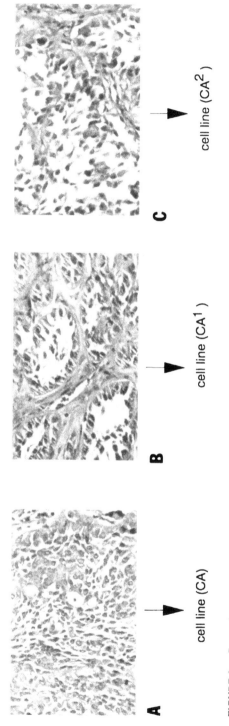

FIGURE 3. Comparison of the histopathology of tumors derived from MMS-converted human AIGNT cells serially passaged in nude mice. SCC-83-01-82 cells were treated in monolayer culture with MMS. The cells (5×10^6) were injected subcutaneously into nude mice and the tumor MMS-SCC (T_1) developed. Portions of the MMS-SCC (T_1) tumor were subpassaged *in vivo* by implantation of tumor cells subcutaneously in another nude mouse. The tumors which developed were similarly serially passaged. (A) Tumor (MMS-SCC [T_1], developed from MMS-treated SCC-83-01-82 cells, which was interpreted as a moderately differentiated squamous cell carcinoma; (B) tumor (T_2, developed *in vivo* subpassage of portions of the MMS-SCC (T_1) tumor, which was interpreted as an undifferentiated adenocarcinoma; (C) tumor (T_3), developed from *in vivo* subpassage of portions of the SCC-T_2 tumor, which was interpreted as a poorly differentiated squamous cell carcinoma.

TABLE 5
Characterization of Tumorigenic Potential of Different Tumor-Derived Cell Lines

Cell line	Number of mice receiving injections/ number of mice with tumors	Tumor latent period
SCC-83-01-82	8/0	NA
MMS-83-01-82	11/5	3 4 months
SCC-83-01-82CA	7/7	5—7 d
SCC-83-01-82CA$_1$	4/4	5—7 d
SCC-83-01-82CA$_2$	2/2	10—15 d
SCC-83-01-82CA*	6/0	NA

Note: All cell lines were grown in monolayer culture to amplify the numbers. For all the cell lines, 5×10^6 cells were injected subcutaneously into nude mice and the area evaluated for the development of tumors by weekly observation. The development of a mass ≥ 2.0 cm in diameter was characterized as a progressively growing tumor. The ≥ 2.0-cm tumors were excised and examined microscopically to determine the final histologic typing. MMS treatments of the AIGNT cell line SCC-83-01-82 were done as previously described. SCC-83-01-82-CA cells were derived from a T_1 tumor that developed following the injection of a nude mouse with MMS-treated SCC-83-01-82 cells. SCC-83-01-82CA$_1$ cells were derived from a T_2 tumor which developed from *in vivo* passage of T_1 tumor fragments. SCC-83-01-82CA$_2$ cells were derived from a T_3 tumor which developed from *in vivo* passage of T_2 tumor fragments. SCC-83-01-82CA* was a cell line derived as previously described which was maintained in a continuous growth phase for 6 months.[10]

in the development of cell lines that were capable of producing tumors in nude mice. However, prolonged passage of these cells *in vitro* resulted in the cell line losing the tumorigenic capacity (Table 5).[10] Thus, the cell line had reverted from the AIGT phenotype back to an AIGNT phenotype. This reversion was back to the originally selected AIGNT phenotype. The reverted AIGNT cell line could be converted to AIGT by following the same MMS exposure protocol (data not shown). Thus, AIGNT cell line SCC-83-01-82 had a plasticity of phenotype that could be modulated by experimental treatments. Moreover, cloning of these cells from the AIGT phenotype (CA-CA$_3$) indicated that these MMS-converted CA phenotypes exhibited plasticity.

E. MOLECULAR CHARACTERIZATION OF TUMOR CELL PHENOTYPES

The *in vitro* and *in vivo* patterns of growth of the different cell lines permitted the designation of the different phenotypic groups. Without some type of intervention, the phenotype of the different cell lines is maintained. However, the data concerning the change in phenotype of the AIGNT cell line following exposure to MMS suggested that molecular changes existed between the different phenotypes. We characterized the pattern of expression of genes, that have been shown to be associated with cell transformation, in the different tumor cell phenotypes and in xenogeneic tumors that developed

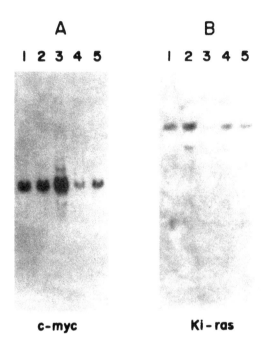

FIGURE 4. Characterization of the c-*myc* and K-*ras* mRNA levels in SCC-83-01-82 cells in the AIGNT stage, after MMS treatment, and in the AIGT stage. (A) *Myc* hybridization to RNA isolated from different cell lines; (B) K-*ras* hybridization to RNA isolated from different cell lines. All filters contained identical RNA preparations and concentrations. Lane 1, untreated SCC-83-01-82 cells; lane 2, MMS-treated SCC-83-01-82; lane 3, SCC-83-01-82-CA cells derived from a T_1 tumor from MMS-treated SCC-83-01-82 cells, lane 4, SCC-83-01-82CA$_1$ cells derived from a T_2 tumor which developed from *in vivo* passage of T_1 tumor fragments; lane 5, SCC-83-01-82CA$_2$ cells derived from a T_3 tumor which developed from *in vivo* passage of T_2 tumor fragments.

from some of the cell lines. The expression of these genes was examined by Northern analysis to detect the level of expression, by *in situ* hybridization to detect the localization of the message and relative number of cells expressing the gene, and by PCR analysis of the gene to detect genomic mutations that have been associated with the activation and transformation ability of these genes.[8-11]

Northern blot analysis was used to determine whether conversion of the nontumorigenic SCC-83-01-82 cell line to tumorigenicity after MMS treatment involved altered levels of expression of c-*myc*, H-*ras*, or K-*ras*.[10] No consistent change in the level of expression for any of the three genes could be detected after MMS treatment of SCC-83-01-82 (Figure 4; H-*ras* data not shown).[10,11] The MMS-induced conversion was not associated with an increase in the level of expression of any of these three genes from the levels present in the AIGNT cell line. Analysis of cell lines derived from tumors that developed from either MMS-treated AIGNT cells or from *in vivo* subpassaged tumors also did not have results consistent with continued overexpression of any of these three genes. In fact, the later serially subpassaged tumors and their cell lines (T_2, T_3, CA_1, and CA_2) had decreased levels of expression of these genes (see Table 4). This result suggested that these cells had undergone a further set of molecular changes that did not require the expression of these genes for continued tumorigenicity. These results imply that the AIGT phenotype contains a set of molecular phenotypes which are associated with differing biologic behaviors.

Localization of the oncogene mRNA in tumors was accomplished by *in situ* hybridization. This technique permitted both spatial localization of the particular mRNA in the cells in the tumor and an assessment of the number of cells in monolayer culture expressing the gene. Examinations of the tumors which developed in the nude mice following injection of the cells revealed a variability in the level of expression between different tumor cell populations. A nonuniform pattern of distribution of cells with high levels of expression of either H-*ras* or c-*myc* was present (Figure 5).[10,11,20] The cells with the highest levels of either c-*myc* or H-*ras* were located primarily in the peripheral regions of the tumors, those areas with the most invasive and proliferative groups of cells. The number of cells expressing these genes and the level of expression in the individual cells decreased in the later-passage tumors (see Table 4). These results confirm the results from Northern blot hybridization that the pattern of expression changes in the poorly differentiated SCC and the cell lines derived from them. These results further suggest that the AIGT phenotype contains multiple subphenotypes, each with a differing pattern of gene expression. These subphenotypes probably represent further molecular changes associated with the progression of the tumorigenic phenotype.

PCR analysis of the oncogenes examined was used to detect the presence of genomic mutations, in specific codons of these genes, which have been associated with the activation of their transforming ability. PCR was used to specifically amplify the genomic regions of interest, and the DNA sequence of these regions was subsequently determined. The regions examined were codons 12, 13, and 61 for the H-*ras*, K-*ras*, and N-*ras* gene (Table 6). A mutation was detected in the 12th codon of the SCC, CA, CA_1, CA_2, and CA_3 cell lines, but not in DNA amplified from normal cells (Figure 6). The other examined genomic regions did not contain mutations in the DNA sequence. The mutation in the 12th codon, which was detected in the five cell lines, results in a change in the glycine coding triplet GGC to GTC which causes a valine to be inserted at that position. This mutation is identical to the H-*ras* mutation which has previously been characterized in the T_{24} bladder carcinoma cell line.[27] The PCR/DNA sequencing analyses did not show any further mutations in the cell lines which have differing tumorigenic behaviors, and thus the molecular changes responsible for the phenotypic alterations observed in these cells must reside in a different genetic element. The results also did not show a direct effect of MMS in the generation of mutations in

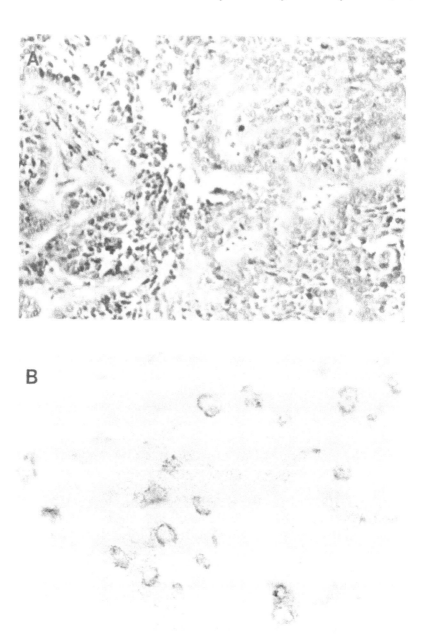

FIGURE 5.   Localization of oncogene mRNA by *in situ* hybridization.[10,11,20] 8-μm thick frozen sections of a nude mouse tumor were hybridized with biotinylated cDNA probes for either c-*myc*, H-*ras*, or Type I keratin. The RNA-DNA hybrids were detected with avidin-horseradish peroxidase-conjugated complex and DAB-H$_2$O$_2$. The sections were photographed without counterstaining to demonstrate the pattern of *in situ* hybridization of each of the three cDNA probes. (A) H & E-stained tumor section (original magnification × 40); (B) *in situ* hybridization with keratin cDNA probe (original magnification × 25); (C) *in situ* hybridization with c-*myc* cDNA probe (original magnification × 40); (D) *in situ* hybridization with H-*ras* cDNA probe (original magnification × 25).

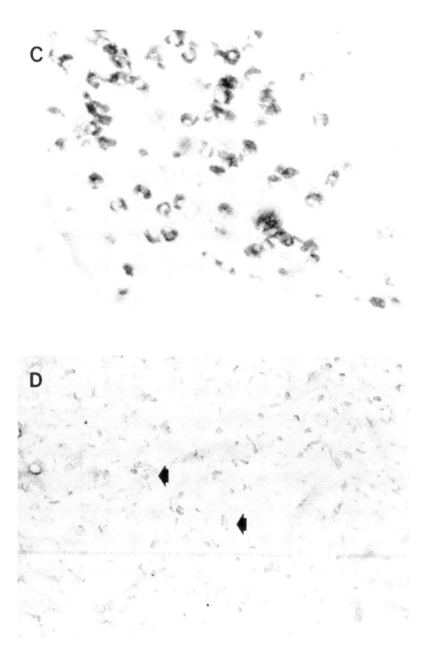

FIGURE 5 (continued)

TABLE 6
PCR Analysis of Mutations in Specific Codons of the H-*ras* and K-*ras* Genes

Amplimer set/codon	Amplimer sequence	Cell line examined					
		Normal	SCC	CA	CA$_1$	CA$_2$	CA$_3$
H-*ras*/12,13	5'ATG ACG GAA TAT AAG CTG GT 3'CGC CAG GCT CAC CTC TAT A	−	+	+	+	+	+
H-*ras*/61	5'AG GTG GTC ATT GAT GGG GAG 3'AG GAA GCC CTC CCC GGT GCG	−	−	−	ND	ND	ND
K-*ras*/12,13	5'ATG ACT GAA TAT AAA CTT GT 3'CTC TAT TGT TGG ATC ATA TT	−	−	−	−	−	−
K-*ras*/61	5'AA GTA GTA ATT GAT GGA GAA 3'AG AAA GCC CTC CCC AGT CCT	−	−	−	−	ND	ND
N-*ras*/12,13	5'ATG ACT GAG TAC AAA CTG GT 3'CTC TAT GGT GGG ATC ATA TT	−	−	−	−	ND	ND
N-*ras*/61	5'CAA GTG GTT ATA GAT GGT GA 3'AG GAA GCC TTC GCC TGT CCT	−	−	−	−	ND	ND

Note: The amplimer sequences are positioned with the translation start sites as 1 and with introns excluded. ND indicates experiments not done and (−) indicates no mutation in the tested region. Cell lines as designated in Table 5. PCR amplification and DNA sequencing as described in Section II.

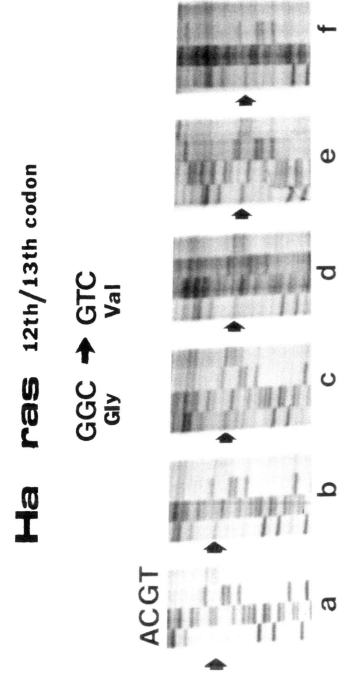

FIGURE 6. Autoradiogram of a polyacrylamide–urea gel comparing sequences of the Ha-*ras* 12th codon region. DNA templates for the sequencing were prepared by conventional PCR amplification of genomic DNA. Asymmetric PCR was carried out using the amplified DNA with specific primers for the codon region (Clontech). Sequencing reactions were labeled by incorporation of [[³⁵S]thio]dATP and one of the primers used for the sequencing reactions. The products were resolved on a sequencing gel. The reaction sets were loaded from left lane to right lane in sets of four from a to f in the sequence A, C, G, T nucleotides. (a) Normal human fibroblast cells; (b) CA clone 1; (c) CA clone 2; (d) CA clone 3; (e) CA cells; (f) SCC AIGNT cells.

these genes, which implied a different mechanism of action of MMS in the AIGNT-AIGT conversion process. Current studies are continuing to characterize the molecular differences between the different tumor cell phenotypes.

IV. DISCUSSION

Although most human tumors are assumed to be clonal in origin, each tumor contains subpopulations of cells with differences in cellular differentiation as determined by histopathology. These subpopulations of cells probably represent different stages in a continuum of potential transformed cell morphologies.[1-3] Thus, individual subpopulations of cells may possess different degrees of malignant vigor. The molecular differences between these different tumor cell populations have not been completely characterized.[28]

Characterization of the sequential changes that occur during the development of tumors has been made in a few systems.[29-34] These studies have primarily concentrated on examining the molecular differences between entire lesions that appear to represent progressively altered tissue with increasing potentials for malignant tumor development. We have used a different approach and concentrated on examining the difference between isolated cell lines with different defined human tumor cell phenotypes.[8-11] These phenotypes have been isolated from human tumors on the basis of their differences in growth pattern both *in vitro* and *in vivo*.[35-40] The plasticity of the AIGNT phenotype has permitted examinations of molecular changes that may be associated with the conversion of these cells to an AIGT phenotype. The ability to observe these molecular events in a defined environment may help define the changes in gene expression that are involved in the sequence of events giving rise to the multiple cellular phenotypes present in human tumors.[41-45]

The MMS-induced conversion of an AIGNT cell line to an AIGT phenotype represents one model for examining the molecular events associated with a transition in the biologic behavior of tumor cells. Both the AIGNT and AIGT-converted cells expressed a similar set of oncogenes which have been previously associated with malignant transformation. The conversion process was not associated with changes in the level of expression of any of this set of oncogenes. Selection of additional cell lines from the nude mouse tumors did show that the expression of this set of oncogenes could change. Interestingly, the more aggressive AIGT cell lines isolated (CA-CA$_2$) had decreased expression of H-*ras*. Thus, AIGT cell lines capable of rapid tumor development in a xenogeneic host did not necessarily express a set of oncogenes that were common to the other AIGNT and MMS-converted AIGT cell lines. These results suggest that the AIGT phenotype actually contains a number of molecular phenotypes which may each exhibit different degrees of malignant vigor. These phenotypes may represent the development of tumor cell subpopulations with the capacity for either local invasion or metastasis, features with the most dire consequences in human tumors.

The plasticity of the tumor cell phenotypes represents a critical area of examination. The AIGNT cell line used in these studies could be converted to an AIGT phenotype by MMS treatment. This tumorigenic conversion could be directly inhibited by treatment of the cells in monolayer culture with benzamide. Thus, the MMS-induced changes were either reversed by the benzamide treatment or directly inhibited by benzamide-induced molecular events. The molecular effects of both MMS and benzamide are currently being examined.

The plasticity of the AIGNT phenotype was also evident by the pattern of growth of MMS-converted cells *in vivo*. The AIGT phenotypic conversion, due to MMS treatment of AIGNT cells, persisted if the tumor cells were maintained by passage *in vivo*. This MMS-acquired AIGT phenotype was also resistant to treatment with benzamide *in vivo*. However, the MMS-induced AIGT phenotype did revert to an AIGNT phenotype with prolonged monolayer culture of cell lines derived from the tumors. The interpretation of these results was that the *in vitro* culture conditions caused a reversion of the cell line to an AIGNT phenotype. The cell lines responded differently to epigenetic effectors present *in vitro* and *in vivo*. Thus, the cellular environment had a role in the development of a particular phenotype.[46] The plasticity of the tumor cell phenotypes represented a combination of heritable effects resulting from both the initial transforming event and chemical exposure and epigenetic effects from the environment. In both cases, transient changes in gene expression could be expected to play a role in the phenotypic fate of the cells.

The observations that we have made contribute to further understanding of the molecular changes that occur during the progressive development of a malignant phenotype. The plasticity of different tumor cell phenotypes represents a condition that may be used to therapeutic advantage if tumor cells could be induced to follow a pattern of terminal differentiation.[47] The AIGNT phenotype model system we have developed presents a method to analyze molecular events associated with specific changes in tumor cell behavior. Further studies will examine specific patterns of gene expression associated with the transitions between different phenotypic groups.

ACKNOWLEDGMENTS

We thank Dr. Hakjoo Lee and Dr. Ju-Cheng Chen for their contributions to this work. The work was supported in part by the National Institutes of Health-National Cancer Institute (NIH-NCI) R01-CA25907-09 (for George E. Milo and Charles F. Shuler) and NIH-NCI P30-CA16058-15 (for the Ohio State University Comprehensive Cancer Center).

REFERENCES

1. **Cairns, J.**, The origin of human cancers, *Nature,* 289, 353, 1981.
2. **Fidler, I. J. and Hart, I. R.**, Biologic diversity in metastatic neoplasms — origins and implications, *Science,* 217, 998, 1982.
3. **Heppner, G. H.**, Tumor heterogeneity, *Cancer Res.,* 44, 2259, 1984.
4. **Milo, G. E., Oldham, J., Zimmerman, R., Hatch, G., and Weisbrode, S.**, Characterization of human cells transformed by chemical and physical carcinogens in vitro, *In Vitro,* 17, 719, 1981.
5. **Milo, G. E., Casto, B., and Ferrone, S.**, Comparison of features of carcinogen-transformed human cells in vitro with sarcoma derived cells, *Mutat. Res.,* 199, 387, 1987.
6. **Rose, P., Koolemans-Beynen, A., Boutselis, J. G., Minton, J. P., and Milo, G. E.**, An improved tumor stem cell assay in ovarian cancer, *Am. J. Obstet. Gynecol.,* 156, 730, 1987.
7. **Milo, G. E., Yohn, J., Schuller, D. E., Noyes, I., and Lehman, T.**, Comparative stages of expression of human squamous carcinoma cells and carcinogen transformed keratinocytes, *J. Invest. Dermatol.,* 92, 848, 1989.
8. **Milo, G. E., Shuler, C. F., and Stoner, G.**, Cell phenotype specific conversion of premalignant human cells to tumorigenic cells by methylmethane sulfonate and methylnitrosoguanidine, *Cell Biol. Toxicol.,* submitted.
9. **Chen, J.-C., Shuler, C. F., Zhang, C.-X., Schuller, D. E., and Milo, G. E.**, Histopathologic comparison between human oral squamous cell carcinomas and their xenografts in nude mice, *Oral Surg. Oral Med. Oral Pathol.,* 71, 457, 1991.
10. **Milo, G. E., Shuler, C. F., Kurian, P., French, B. T., Mannix, D. G., Noyes, I., Hollering, J., Sital, N., Schuller, D. E., and Trewyn, R. W.**, Nontumorigenic squamous cell carcinoma line converted to tumorigenicity with methylmethane sulfonate without activation of HRAS or MYC, *Proc. Natl. Acad. Sci. U.S.A.,* 87, 1268, 1990.
11. **Shuler, C. F., Kurian, P., French, B. T., Noyes, I., Sital, N., Hollering, J., Trewyn, R. W., Schuller, D. E., and Milo, G. E.**, Noncorrelative c-*myc* and *ras* oncogene expression in squamous cell carcinoma cells with tumorigenic potential, *Terat. Carcinog. Mutagen.,* 10, 53, 1990.
12. **Donahoe, J., Noyes, I., Milo, G. E., and Weisbrode, S.**, A comparison of expression of neoplastic potential of carcinogen-induced transformed human fibroblasts in nude mice and chick embryonic skin, *In Vitro,* 18, 429, 1982.
13. **Rygaard, J. and Poulsen, C. O.**, Heterotransplantation of human malignant tumor to ''nude'' mice, *Acta Pathol. Microbiol. Scand.,* 77, 758, 1969.
14. **Kerbel, R. S., Frost, P., Liteplo, R., Carlow, D. A., and Elliot, B. E.**, Possible epigenetic mechanisms of tumor progression: induction of high frequency heritable but phenotypically unstable changes in the tumorigenic and metastatic properties of tumor cell populations by 5-azacytidine treatment, *J. Cell Physiol. Suppl.,* 3, 87, 1984.
15. **Milo, G. E. and Lee, H.**, Molecular control of expression of plasticity of tumorigenic/ metastatic phenotypes, in *Neoplastic Transformation in Human Cell Systems In Vitro: Mechanisms of Carcinogenesis,* Rhim, J. S. and Dritschilo, J., Eds., Humana Press, Clifton, NJ, in press.
16. **Rigby, P. W. J., Dieckmann, M., Rhodes, C., and Berg, P.**, Labeling deoxyribionucleic acid to high specific activity in vitro by nick translation with DNA polymerase, *J. Mol. Biol.,* 113, 327, 1977.
17. **Singer, R. H., Larence, J. B., and Villnave, C.**, Optimization of in situ hybridization using isotopic and non-isotopic detection methods, *BioTechniques,* 4, 230, 1986.
18. **Angerer, R. C., Cox, K. H., and Angerer, L. M.**, *In situ* hybridization to cellular RNAs, *Genet. Eng.,* 7, 43, 1985.

19. **Brigatic, D. J., Myerson, D., Leary, J. J., Spalholz, B., Travis, S. Z., Fong, C. K. Y., Ksiung, G. D., and Ward, D. C.,** Detection of viral genomes in cultured cells and paraffin-embedded tissue sections using biotin-labeled hybridization probes, *Virology,* 125, 32, 1983.

20. **Hoellering, J. and Shuler, C. F.,** Localization of H-*ras* mRNA in oral squamous cell carcinomas, *J. Oral Pathol. Med.,* 18, 74, 1989.

21. **Saiki, R. K., Scharf, S., Faloona, F., Mullis, K. B., Horn, G. T., Ehrlich, H. A., and Arnheim, N.,** Enzymatic amplification of B-globin genomic sequences and restriction site analysis for diagnosis of sickle cell anemia, *Science,* 230, 1350, 1985.

22. **Lyons, J.,** Analysis of *ras* gene point mutations by PCR and oligonucleotide hybridization, in *PCR Protocols,* Innis, M. A., Gelfand, D. H., Sninsky, J. J., and White, T. J., Eds., Academic Press, New York, 1990, 386.

23. **Sanger, F., Nicklen, S., and Coulson, A. R.,** DNA sequencing with chain-terminating inhibitors, *Proc. Natl. Acad. Sci. U.S.A.,* 74, 5463, 1977.

24. Sequenase, *Step by Step Protocol for DNA Sequencing with Sequenase,* 5th ed., U.S. Biochemical Corp., 1989.

25. **Chrigwin, J. M., Przybyla, A. E., MacDonald, R. J., and Rutter, W. J.,** Isolation of biologically active ribonucleic acid from sources enriched in ribonuclease, *Biochemistry,* 18, 5294, 1979.

26. **Sambrook, J., Fritsch, E. F., and Maniatis, T.,** in *Molecular Cloning A Laboratory Manual,* 2nd ed., Cold Spring Harbor Laboratory Press, Cold Spring Harbor, NY, 1989.

27. **Senger, D. R., Peruzzi, C. A., and Ali, I. U.,** T24 human bladder carcinoma cells with activated Ha-ras protooncogene: nontumorigenic cells susceptible to malignant transformation with carcinogen, *Proc. Natl. Acad. Sci. U.S.A.,* 85, 5107, 1988.

28. **Milo, G. E., Casto, B., and Shuler, C. F., Eds.,** *Transformation of Human Epithelial Cells: Molecular and Oncogenetic Mechanisms,* CRC Press, Boca Raton, FL, in preparation.

29. **Nowell, P. C.,** Mechanisms of tumor progression, *Cancer Res.,* 46, 2203, 1986.

30. **Newbold, R. F.,** Multistep malignant transformation of mammalian cell by carcinogens: induction of immortality as a key event, *Carcinogenesis,* 9, 17, 1985.

31. **Sukumar, S.,** An experimental analysis of cancer: role of *ras* oncogenes in multistep carcinogenesis, *Cancer Cells,* 2, 199, 1990.

32. **Thompson, T. C., Southgate, J., Kitchener, G., and Land, H.,** Multistage carcinogenesis induced by *ras* and *myc* oncogenes in a reconstituted organ, *Cell,* 56, 917, 1989.

33. **Fearon, E. R. and Vogelstein, B.,** A genetic model for colorectal tumorigenesis, *Cell,* 61, 759, 1990.

34. **Vogelstein, B., Fearon, E. R., Kern, S. E., Hamilton, S. R., Preisinger, A. C., Nakamura, Y., and White, R.,** Allelotype of colorectal carcinomas, *Science,* 244, 207, 1989.

35. **Bishop, J. M.,** Molecular themes in oncogenesis, *Cell,* 64, 235, 1991.

36. **Hunter, T.,** Cooperation between oncogenes, *Cell,* 64, 249, 1991.

37. **Barbacid, M.,** Ras genes, *Annu. Rev. Biochem.,* 56, 779, 1987.

38. **Cole, M. D.,** The *myc* oncogene: its role in transformation and differentiation, *Annu. Rev. Genet.,* 20, 361, 1986.

39. **Bos, J. L.,** The *ras* family and human carcinogenesis, *Mutat. Res.,* 195, 255, 1988.

40. **Ruley, H. E.,** Transforming collaborations between *ras* and nuclear oncogenes, *Cancer Cells,* 2, 258, 1991.

41. **Liotta, L. A., Steeg, P. S., and Stetler-Stevenson, W. G.,** Cancer metastasis and angiogenesis: an imbalance of positive and negative regulation, *Cell,* 64, 327, 1991.

42. **Greenberg, A. H., Egen, S. E., and Wright, J. A.,** Oncogenes and metastatic progression, *Invasion Metastasis,* 9, 360, 1989.

43. **Kerbel, R. S.,** Growth dominance of the metastatic cancer cell: cellular and molecular aspects, *Adv. Cancer Res.,* 55, 87, 1990.

44. **Nicolson, G. L.**, Tumor cell instability, diversification and progression to the metastatic phenotype: from oncogene to oncofetal expression, *Cancer Res.*, 47, 1473, 1987.
45. **Fidler, I. J. and Radinsky, R.**, Genetic control of cancer metastasis, *J. Natl. Cancer Inst.*, 82. 166, 1990.
46. **Dotto, G. P., Weinberg, R. A., and Ariza, A.**, Malignant transformation of mouse primary keratinocytes by HaSV and its modulation by surrounding normal cells, *Proc. Natl. Acad. Sci. U.S.A.*, 85, 6389, 1988.
47. **Harris, H.**, The role of differentiation in the suppression of malignancy, *J. Cell Sci.*, 97, 5, 1990.

Chapter 10

ONCOGENE AND TUMOR SUPPRESSOR GENE INVOLVE-MENT IN HUMAN LUNG CARCINOGENESIS

Teresa A. Lehman and Curtis C. Harris

TABLE OF CONTENTS

I. INTRODUCTION

Carcinogenesis is a multistage process which occurs due to an accumulation of genetic and epigenetic changes that dysregulate molecular control of cell growth. The genetic changes can be the activation of protooncogenes and/or the inactivation of tumor suppressor genes that can initiate tumorigenesis as well as enhance its progression. For example, Ki-*ras* activation in colorectal carcinoma by base substitution is considered an early event,[1] whereas gene amplification of N-*myc* has been associated with progression of human neuroblastoma.[2] To date, only three putative tumor suppressor genes have been well characterized. In the inherited form of retinoblastoma, the retinoblastoma gene Rb-1 has been found to be inactivated by mutation, including deletions in retinoblastomas and other human tumors,[3-5] including small-cell lung carcinoma (SCLC).[6-8] The p53 gene is mutated in many types of cancer, including lung carcinomas, and may be involved in tumor progression.[9,10] The most recently identified putative tumor suppressor gene is DCC (deleted in colorectal carcinoma), which is located on chromosome 18q.[11]

II. ONCOGENES

The strategy which we have taken for investigating the role of oncogenes in the neoplastic transformation of normal human bronchial epithelial cells is shown in Table 1. Seven families of activated protooncogenes — *ras*,[12-14] *raf*,[15,16] *jun*,[17] erb-B2 (*neu*),[18] *myb*,[19] *myc*,[20-22] and *fms*[23] — have been associated with human lung cancer. Since association does not necessarily indicate causation, the actual functional role of these oncogenes in lung carcinogenesis is being studied *in vitro* by introducing these genes, singly or in combination, into normal human bronchial epithelial cells (NHBE) and SV40 T-antigen "immortalized" bronchial epithelial cells. Since the NHBE cells are the presumed progenitor cells for bronchogenic carcinoma, we have optimized their growth in culture by creating a chemically defined medium.[24] This medium is free of serum and transforming growth factor-β_1 (TGF-β_1), which will inhibit cell growth and induce terminal squamous differentiation in these cells at clonal density.[25]

To study the functional involvement of Ha-*ras* in human lung carcinoma, we have transferred vHa-*ras* into NHBE cells by protoplast fusion.[26] These cells sustained a cascade of genotypic and phenotypic changes that included decreased responsiveness to induction of terminal squamous differentiation, increased responsiveness to serum mitogens, increased life-span, aneuploidy, and, rarely, "immortality" and tumorigenicity in athymic nude mice (see Table 2). Therefore, neoplastic transformation of NHBE cells by Ha-*ras* is a rare event. Both the occurrence of frequent chromosomal aberrations and the lengthy cell crisis period of these transfected cells suggest that one or more unidentified events, in addition to the introduction of Ha-*ras*, may be involved in the development of the neoplastic phenotype.

TABLE 1
Strategy for Studying Neoplastic Transformation of Human Bronchial Epithelial Cells by Activated Protooncogenes

A. Select activated protooncogenes associated with human lung cancer.
B. Transfer activated protooncogenes into the progenitor epithelial cells of bronchogenic carcinoma.
C. Select preneoplastic and neoplastic cells from putative suppressive normal cells.
D. Determine tumorigenic potential in athymic nude mice.
E. Investigate dysregulation in molecular controls of growth and terminal differentiation.

TABLE 2
Progressive Phenotypic and Genotypic Changes in Normal Human Bronchial Epithelial Cells Transfected with vHa-*ras*

A. Decreased response to inducers of terminal squamous differentiation
B. Increased response to serum mitogens
C. Increased frequency of chromosomal aberrations and aneuploidy
D. Increased cell population doublings
E. Cell "crisis"
F. Continuous cell line
G. Tumorigenicity
H. Increased *ras* p21 expression in tumor cells
I. Metastasis

Although normal human cells in culture are relatively resistant to neoplastic transformation events,[27-29] several studies have indicated that immortalization is the rate-limiting step in the multistage process of *in vitro* human cell carcinogenesis.[30-32] In order to develop an immortalized cell system for studies of carcinogenesis, we have infected NHBE cells with the SV40 large T-antigen gene.[33] Unlike the NHBE cells, these SV40 T-antigen-containing cells, e.g., the BEAS-2B cell line, became immortalized. An attractive feature of these cells for use in carcinogenesis assays is the fact that they are nontumorigenic in early passage. In addition, these cells are aneuploid and undergo squamous differentiation in response to serum or TGF-β_1.[25] This is illustrated in Figure 1. Recently, HPV16 and 18 "immortalized" bronchial epithelial cells have recently been produced for similar studies.[34]

Many human lung adenocarcinomas have been shown to contain activated *ras* genes, which are thought to be involved in both the early and late stages of carcinogenesis.[1,12,35-38] The activated *ras* gene is most frequently Ki-*ras*,[39] but activated N-*ras* and Ha-*ras* have also been observed in lung cancer cell lines.[14] In this laboratory, the immortalized BEAS-2B cell line[33] has been used to define conditions under which *ras* and other oncogenes reproducibly cause neoplastic transformation.

Infection of BEAS-2B cells with a recombinant retrovirus containing vHa-*ras* produced cells (BZR) which were tumorigenic in athymic nude mice.[32] Tumor analysis revealed cells of human origin with the isoenzyme phenotype

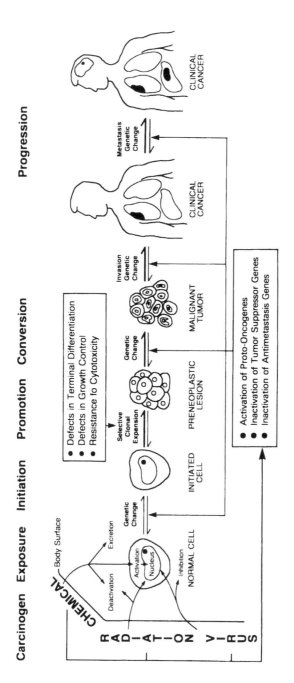

FIGURE 1. Multistep human epithelial cell carcinogenesis *in vitro*. Immortalization appears to be the rate-limiting step in *in vitro* human cell carcinogenesis.

and marker chromosomes of BEAS-2B cells. In addition, cell lines developed from the BZR tumors (BZRT33 and others) expressed abundant 21-kDa protein immunoreactive to antibodies specific for the codon 12 mutation present in the vHa-*ras* retroviral vector.[32] In contrast to the cellular *ras* oncogene, this 21-kDa protein was autophosphorylated, indicating expression of the vHa-*ras* gene as opposed to an endogenous *ras* gene. BEAS-2B, BZR, and BZRT33 cells were also examined for their invasiveness, metastatic potential, and ability to repopulate deepithelialized rat tracheal xenotransplants. Injection of these three cell lines into athymic nude mice revealed that BEAS-2B was not tumorigenic, BZR cells induced tumors with a latency period of 1 to 3 weeks, and BZRT33 induced tumors in less than 1 week.[32] The incidence of spontaneous metastasis to the lung following subcutaneous injection was negative for BEAS-2B (0%), intermediate for BZR (33%), and extensive for BZRT33 (100%).[32] The *in vivo* growth and invasiveness of normal and immortalized cells were studied by several procedures.

Immortalized BEAS-2B cells were able to reconstitute a mucus-producing columnar epithelium in deepithelialized rat tracheas that were transplanted subcutaneously into athymic nude mice.[40] BZR cells, obtained by transfer of vHa-*ras* into BEAS-2B cells, were tumorigenic in this xenotransplantation model, and the tumor-derived cell lines (e.g., BZRT33 and BZRT35 cells), which have increased ploidy and increased expression of the vHa-*ras* p21 protein, were more malignant than the BZR cells. This increasing malignancy in the tumor-derived cell lines correlated with increased type IV collagenase enzyme activity and mRNA expression.[40] These progressive changes are associated with a malignant phenotype, which is further enhanced by *in vivo* passaging.

The presence of an activated c-Ki-gene in human lung carcinomas has been well documented.[12,41-45] We have investigated the role of Ki-*ras* in the multistep neoplastic transformation of human bronchial epithelial cells. The vKi-*ras* oncogene used for these transfections contained mutations at codons 12 and 59. The mutation at codon 12 has also been observed in the lung carcinoma cell line A549.[46] Transfer of this oncogene into BEAS-2B by either infection or transfection resulted in neoplastic transformation.[47] The cells produced were not affected by TGF-β_1, and they were mitogenically stimulated by serum.[48] Thus, one of the earliest changes which occurs in lung carcinogenesis, decreased responsiveness to induction of terminally squamous differentiation, has occurred in the Ki-*ras*-transfected BEAS-2B cells. Tumors induced by the transfection of vKi-*ras* had adenocarcinomatous elements.[47] This is an interesting observation since the Ki-*ras* oncogene is most frequently found to be activated in human lung cancers, and most of these are adenocarcinomas.[12,39,49]

Abnormalities in the *raf, myc,* and *ras* protooncogene families have been associated with both human small-cell[6,16,20,22,50,51] and non-small-cell lung carcinomas.[12,49,52,53] We have assayed the functional role of c-*raf*-1 and c-*myc*

TABLE 3
Strategy for Identifying and Studying Tumor Suppressor Genes in Lung Carcinogenesis

A. Identify chromosomal location of putative tumor suppressor genes
   1. Allelic deletion analysis of tumor DNA vs. germ-line DNA
   2. Monochromosome cell hybrids
B. Genetic analysis of somatic cell hybrids
C. Isolate genes by subtraction library approach
   1. Tumorigenic vs. nontumorigenic hybrids
   2. Terminal squamous differentiation-resistant vs. differentiation-sensitive cells
D. Isolate genes by insertional mutagenesis approach
E. Determine structure and function of isolated genes
   1. p53
   2. Rb-1
   3. Nm23
   4. DCC

protooncogenes in lung carcinogenesis by introducing these genes, both alone and in combination, into human bronchial epithelial BEAS-2B cells.[54] Two retroviral recombinants, p-Zip-*raf* and p-Zip-*myc*, containing the complete coding sequences of the human c-*raf*-1 and the murine c-*myc* genes, respectively, were constructed and transfected into BEAS-2B cells. BEAS-2B cells transfected with Zip-*raf* or Zip-*myc* alone were nontumorigenic after 12 months, but BEAS-2B cells transfected with Zip-*raf* and Zip-*myc* together formed large-cell carcinomas in athymic nude mice in 4 to 21 weeks.[54] Carcinomas induced by the combination of c-*raf*-1 and c-*myc* were of human epithelial origin and exhibited specific surface antigens and several neuroendocrine markers. An increase in the mRNA levels of neuron-specific enolase was detected in BEAS-2B cells containing c-*raf*-1 and c-*myc* genes, suggesting an association between transformation and the expression of several neuroendocrine markers.[54,55]

III. TUMOR SUPPRESSION

The primary indication for the existence of the dominantly acting tumor suppressor genes originates from epidemiological studies.[3] Further evidence comes from the analysis of genetic loci exhibiting DNA restriction fragment-length polymorphisms (RFLPs) showing reduction to homozygosity of chromosome 13 in retinoblastoma and osteosarcoma,[5,56,57] chromosome 11 in Wilms' tumor[58,59] and bladder carinoma,[60] chromosome 17p in colorectal, lung, and brain tumors,[9] and chromosome 18q in colorectal carcinoma.[11] Several of these studies have been corroborated by genetic studies using the technique of somatic cell hybridization.[61,62] Our strategy for identifying tumor suppressor genes involved in human lung cancer involves several approaches, which are illustrated in Table 3.

A. LOSS OF HETEROZYGOSITY

Since the location of the tumor suppressor genes may be unknown and since these genes may have different functions, a well-defined and comprehensive approach is required. An initial approach is allelic DNA sequence deletion analysis, which identifies the chromosomal regions that may harbor the tumor suppressor genes. The loss of heterozygosity (LOH) of RFLP has been used to investigate the loss of allelic DNA sequences on specific chromosomes in several types of hereditary and sporadic tumors.[56,58-60,63-67] RFLP analysis came into prominence when the analysis of loci on 13q in hereditary retinoblastoma revealed the loss of genes on 13q. This eventually led to the identification of the Rb-1 gene on chromosome 13q. Recent RFLP analyses of 11p have detected the loss of alleles in Wilms' tumor[63,64] and also in tumors associated with Beckwith-Wiedemann syndrome.[68]

Many RFLP studies of human lung cancer have focused on small-cell carcinoma.[69,70] A small number of non-small-cell carcinomas have been studied by DNA sequence deletion analysis.[69-71] Recently, we have concluded an analysis of non-small-cell lung carcinoma (non-SCLC) for allelic DNA sequence losses on six different chromosomes at 13 different genetic loci. This study was conducted on tumors of varied histological types, including squamous cell carcinoma, adenocarcinoma of the lung, and large-cell carcinoma of the lung. This analysis allowed the comparison of the allelic DNA sequence losses in different histological classes of tumors.[72] Interestingly, in squamous cell carcinoma, consistent LOH was found at 17p13 using the D17S1 probe, while consistent LOH at this locus in adenocarcinomas and large-cell carcinomas was not detected. Frequent LOH at this locus has also been associated with colorectal[73-75] and small-cell carcinoma of the lung.[69,71,76,77]

LOH on chromosome 3 has been reported in small-cell carcinomas.[69-71,76-78] It has been speculated that this region contains a putative tumor suppressor gene for small-cell carcinomas of the lung.[69] Our study of LOH for markers on chromosome 3 in approximtely 60% of the tumors showed agreement with other reports that use DNA-RFLP to examine genetic loci on chromosome 3 in non-SCLC. However, LOH is substantially less than 100%, which is not in agreement with one report[78] in the literature.

We have studied chromosome 11 extensively for loss of alleles because it has been speculated to have at least one, if not more, tumor suppressor gene. Six different loci on this chromosome have been studied. LOH was observed in 45% of the squamous cell carcinomas and adenocarcinomas studied.[72] LOH was most frequently observed at the HBG2, insulin, and cHa-*ras* loci in both types of cancers. From these data, it was possible to establish two commonly deleted regions in lung cancer for this chromsome, namely, 11pter-p15.5 and 11p13-11q13 (see Figure 2). These findings are consistent with observations that describe two separate regions on chromosome 11 that may harbor tumor suppressor genes that correspond to 11p13 in Wilms' tumor and 11pter-11p15.5 in rhabdomyosarcoma.[58,79]

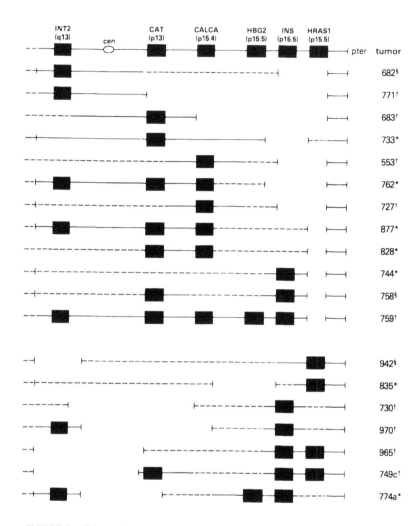

FIGURE 2.   Schematic representation of shortest region of overlapping DNA sequence deletion analysis on chromosome 11 for nine squamous cell carcinomas (†), seven adenocarcinomas (*), and three large-cell carcinomas (§) of the lung. Solid lines show intact genetic loci, dashed lines show regions for which no information is available, and gaps show regions of gene deletion.

The LOH results obtained for non-SCLC show differences in the genetic deletions observed in various histological types of lung cancers; mitotic recombination was a rare cause of LOH. Interestingly, in squamous cell carcinoma, coincidental loss of heterozygosity for several chromosomes was observed. For example, in eight of nine cases allelic DNA sequence deletion was observed for both chromosomes 11 and 17 where the analyses were informative for both chromosomes. Similarly, allelic DNA sequence deletions occurred for chromosomes 3 and 17 in three of five informative cases. Other

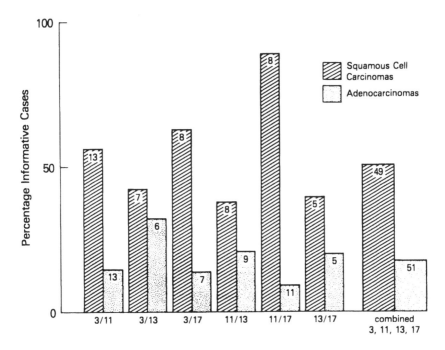

FIGURE 3. Coincident loss of heterozygosity for numbered chromosomes. For each of either 23 squamous cell or 23 adenocarcinomas of the lung, coincident loss of heterozygosity is shown for different combinations of chromosomes 3, 11, 13, and 17. Numbers in bars indicate number of cases informative for the chromosome indicated.

combinations of coincident loss in squamous and adenocarcinomas of the lung are shown in Figure 3.

The genetic changes observed in these tumors may also be involved in the pathogenesis of lung cancer in combination with other tumor suppressor genes. From this study, we can conclude that loss of putative tumor suppressor genes identified in other cancers may have a role, independently or in combination, in the development of non-SCLC.

B. MONOCHROMOSOME-CELL FUSION

In several cases where LOH studies suggest that deletion of a particular chromosomal region is associated with development of tumorigenicity, the technique of monochromosome fusion[80,81] can be employed to investigate this hypothesis. Using this technique, a single normal human chromosome is introduced into recipient tumorigenic cells. The tumorigenic potential of the microcell hybrid is assessed in athymic nude mice, and in many cases, suppression of the tumorigenic phenotype is observed. For example, Stanbridge and co-workers have shown that introduction of a normal human chromosome 11 into Wilms' tumor cells suppresses the tumorigenicity of these cells.[82] Likewise, Barrett, Oshimura, and co-workers have observed similar suppression in cervical carcinoma cell lines,[83,84] as have Stanbridge et al., using HeLa cells.[85] In addition, rhabdomyosarcoma cells, which are associated with 11p15 sequence deletions, were suppressed by the monochromosome transfer of a normal human chromosome 11.[84] Therefore, the malignant growth of these three different types of tumor cells appears to be dependent on the absence of a gene or genes normally present on chromosome 11.

Tumor suppression has also been documented for other human chromosomes and malignant cells. Chromosome 3 has been shown to decrease tumorigenicity in renal cells in which RFLP analysis has suggested that 3p deletions may be important in the development of the disease.[84]

Nagle and co-workers[86] have introduced a normal human chromosome 6 into two human malignant melanoma cell lines. In addition to reversion of the transformed *in vitro* phenotype and decreased soft-agar cloning efficiency, tumorigenicity of the hybrid cells was intially suppressed. Interestingly, all animals later developed tumors. However, cytogenetic and RFLP analysis of the tumors revealed a loss of the introduced chromosome 6 from the melanoma cell hybrids.

C. CELL-CELL HYBRIDS

Tumor suppression was first demonstrated by Harris and co-workers,[87] who produced murine cell hybrids between cells of high and low tumorigenic potential. The tumorigenicity of these hybirds was transiently suppressed, but as the hybrid clones were propagated in culture, tumorigenic segregants rapidly developed. As chromosomes in the hybrid cells were lost, the tumorigenicity of the hybrids increased to that of the parent cell of high tumorigenic potential.[88]

Genetic analysis of somatic cell hybrids between tumorigenic and normal human cells has shown that suppressor activity of the normal cell is functionally dominant over the tumorigenic cell. Hybrids formed from the human cervical carcinoma cell line, HeLa, and normal human fibroblasts[89] or normal human epidermal keratinocytes[90] showed suppressed tumorigenicity, as did hybrids between EJ bladder carcinoma cells containing a mutant cHa-*ras* and normal human fibroblasts.[91] However, studies which examine the tumorigenicity of a cancerous cell type hybridized with its normal epithelial progenitor cell have not been performed. We therefore created cell-cell hybrids between the cancer cell line HuT292DM and NHBE, SV40 T-antigen-immortalized nontumorigenic human bronchus cells (BEAS-2B),[33] or a weakly tumorigenic cell line derived from BEAS-2B which has a 3p deletion (B39TL) following growth in nude mice.[137] Hybrids formed between NHBE and HuT292DM cells had a limited doubling potential in culture and senesced after 40 to 43 population doublings. Therefore, tumorigenicity assays could not be performed with these hybrids due to an insufficient number of cells.

TABLE 4
Suppression of Tumorigenicity in Somatic Cell Hybrids Between a Human Lung Cancer Cell Line and Immortalized Bronchial Epithelial Cells

Cell line	Number of injected mice	Tumors[a]/number of injected mice	Latency[b] (days)	% totally suppressed	Number regressed[c]
BEAS-2B	15	0/12	>294	100	0
B39TL	15	7/14	148	50	3
HuT292DM	20	19/19	27	0	0
BEAS-2B × Hu-T292DM[d]	55	13/54	88[e]	76	1
B39TL × Hu-T292DM[f]	30	13/28	83[g]	54	3

[a]   A nonregressing nodule ≥1.0 cm in the largest dimension. Mice surviving less than 3 months without tumors have been excluded.
[b]   Mean number of days to reach scorable size.
[c]   Not scored as tumors.
[d]   Pooled data for 11 hybrid lines, each injected into five mice.
[e]   Mean tumor latency in the remaining 24% of the hybrids that produced tumors.
[f]   Pooled data from six hybrid lines, each injected into five mice.
[g]   Mean tumor latency in the remaining 46% of the hybrids that produced tumors.

In contrast to NHBE and HuT292DM cell hybrids, hybrids of BEAS-2B and HuT292DM cells have an indefinite life span in culture.[92] Tumor incidence in the parental line HuT292DM was 100% with a mean latency of 27 d, 50% in B39TL with a mean latency of 148 d, and 0% in BEAS-2B after approximately 1 year. Hybrids of BEAS-2B and HuT292DM cells yielded total suppression of tumorigenicity in 76% of the mice injected, while the immortalized, weakly tumorigenic B39TL as a parent yielded only 54% suppression of tumorigenicity of HuT292DM. Tumorigenicity of the B39TL × HuT292DM cell hybrids is comparable to the tumorigenicity of the parent B39TL at 50% (7/14). In addition, latency of tumor development in BEAS-2D × HuT292DM cell hybrids was extended two- to three-fold over that of the parent HuT292DM. These data are presented in Table 4.

Cell lines were isolated from tumors arising from the BEAS-2B × HuT292DM cell hybrids and the B39TL × HuT292DM hybrids. Upon reinjection of these lines into athymic nude mice, tumors were produced with latency periods comparable to the parent HuT292DM cells. These data suggest that reversion to tumor-forming ability may occur due to loss of one or more chromosomes that harbor tumor suppressor genes.

Karyotype analysis of parental lines, cell-cell hybrids, and hybrid-derived tumor cell lines was performed, and the results are shown in Table 5. The parental lines are hypodiploid, while the hybrid lines are hypotriploid to hypotetraploid. The hybrid lines contained all the marker chromosomes of

TABLE 5
Chromosomal Characteristics of Hybrids and Parental Human
Bronchial Epithelial Cell Lines

Cell line	Ploidy[a]	Marker chromosomes				Y chromosome
		HuT292 DM	BEAS-2B	B39 TL	New	
Hybrids						
HuT292DM × BEAS-2B-1, P14	75—85 (92)[b]	13	9		6	Present
HuT292DM × BEAS-2B-2, P10	75—90 (95)	13	8			Present
HuT292DM × B39TL-1, P12	80—90 (94)	12		8	3	Present
HuT292DM × B39TL-2, P10	75—90 (98)	15		7	3	Present
Tumor lines						
HuT292DM × B39TL-T, P4	65—85 (93)	7		9	10	Absent
HuT292DM × BEAS-2B-T, P4	68—78 (92)	7	2		5	Present
Parental lines						
HuT292DM	43—45 (96)	7				Absent
BEAS-2B, P27	44—48 (85)		6			Present
B39TL, P3	40—47 (90)		4	7		Present

a   Range of chromosome numbers (% in range) based on counts of 100 metaphases per cell line.
b   The remaining metaphases had 120 to 150 chromosomes.

both parents. In addition, new marker chromosomes were present in the hybrid tumor-derived cell lines as well as a loss of the Y chromosome from B39TL in the B39TL × HuT292DM hybrid tumor cell line. Karyotype analysis of the hybrid tumor cell lines revealed varied chromosome counts, mostly in the triploid range, suggesting a loss of chromosomes from the hypotriploid to hypotetraploid range observed in the hybrids.

From these experiments, we can conclude that nontumorigenic or weakly tumorigenic parents in a cell-cell hybrid with tumorigenic cells will dominantly control culture longevity and tumorigenicity of the more tumorigenic parent. Further, genes other than those involved in senescence can exhibit tumor suppressor activity.

IV. RETINOBLASTOMA SUSCEPTIBILITY GENE (Rb-1)

Retinoblastoma is a childhood cancer that occurs in familial and spontaneous forms. In 1971, Knudson[57] proposed that this retinal cancer is caused by two mutational events. In the familial form, a germ-line mutation predisposes the individual to retinoblastoma and a second mutation is acquired somatically, leading to tumor development.[57] In the spontaneous form of retinoblastoma, both mutations are somatic in origin. Further, those with the hereditary form are at risk for developing secondary cancers later in life. These second cancers are of unusual types such as osteosarcoma and fibrosarcoma. Individuals with the nonhereditary form are at no increased risk for other cancers. The evidence that one of these mutations creates an inactive allele was provided by the loss of genetic material on chromosome 13q14 in retinoblastomas.[93] This also suggested that this region harbors a gene, Rb-1, that serves as the first target for inactivation by these mutations. The second of Knudson's hypothesized target genes was soon identified to be the other copy of the intact Rb-1 gene. This was recognized by studying a closely linked marker gene, esterase D, on chromosome 13. Loss of heterozygosity studies revealed that the esterase D gene was heterozygous in normal tissue of a retinoblastoma patient, but in the tumor cells, it was reduced to a homozygous state. This implied that in tumor cells, the intact Rb-1 gene was replaced by a copy of the mutated allele. This demonstrated that both copies of the Rb-1 gene need to be lost or inactivated for tumor development. Using RFLP techniques and chromosome walking, the Rb-1 gene has been isolated and cloned.[5,94,95] It has also been shown that the Rb-1 protein is present in normal retinoblasts but absent in retinoblastomas.

All of the evidence collected to date suggests that the Rb-1 protein acts as a negative regulator of cell proliferation.[96-98] If this is true, the Rb-1 protein must be posttranslationally regulated. It has been shown that the phosphorylation level of Rb-1 changes rapidly, suggesting that specific kinases and phosphatases are involved. In addition, phosphorylation of Rb-1 is linked to the cell cycle.[99-102] Although synthesis of the Rb-1 protein is relatively constant throughout the cell cycle, phosphorylated Rb-1 protein can be detected in cells in late G_1 and the S-phase, while cells in G_0 and early G_1 are less phosphorylated. The state of phosphorylation of the Rb-1 protein may act as a "gate" to allow cells to enter the S-phase and proliferate. In contrast, unphosphorylated Rb-1 protein may inhibit cell proliferation and enhance differentiation.

Further evidence that the unphosphorylated form of Rb-1 protein inhibits cell proliferation comes from work by Ludlow,[103] who demonstrated that SV40 T-antigen binds only to the unphosphorylated form of the Rb-1 protein. This binding may functionally inactivate the unphosphorylated form of Rb-1 by removing its regulatory effects on the cell cycle and promoting cell proliferation. The functional inactivation of Rb-1 by SV40 T-binding may correspond to the "second hit" of Knudson's hypothesis, thereby increasing the neoplastic potential of these infected cells. This regulation may be a key step in the modulation of cell growth mediated by the Rb-1 protein.

Several studies have shown that nuclear viral oncogene products from adenovirus E1a,[104,105] SV40 T-antigen[103,106] (as discussed above), and HPV16 E7[107,108] bind to the Rb-1 protein. The importance of these interactions has not been conclusively demonstrated thus far. However, mutations in the Rb-1 binding regions of the viral protein E1a[105] and SV40 T-antigen[106] prevent the association of viral oncogene and Rb-1 gene products. This has been hypothesized to prevent entry of the virus-infected cells into the S-phase of the cell cycle, thus preventing viral DNA replication. When viral DNA replication is prevented, the oncogenic effects of the virus are not expressed, and Rb-1 acts as a suppressor of cellular transformation.

Several different abnormalities have been observed in the Rb-1 gene and its product in retinoblastoma,[5,95] osteosarcoma,[5] SCLCs,[6,7] and breast[109,110] and bladder carcinomas.[111] These abnormalities include point mutations altering the splicing patterns of mRNA, small deletions or duplications, truncations of the protein, and abnormal levels of the Rb-1 transcript. In 50% of human retinoblastoma tumors, point mutations which either alter the splicing pattern or generate small deletions or duplications in the gene were observed.[112,113]

Inactivation of the Rb-1 gene may be involved in the development of lung cancers as well, especially in the case of SCLC. In 60% of the SCLCs studied, no detectable Rb-1 transcript was observed, while 10% of the non-SCLCs had abnormal or absent Rb-1 transcripts.[6,7] All SCLCs examined for Rb-1 protein were found to be negative.[7] One of four pulmonary carcinoids examined had Rb-1 structural abnormalities, while three expressed no Rb-1 mRNA.[6]

One recent demonstration of tumor suppression by Rb-1 was shown by introducing the Rb-1 gene into a tumorigenic cell line which lacks the gene and then examining changes in growth and tumorigenic potential. Lee and co-workers[98] have shown that introduction of the cDNA from Rb-1 into a retinoblastoma cell line which lacks the Rb-1 protein as well as an osteogenic sarcoma line expressing a truncated Rb-1 protein reduced the rate of growth in culture and the ability to grow in an anchorage-independent manner. Furthermore, the tumorigenic potential of the retinoblastoma and osteogenic sarcoma cell lines was decreased in the cells which now contained the Rb-1 gene.[98] However, introduction of the same Rb-1 gene construct into the human prostate cell line DU145, which has a 35-amino acid in-frame deletion, did not significantly alter its growth rate in culture.[114] Unlike the retinoblastoma and osteosarcoma cell lines containing Rb-1, the tumorigenicity was not inhibited to the same extent, but the tumor sizes were greatly reduced in the mice injected with the prostate cell line containing the Rb-1 gene.[114]

V. p53

Phosphoprotein p53 is a nuclear protein which is present in high amounts in transformed human[115] and mouse cells.[116] Although no specific function has been assigned to this protein, antibody injections into dividing cells have implicated p53 in cell cycle regulation.[117] Initial studies in rat embryo fibroblasts have shown that p53 can cooperate with *ras* in neoplastic transformation.[118] Recently, it has been shown that the p53 gene used in this and other studies was a mutated, and not a wild-type gene.[119] In fact, it has been demonstrated that wild-type p53 does not cooperate with *ras,* but suppresses focus formation when cotransfected with *ras* in this assay.[120,121]

One of the best-characterized features of p53 is its ability to form complexes with other proteins. p53 was first identified in a complex with SV40 T-antigen.[116,122] Since that time, it has been found associated with adenovirus E1b in transformed rodent cells[123] and HPV16 and 18 E6.[124,125] While binding of p53 to SV40 T-antigen and adenovirus E1b appears to increase the normally short half-life of p53, binding to HPV16 and HPV18 promotes degradation of p53.[125] In addition, p53 complexes with itself to form homooligomeric structures.[126,127]

In situ hybridization analysis has assigned the p53 gene to the short arm of human chromosome 17, banding region 13.[128] As discussed above, several recent RFLP studies in human lung carcinoma, breast carcinoma, colorectal carcinoma, and brain tumors have shown LOH in this region of the chromosome. This finding led to the hypothesis that this region harbors a tumor suppressor gene. Vogelstein and co-workers have shown that in two colorectal carcinomas, one of the 17p alleles is lost and the p53 gene on the other is mutated, while normal tissue surrounding the tumor has the p53 wild-type sequence.[9] This finding sparked speculation that progression of these tumors occurs through a dominant negative effect mediated by the presence of mutant p53 or complete loss of wild-type p53.[9,120,129] A dominant negative effect may occur when pseudohomodimers of wild-type and mutated p53 are formed which functionally inactivate the wild-type p53.[130]

To further explore the possible dominant negative effect of mutant p53, Bernstein et al.[131] have generated independent lines of transgenic mice carrying genomic clones of a mutant p53 gene. These mice expressed high levels of mutant p53 in a wide variety of tissues, and have a greatly elevated predisposition to malignancies, particularly osteosarcomas, lung adenocarcinomas, and lymphomas.[131] Both alleles of the p53 gene used to develop the transgenic mice have sustained mutation in the coding region. The elevated tumor incidence in mice could be due to a dominant negative effect of functionally inactive transgenic protein inhibiting normal endogenous wild-type p53 protein.

Previous studies in rodent systems have shown that mutant p53 binds to cellular heat-shock protein 70 (hsp70).[130] Immunoprecipitation using hsp antisera or p53 antibodies has clearly demonstrated this complex formation, which results from conformational changes in the p53 due to mutation. The association of p53 with hsp70 in a human system has recently been demonstrated.[132] Cell lysates of a human osteosarcoma cell line, HOS-SL, were immunoprecipitated with anti-hsp70 and anti-p53 antibodies, and coimmunoprecipitation of p53 and hsp70 was observed. Subsequent cloning and

TABLE 6
Comparison of Characteristics, Activities, and Functions of Rb-1 and p53 Proteins

Rb-1	p53
DNA binding activity	DNA binding activity
Nuclear phosphoprotein	Nuclear phosphoprotein
Binds SV40 T-antigen	Binds SV40 T-antigen
Binds adenovirus E1a	Binds adenovirus E1b
Binds HPV-16 E7	Binds HPV-16 E6
Regulates G_1-S transition in normal cells	Regulates G_1-S transition in normal cells
Mutation in 95% small-cell lung carcinoma (SCLC) and 15% non-SCLC	Mutation in 80% SCLC and 60% non-SCLC

sequencing of the p53 gene has revealed a mutation in codon 156 of the p53 gene.[133] We are currently using coimmunoprecipitation with hsp70 as a rapid method of screening cell lines for mutations in the p53 gene.

Recently, there have been many reports regarding p53 mutations detected in human tumors and cell lines.[9,10,134,135] A wide range of abnormalities of the p53 gene, its RNA, and protein products have been reported in human lung cancer cell lines. A panel of human SCLC and non-SCLC cell lines have been examined as well as samples from normal lung obtaiend at the time of surgical resection. Of the 30 lung cell lines examined, 1 had a DNA rearrangement, 4 had abnormally sized p53 mRNA, 4 had decreased levels of p53 mRNA, 2 had only trace amounts of p53 mRNA, and 10 had point mutations.[134]

Using the LOH studies on chromosome 17 as background, 21 tumors of various histological types (colorectal, lung, breast, and brain) have been analyzed for mutations in the p53 gene.[10] Fifteen of the tumors contained a single missense mutation, two contained two missense mutations, one tumor had a frame-shift mutation, and in three tumors, no p53 mutations were detected. The mutations identified in this study were clustered in four regions, ''hot-spots'' of the p53 gene. These regions, exons 5, 6, and 7, are the most highly conserved among species.[136] Although more data are needed, these initial results suggest that these regions of the p53 gene may be especially important in mediation of the normal function of the p53 gene product. Normal cells from tissue surrounding these tumors were also analyzed for p53 mutations, and none were found.

Wild-type p53 as a putative tumor suppressor gene has many properties in common with Rb-1, the only other known tumor suppressor gene. A comparison of the characteristics and activities of the Rb-1 and p53 gene products is shown in Table 6. Both these genes encode nuclear phosphoproteins which bind DNA and have a possible regulatory function in the cell cycle. Most notably, both of these proteins form complexes with oncoproteins of DNA tumor viruses. The binding regions of these oncoproteins to Rb-1

SCHEMATIC REPRESENTATION OF THE INTERACTION
BETWEEN THE RETINOBLASTOMA PROTEIN, p53
VIRAL PROTEINS AND HEAT SHOCK PROTEIN 70

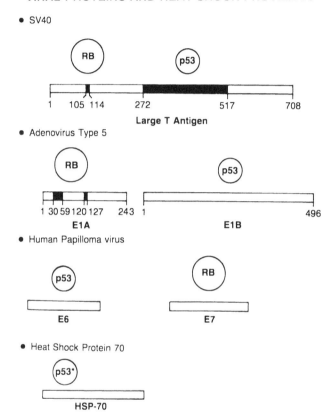

FIGURE 4. Schematic representation of the interaction between the
Rb-1 protein, p53, viral proteins, and hsp70.

and p53 are shown in Figure 4. As was discussed earlier in the case of Rb-1,
these nuclear oncoproteins participate in transformation through at least one
common mechanism, namely, binding to, and thereby inactivating, Rb-1 and/
or p53. Since p53 is believed to be involved in the transition of cells from
G_1 to the S-phase of the cell cycle by binding to the p53 protein, SV40 T-
antigen would inactivate this function of p53 in the cell cycle, promote the
replication of viral DNA, and cause transformation.

We are examining the status of p53 in primary lung tumors compared to
surrounding normal tissue, and in lung carcinoma cell lines. Several different
approaches are being taken. In the first approach, we are sequencing all coding
regions of the gene using intron primers to amplify the DNA. The PCR product
is sequenced and examined for mutations.

The second approach takes advantage of the association of mutated p53 and heat-shock proteins.[137] Using antibodies against both p53 and heat-shock proteins, immunoprecipitations of the various cell lines are performed and the presence of mutated p53 is detected by the coimmunoprecipitation of the p53-hsp complex. As lung cancer cell lines which contain a mutated p53 are identified, they are transfected with a variety of plasmids containing wild-type p53 either constitutively or inducibly expressed. In addition, NHBE and T-antigen-immortalized BEAS-2B cells are also transfected with wild-type and mutated p53 in constitutive or inducible expression vectors. The tumorigenicity of these transfected cells, as well as growth characteristics, will be determined. These experiments are designed to provide data that may give us some insight into the biological effects of mutated and wild-type p53 in lung cells.

REFERENCES

1. Vogelstein, B., Fearon, E. R., Hamilton, S. R., Kern, S. E., Preisinger, A. C., Leppert, M., Nakamura, Y., White, R., Smits, A. M., and Bos, J. L., Genetic alterations during colorectal-tumor development, *N. Engl. J. Med.*, 319, 525, 1988.

2. Brodeur, G. M., Hayes, F. A., Green, A. A., Casper, J. T., Wasson, J., Wallach, S., and Seeger, R. C., Consistent N-*myc* copy number in simultaneous or consecutive neuroblastoma samples from sixty individual patients, *Cancer Res.*, 47, 4248, 1987.

3. Knudson, A. G., Jr., Hereditary cancer, oncogenes, and antioncogenes, *Cancer Res.*, 45, 1437, 1985.

4. Hansen, M. F., Koufos, A., Gallie, B. L., Phillips, R. A., Fodstad, O., Brogger, A., Gedde-Dahl, T., and Cavenee, W. K., Osteosarcoma and retinoblastoma: a shared chromosomal mechanism revealing recessive predisposition, *Proc. Natl. Acad. Sci. U.S.A.*, 82, 6216, 1985.

5. Friend, S. H., Bernards, R., Rogelj, S., Weinberg, R. A., Rapaport, J. M., Albert, D. M., and Dryja, T. P., A human DNA segment with properties of the gene that predisposes to retinoblastoma and osteosarcoma, *Nature*, 323, 643, 1986.

6. Harbour, J. W., Lai, S. L., Whang-Peng, J., Gazdar, A. F., Minna, J. D., and Kaye, F. J., Abnormalities in structure and expression of the human retinoblastoma gene in SCLC, *Science*, 241, 353, 1988.

7. Yokota, J., Akiyama, T., Fung, Y. K., Benedict, W. F., Namba, Y., Hanaoka, M., Wada, M., Terasaki, T., Shimosato, Y., Sugimura, T., and Terada, M., Altered expression of the retinoblastoma (RB) gene in small-cell carcinoma of the lung, *Oncogene*, 3, 471, 1988.

8. Saksela, K., Makela, T. P., and Alitalo, K., Oncogene expression in small-cell lung cancer cell lines and testicular germ-cell tumor: activation of the N-*myc* gene and decreased RB mRNA, *Int. J. Cancer*, 44, 182, 1989.

9. Baker, S. J., Fearon, E. R., Nigro, J. M., Hamilton, S. R., Preisinger, A. C., Jessup, J. M., vanTuinen, P., Ledbetter, D. H., Barker, D. F., and Nakamura, Y., Chromosome 17 deletions and p53 gene mutations in colorectal carcinomas, *Science*, 244, 217, 1989.

10. Nigro, J. M., Baker, S. J., Preisinger, A. C., Jessup, J. M., Hostetter, R., Cleary, K., Bigner, S. H., Davidson, N., Baylin, S., Devilee, P., Glover, T., Collins, F. S., Weston, A., Modali, R., Harris, C. C., and Vogelstein, B., Mutations in the p53 gene occur in diverse human tumor types, *Nature*, 342, 705, 1989.

11. Fearon, E. R., Cho, K. R., Nigro, J. M., Kern, S. E., Simons, J. W., Ruppert, J. M., Hamilton, S. R., Preisinger, A. C., Thomas, G., Kinzler, K. W., and Vogelstein, B., Identification of a chromosome 18q gene that is altered in colorectal cancers, *Science*, 247, 49, 1990.

12. Rodenhuis, S., Van de Wetering, M. L., Mooi, W. J., Evers, S. G., van Zandwijk, N., and Bos, J. L., Mutational activation of the K-*ras* oncogene. A possible pathogenetic factor in adenocarcinoma of the lung, *N. Engl. J. Med.*, 317, 929, 1987.

13. Yuasa, Y., Srivastava, S. K., Dunn, C. Y., Rhim, J. S., Reddy, E. P., and Aaronson, S. A., Acquisition of transforming properties by alternative point mutations within c-*bas/has* human proto-oncogene, *Nature*, 303, 775, 1983.

14. Yuasa, Y., Gol, R. A., Chang, A., Chiu, I. M., Reddy, E. P., Tronick, S. R., and Aaronson, S. A., Mechanism of activation of an N-*ras* oncogene of SW-1271 human lung carcinoma cells, *Proc. Natl. Acad. Sci. U.S.A.*, 81, 3670, 1984.

15. Rapp, U. R., Huleihel, M., Pawson, T., Linnoila, I., Minna, J. D., Heidecker, G., Cleveland, J. L., Beck, T., Forchhammer, J., and Storm, S. M., Role of *raf* oncogenes in lung carcinogenesis, *Lung Center*, 4, 162, 1988.

16. Graziano, S. L., Cowan, B. Y., Carney, D. N., Bryke, C. R., Mitter, N. S., Johnson, B. E., Mark, G. E., Planas, A. T., Catino, J. J., Comis, R. L., and Poiesz, B. J., Small cell lung cancer cell line derived from a primary tumor with a characteristic deletion of 3p, *Cancer Res.*, 47, 2148, 1987.

17. Schuette, J., Nau, M., Birrer, M., Thomas, F., Gazdar, A., and Minna, J., Constitutive expression of multiple mRNA forms of the c-*jun* oncogene in human lung cancer cell lines, *Proc. Am. Assoc. Cancer Res.*, 29, 1808, 1988.

18. Weiner, D. B., Nordberg, J., Robinson, R., Nowell, P. C., Gazdar, A. F., Greene, M. I., Williams, W. V., Cohen, J. A., and Kern, J. A., Expression of the *neu* gene encoded protein (P185^{neu}) in human non-small cell carcinomas of the lung, *Cancer Res.*, 50, 421, 1990.

19. Griffin, C. A. and Baylin, S. B., Expression of the c-*myb* oncogene in human small cell lung carinoma, *Cancer Res.*, 45, 272, 1985.

20. Little, C. D., Nau, M. M., Carney, D.N., Gazdar, A. F., and Minna, J. D., Amplification and expression of the c-*myc* oncogene in human lung cancer cell lines, *Nature*, 306, 194, 1983.

21. Nau, M. M., Brooks, B. J., Jr., Carney, D. N., Gazdar, A. F., Battey, J. F., Sausville, E. A., and Minna, J. D., Human small-cell lung cancers show amplification and expression of the N-*myc* gene, *Proc. Natl. Acad. Sci. U.S.A.*, 83, 1092, 1986.

22. Nau, M. M., Brooks, B. J., Jr., Battey, J. F., Sausville, E., Gazdar, A. F., Kirsch, I. R., McBride, O. W., Bertness, V., Hollis, G. F., and Minna, J. D., L-*myc*, a new *myc*-related gene amplified and expressed in human small cell lung cancer, *Nature*, 318, 69, 1985.

23. Kiefer, P. E., Bepler, G., Kubasch, M., and Havemann, K., Amplification and expression of protooncogenes in human small cell lung cancer cell lines, *Cancer Res.*, 47, 6236, 1987.

24. Lechner, J. F. and LaVeck, M. A., A serum-free method for culturing normal human bronchial epithelial cells at clonal density, *J. Tissue Culture Methods*, 9, 43, 1985.

25. Ke, Y., Reddel, R. R., Gerwin, B. I., Miyashita, M., McMenamin, M. G., Lechner, J. F., and Harris, C. C., Human bronchial epithelial cells with integrated SV40 virus T antigen genes retain the ability to undergo squamous differentiation, *Differentiation*, 38, 60, 1988.

26. **Yoakum, G. H., Lechner, J. F., Gabrielson, E. W., Korba, B. E., Malan-Shibley, L., Willey, J. C., Valerio, M. G., Shamsuddin, A. K. M., Trump, B. F., and Harris, C. C.,** Transformation of human bronchial epithelial cells transfected by Harvey *ras* oncogene. *Science,* 227, 1174, 1985.

27. **DiPaolo, J. A.,** Relative difficulties in transforming human and animal cells *in vitro, J. Natl. Cancer Inst.,* 70, 3, 1983.

28. **DiPaolo, J. A., DeMarinis, A. J., and Doniger, J.** Asbestos and benzo(a)pyrene synergism in the transformation of Syrian hamster embryo cells, *Pharmacology,* 27, 65, 1983.

29. **Harris, C. C.,** Human tissues and cells in carcinogenesis research, *Cancer Res.,* 47, 1, 1987.

30. **Rhim, J. S., Trimmer, R., Arnstein, P., and Huebner, R. J.,** Neoplastic transformation of chimpanzee cells induced by adenovirus type 12-simian virus 40 hybrid virus, *Proc. Natl. Acad. Sci. U.S.A.,* 78, 313, 1981.

31. **Namba, M., Nishitani, K., Fukushima, F., Kimoto, T., and Nose, K.,** Multistep process of neoplastic transformation of normal human fibroblasts by ^{60}Co gamma rays and Harvey sarcoma viruses, *Int. J. Cancer,* 37, 419, 1986.

32. **Amstad, P., Reddel, R. R., Pfeifer, A., Malan-Shibley, L., Mark, G. E., and Harris, C. C.,** Neoplastic transformation of a human bronchial epithelial cell line by a recombinant retrovirus encoding viral harvey *ras, Mol. Carcinogen.,* 1, 151, 1988.

33. **Reddel, R. R., Ke, Y., Gerwin, B. I., McMenamin, M. G., Lechner, J. F., Su, R. T., Brash, D. E., Park, J. B., Rhim, J. S., and Harris, C. C.,** Transformation of human bronchial epithelial cells by infection with SV40 or adenovirus-12 SV40 hybrid virus, or transfection via strontium phosphate coprecipitation with a plasmid containing SV40 early region genes, *Cancer Res.,* 48, 1904, 1988.

34. **Willey, J. C., Sleemi, A., and Harris, C. C.,** Transformation of normal human bronchial epithelial cells by human papillomavirus types 16 and 18 DNA, *J. Cell Biochem.,* 14C, 1, 1990.

35. **Guerrero, I., Calzada, P., Mayer, A., and Pellicer, A.,** A molecular approach to leukemogenesis: mouse lymphomas contain an activated c-*ras* oncogene, *Proc. Natl. Acad. Sci. U.S.A.,* 81, 202, 1984.

36. **Zarbl, H., Sukumar, S., Arthur, A. V., Martin-Zanca, D., and Barbacid, M.,** Direct mutagenesis of Ha-*ras*-1 oncogenes by *N*-nitroso-*N*-methylurea during initiation of mammary carcinogenesis in rats, *Nature,* 315, 382, 1985.

37. **Bondy, G. P., Wilson, S., and Chambers, A. F.,** Experimental metastatic ability of H-*ras*-transformed NIH3T3 cells, *Cancer Res.,*45, 6005, 1985.

38. **Kasid, A., Lippman, M. E., Papageorge, A. G., Lowy, D. R., and Gelmann, E. P.,** Transfection of v-*ras*H DNA into MCF-7 human breast cancer cells bypasses dependence on estrogen for tumorigenicity, *Science,* 228, 725, 1985.

39. **Slebos, R. J., Kibbelaar, R. E., Dalesio, O., Kooistra, A., Stam, J., Meijer, C. J., Wagenaar, S. S., Vanderschueren, R. G., van Zandwijk, N., Mooi, W. J., Bos, J. L., and Rodenhuis, S.,** K-*ras* oncogene activation as a prognostic marker in adenocarcinoma of the lung, *N. Engl. J. Med.,* 323, 561, 1990.

40. **Bonfil, R. D., Reddel, R. R., Ura, H., Reich, R., Fridman, R., Harris, C. C., and Klein-Szanto, A. J. P.,** Invasive and metastatic potential of a v-Ha-*ras*-transformed human bronchial epithelial cell line, *J. Natl. Cancer Inst.,* 81, 587, 1989.

41. **Der, C. J., Krontiris, T. G., and Cooper, G. M.,** Transforming genes of human bladder and lung carcinoma cell lines are homologous to the *ras* genes of Harvey and Kirsten sarcoma viruses, *Proc. Natl. Acad. Sci. U.S.A.,* 79, 3637, 1982.

42. **Pulciani, S., Santos, E., Lauver, A. V., Long, L. K., Aaronson, S. A., and Barbacid, M.,** Oncogenes in solid human tumours, *Nature,* 300, 539, 1982.

43. **Capon, D. J., Seeburg, P. H., McGrath, J. P., Hayflick, J. S., Edman, U., Levinson, A. D., and Goeddel, D. V.,** Activation if Ki-*ras*2 gene in human colon and lung carcinomas by two different point mutations, *Nature,* 304, 507, 1983.

44. Shimizu, K., Birnbaum, D., Ruley, M. A., Fasano, O., Suard, Y., Edlund, L., Taparowsky, E., Goldfarb, M., and Wigler, M., Structure of the Ki-*ras* gene of the human lung carcinoma cell line Calu-1, *Nature*, 304, 497, 1983.

45. Santos, E., Martin-Zanca, D., Reddy, E. P., Pierotti, M. A., Della Porta, G., and Barbacid, M., Malignant activation of a K-*ras* oncogene in lung carcinoma but not in normal tissue of the same patient, *Science*, 223, 661, 1984.

46. Valenzuela, D. M. and Groffen, J., Four human carcinoma cell lines with novel mutations in position 12 of c-K-*ras* oncogene, *Nucleic Acids Res.*, 14, 843, 1986.

47. Reddel, R. R., Key, Y., Kaighn, M. E., Malan-Shibley, L., Lechner, J. F., Rhim, J. S., and Harris, C. C., Human bronchial epithelial cells neoplastically transformed by v-Ki-*ras*: altered response to inducers of terminal squamous differentiation, *Oncogene Res.*, 3, 401, 1988.

48. Masui, T., Wakefield, L. M., Lechner, J. F., LaVeck, M. A., Sporn, M. B., and Harris, C. C., Type beta transforming growth factor is the primary differentiation-inducing serum factor for normal human bronchial epithelial cells, *Proc. Natl. Acad. Sci. U.S.A.*, 83, 2438, 1986.

49. Rodenhuis, S., Slebos, R. J., Boot, A. J., Evers, S. G., Mooi, W. J., Wagenaar, S. S., van Bodegom, P. C., and Bos, J. L., Incidence and possible clinical significance of K-*ras* oncogene activation in adenocarcinoma of the human lung, *Cancer Res.*, 48, 5738, 1988.

50. Nakano, H., Yamamoto, F., Neville, C., Evans, D. A., Mizuno, T., and Perucho, M., Isolation of transforming sequences of two human lung carcinomas: structural and functional analysis of the activated c-K-*ras* oncogenes, *Proc. Natl. Acad. Sci. U.S.A.*, 81, 71, 1984.

51. Graziano, S. L., Mark, G. E., Murray, C., Mann, D. L., Ehrlich, G. D., Poiesz, B. J., and Weston, A., DNA restriction fragment length polymorphisms at either end of the c-*raf*-1 locus at 3p25, *Oncogene Res.*, 3, 99, 1988.

52. Cline, M. J. and Battifora, H., Abnormalities of protooncogenes in non-small cell lung cancer. Correlations with tumor type and clinical characteristics [published erratum appears in *Cancer*, 61(5), 1064, 1988], *Cancer*, 60, 2669, 1987.

53. Kurzrock, R., Gallick, G. E., and Gutterman, J. U., Differential expression of p21*ras* gene products among histologic subtypes of fresh primary human lung tumors, *Cancer Res.*, 46, 1530, 1986.

54. Pfeifer, A., Mark, G. E., Malan-Shibley, L., Graziano, S. L., Amstad, P., and Harris, C. C., Cooperation of c-*raf*-1 and c-*myc* protooncogenes in the neoplastic transformation of SV40 T-antigen immortalized human bronchial epithelial cells, *Proc. Natl. Acad. Sci. U.S.A.*, 86, 10075, 1989.

55. Pfeifer, A. M. A., Jones, R. T., Bowder, P. E., Mann, D., Spillare, E., Klein-Szanto, A. J. P., Trump, B. F., and Harris, C. C., Human bronchial epithelial cells transformed by the c-*raf*-1 and c-*myc* protooncogenes induce multidifferentiated carcinomas in nude mice: a model for lung carcinogenesis, *Cancer Res.*, 51, 3793, 1991.

56. Cavenee, W. K., Dryja, T. P., Phillips, R. A., Benedict, W. F., Godbout, R., Gallie, B. L., Murphree, A. L., Strong, L. C., and White, R. L., Expression of recessive alleles by chromosomal mechanisms in retinoblastoma, *Nature*, 305, 779, 1983.

57. Knudson, A. G., Jr., Mutation and cancer: statistical study of retinoblastoma, *Proc. Natl. Acad. Sci. U.S.A.*, 68, 820, 1971.

58. Koufos, A., Hansen, M. F., Lampkin, B. C., Workman, M. L., Copeland, N. G., Jenkins, N. A., and Cavenee, W. K., Loss of alleles at loci on human chromosome 11 during genesis of Wilms' tumour, *Nature*, 309, 170, 1984.

59. Orkin, S. H., Goldman, D. S., and Sallan, S. E., Development of homozygosity for chromosome 11p markers in Wilms' tumour, *Nature*, 309, 172, 1984.

60. Fearon, E. R., Feinberg, A. P., Hamilton, S. H., and Vogelstein, B., Loss of genes on the short arm of chromosome 11 in bladder cancer, *Nature*, 318, 377, 1985.

61. Stanbridge, E. J., Der, C. J., Doersen, C. J., Nishimi, R. Y., Peehl, D. M., Weissman, B. E., and Wilkinson, J. E., Human cell hybrids: analysis of transformation and tumorigenicity, *Science*, 215, 252, 1982.

62. Sager, R., Genetic suppression of tumor formation, *Adv. Cancer Res.*, 44, 43, 1985.

63. Fearon, E. R., Vogelstein, B., and Feinberg, A. P., Somatic deletion and duplication of genes on chromosome 11 in Wilms' tumours, *Nature*, 309, 176, 1984.

64. Koufos, A., Hansen, M. F., Copeland, N. G., Jenkins, N. A., Lampkin, B. C., and Cavenee, W. K., Loss of heterozygosity in three embryonal tumours suggests a common pathogenetic mechanism, *Nature*, 316, 330, 1985.

65. Reeve, A. E., Housiaux, P. J., Gardner, R. J., Chewings, W. E., Grindley, R. M., and Millow, L. J., Loss of a Harvey *ras* allele in sporadic Wilms' tumour, *Nature*, 309, 174, 1984.

66. Zbar, B., Brauch, H., Talmadge, C., and Linehan, M., Loss of alleles of loci on the short arm of chromosome 3 in renal cell carcinoma, *Nature*, 327, 721, 1987.

67. Lundberg, C., Skoog, L., Cavenee, W. K., and Nordenskjold, M., Loss of heterozygosity in human ductal breast tumors indicates a recessive mutations on chromosome 13, *Proc. Natl. Acad. Sci. U.S.A.*, 84, 2372, 1987.

68. Sotelo-Avila, C., Gonzalez-Crussi, F., and Fowler, J. W., Complete and incomplete forms of Beckwith-Wiedemann syndrome: their oncogenic potential, *J. Pediatr.*, 96, 47, 1980.

69. Naylor, S. L., Johnson, B. E., Minna, J. D., and Sakaguchi, A. Y., Loss of heterozygosity of chromosome 3p markers in small-cell lung cancer, *Nature*, 329, 451, 1987.

70. Yokota, J., Wada, M., Shimosato, Y., Terada, M., and Sugimura, T., Loss of heterozygosity on chromosomes 3, 13, and 17 in small-cell carcinoma and on chromosome 3 in adenocarcinoma of the lung, *Proc. Natl. Acad. Sci. U.S.A.*, 84, 9252, 1987.

71. Brauch, H., Johnson, B., Hovis, J., Yano, T., Gazdar, A. F., Pettengill, O. S. Graziano, S. L., Sorenson, G. D., Poiesz, B. J., Minna, J., Linehan, M., and Zbar, B., Molecular analysis of the short arm of chromosome 3 in small-cell and non-small-cell carcinoma of the lung, *N. Engl. J. Med.*, 317, 1109, 1987.

72. Weston, A., Willey, J. C., Modali, R., Sugimura, H., McDowell, E. M., Resau, J., Light, B., Haugen, A., Mann, D. L., Trump, B. F., and Harris, C. C., Differential DNA sequence deletions from chromosomes 3, 11, 13 and 17 in squamous cell carcinoma, large cell carcinoma and adenocarcinoma of the human lung, *Proc. Natl. Acad. Sci. U.S.A.*, 86, 5099, 1989.

73. Fearon, E. R., Hamilton, S. R., and Vogelstein, B., Clonal analysis of human colorectal tumors, *Science*, 238, 193, 1987.

74. Okamoto, M., Sasaki, M., Sugio, K., Sato, C., Iwama, T., Ikeuchi, T., Tonomura, A., Sasazuki, T., and Miyaki, M., Loss of constitutional heterozygosity in colon carcinoma from patients with familial polyposis coli, *Nature*, 331, 273, 1988.

75. Solomon, E., Voss, R., Hall, V., Bodmer, W. F., Jass, J. R., Jeffreys, A. J., Lucibello, F. C., Patel, I., and Rider, S. H., Chromosome 5 allele loss in human colorectal carcinomas, *Nature*, 328, 616, 1987.

76. Dobrovic, A., Houle, B., Belouchi, A., and Bradley, W. E., erbA-related sequence coding for DNA-binding hormone receptor localized to chromosome 3p21-3p25 and deleted in small cell lung carcinoma, *Cancer Res.*, 48, 682, 1988.

77. Johnson, B. E., Sakaguchi, A. Y., Gazdar, A. F., Minna, J. D., Burch, D., Marshall, A., and Naylor, S. L., Restriction fragment length polymorphism studies show consistent loss of chromosome 3p alleles in small cell lung cancer patients' tumors, *J. Clin. Invest.*, 82, 502, 1988.

78. Kok, K., Osinga, J., Carritt, B., Davis, M. B., van der Hout, A. H., van der Veen, A. Y., Landsvater, R. M., de Leij, L. F., Berendsen, H. H., Postmus, P. E., Poppema, S., and Buys, C. H., Deletion of a DNA sequence at the chromosomal region 3p21 in all major types of lung cancer, *Nature*, 330, 578, 1987.

79. **Scrable, H. J., Witte, D. P., Lampkin, B. C., and Cavenee, W. K.**, Chromosomal localization of the human rhabdomyosarcoma locus by mitotic recombination mapping, *Nature*, 329, 645, 1987.

80. **Fournier, R. E. and Ruddle, F. H.**, Microcell-mediated transfer of murine chromosomes into mouse, Chinese hamster, and human somatic cells, *Proc. Natl. Acad. Sci. U.S.A.*, 74, 319, 1977.

81. **Ege, T., Ringertz, N. R., Hamberg, H., and Sidebottom, E.**, Preparation of microcells, *Methods Cell. Biol.*, 15, 339, 1977.

82. **Weissman, B. E., Saxon, P. J., Pasquale, S. R., Jones, G. R., Geiser, A. G., and Stanbridge, E. J.**, Introduction of a normal human chromosome 11 into a Wilms' tumor cell line controls its tumorigenic expression, *Science*, 236, 175, 1987.

83. **Koi, M., Morita, H., Yamada, H., Satoh, H., Barrett, J. C., and Oshimura, M.**, Normal human chromosome 11 suppresses tumorigenicity of human cervical tumor cell line SiHa, *Mol. Carcinogen.*, 2, 12, 1989.

84. **Oshimura, M., Koi, M., Morita, H., Yamada, H., Shimizu, M., and Ono, T.**, Suppression of tumorigenicities of human cancer cell lines following chromosome transfer via microcell fusion, *Proc. Am. Assoc. Cancer Res.*, 30, 786, 1989.

85. **Saxon, P. J., Srivatsan, E. S., and Stanbridge, E. J.**, Introduction of human chromosome 11 via microcell transfer controls tumorigenic expression of HeLa cells, *EMBO J.*, 5, 3461, 1986.

86. **Trent, J. M., Stanbridge, E. J., McBride, H. L., Meese, E. U., Casey, G., Araujo, D. E., Witkowski, C. M., and Nagle, R. B.**, Tumorigenicity in human melanoma cell lines controlled by introduction of human chromosome 6, *Science*, 247, 568, 1990.

87. **Harris, H., Miller, O. J., Klein, G., Worst, P., and Tachibana, T.**, Suppression of malignancy by cell fusion, *Nature*, 223, 363, 1969.

88. **Harris, H.**, The analysis of malignancy by cell fusion: the position in 1988, *Cancer Res.*, 48, 3302, 1988.

89. **Stanbridge, E. J.**, Genetic regulation of tumorigenic expression in somatic cell hybrids, in *Advances in Viral Oncology*, Vol. 6, Klein, G., Ed., Raven Press, New York, 1987, 83.

90. **Peehl, D. M. and Stanbridge, E. J.**, Characterization of human keratinocyte × HeLa somatic cell hybrids, *Int. J. Cancer*, 27, 625, 1981.

91. **Geiser, A. G., Der, C. J., Marshall, C. J., and Stanbridge, E. J.**, Suppression of tumorigenicity with continued expression of the c-Ha-*ras* oncogene in EJ bladder carcinoma-human fibroblast hybrid cells, *Proc. Natl. Acad. Sci. U.S.A.*, 83, 5209, 1986.

92. **Kaighn, M. E., Gabrielson, E. W., Iman, D. S., Pauls, E. A., and Harris, C. C.**, Suppression of tumorigenicity of a human lung carcinoma line by nontumorigenic bronchial epithelial cells in somatic cell hybrids, *Cancer Res.*, 50, 1890, 1990.

93. **Yunis, J. J. and Ramsay, N.**, Retinoblastoma and subband deletion of chromosome 13, *Am. J. Dis. Child.*, 132, 161, 1978.

94. **Lee, W. H., Bookstein, R., Hong, F., Young, L. J., Shew, J. Y., and Lee, E. Y.**, Human retinoblastoma susceptibility gene: cloning, identification, and sequence, *Science*, 235, 1394, 1987.

95. **Fung, Y. K., Murphree, A. L., T'Ang, A., Qian, J., Hinrichs, S. H., and Benedict, W. F.**, Structural evidence for the authenticity of the human retinoblastoma gene, *Science*, 236, 1657, 1987.

96. **Knudson, A. G., Jr.**, A two-mutation model for human cancer, in *Advances in Viral Oncology*, Vol. 7, Klein, G., Ed., Raven Press, New York, 1987, 1.

97. **Benedict, W. F.**, Recessive human cancer susceptibility genes (retinoblastoma and Wilms' loci), in *Advances in Viral Oncology*, Vol. 7, Klein, G., Ed., Raven Press, New York, 1987, 19.

98. **Huang, H. J., Yee, J. K., Shew, J. Y., Chen, P. L., Bookstein, R., Friedmann, T., Lee, E. Y., and Lee, W. H.**, Suppression of the neoplastic phenotype by replacement of the RB gene in human cancer cells, *Science*, 242, 1563, 1988.

99. Mihara, K., Cao, X. R., Yen, A., Chandler, S., Driscoll, B., Murphree, A. L., T'Ang, A., and Fung, Y. K., Cell cycle-dependent regulation of phosphorylation of the human retinoblastoma gene product, *Science*, 246, 1300, 1989.

100. Chen, P. L., Scully, P., Shew, J. Y., Wang, J. Y., and Lee, W. H., Phosphorylation of the retinoblastoma gene product is modulated during the cell cycle and cellular differentiation, *Cell*, 58, 1193, 1989.

101. Buchkovich, K., Duffy, L. A., and Harlow, E., The retinoblastoma protein is phosphorylated during specific phases of the cell cycle, *Cell*, 58, 1097, 1989.

102. DeCaprio, J. A., Ludlow, J. W., Lynch, D., Furukawa, Y., Griffin, J., Piwnica-Worms, H., Huang, C. M., and Livingston, D. M., The product of the retinoblastoma susceptibility gene has properties of a cell cycle regulatory element, *Cell*, 58, 1085, 1989.

103. Ludlow, J. W., DeCaprio, J. A., Huang, C. M., Lee, W. H., Paucha, E., and Livingston, D. M., SV40 large T antigen binds preferentially to an underphosphorylated member of the retinoblastoma susceptibility gene product family, *Cell*, 56, 57, 1989.

104. Whyte, P., Buchkovich, K. J., Horowitz, J. M., Friend, S. H., Raybuck, M., Weinberg, R. A., and Harlow, E., Association between an oncogene and an anti-oncogene: the adenovirus E1A proteins bind to the retinoblastoma gene product, *Nature*, 334, 124, 1988.

105. Egan, C., Bayley, S. T., and Branton, P. E., Binding of the Rb1 protein to E1A products is required for adenovirus transformation, *Oncogene*, 4, 383, 1989.

106. DeCaprio, J. A., Ludlow, J. W., Figge, J., Shew, J. Y., Huang, C. M., Lee, W. H., Marsilio, E., Paucha, E., and Livingston, D. M., SV40 large tumor antigen forms a specific complex with the product of the retinoblastoma susceptibility gene, *Cell*, 54, 275, 1988.

107. Munger, K., Werness, B. A., Dyson, N., Phelps, W. C., Harlow, E., and Howley, P. M., Complex formation of human papillomavirus E7 proteins with the retinoblastoma tumor suppressor gene product, *EMBO J.*, 8, 4099, 1989.

108. Dyson, N., Howley, P. M., Munger, K., and Harlow, E., The human papilloma virus-16 E7 oncoprotein is able to bind to the retinoblastoma gene product, *Science*, 243, 934, 1989.

109. Lee, E. Y., To, H., Shew, J. Y., Bookstein, R., Scully, P., and Lee, W. H., Inactivation of the retinoblastoma susceptibility gene in human breast cancers, *Science*, 241, 218, 1988.

110. T'Ang, A., Varley, J. M., Chakraborty, S., Murphree, A. L., and Fung, Y. K., Structural rearrangement of the retinoblastoma gene in human breast carcinoma, *Science*, 242, 263, 1988.

111. Horowitz, J. M., Yandell, D. W., Park, S. H., Canning, S., Whyte, P., Buchkovich, K., Harlow, E., Weinberg, R. A., and Dryja, T. P., Point mutational inactivation of the retinoblastoma antioncogene, *Science*, 243, 937, 1989.

112. Dunn, J. M., Phillips, R. A., Becker, A. J., and Gallie, B. L., Identification of germline and somatic mutations affecting the retinoblastoma gene, *Science*, 241, 1797, 1988.

113. Dunn, J. M., Phillips, R. A., Zhu, X., Becker, A. J., and Gallie, B. L., Mutations in the RB1 gene and their effects on transcription, *Mol. Cell Biol.*, 9, 4596, 1989.

114. Bookstein, R., Shew, J. Y., Chen, P. L., Scully, P., and Lee, W. H., Suppression of tumorigenicity of human prostate carcinoma cells by replacing a mutated *RB* gene, *Science*, 247, 712, 1990.

115. Crawford, L. V., Pim, D. C., and Lamb, P., The cellular protein p53 in human tumours, *Mol. Biol. Med.*, 2, 261, 1984.

116. Linzer, D. I. and Levine, A. J., Characterization of a 54K dalton cellular SV40 tumor antigen present in SV40-transformed cells and uninfected embryonal carcinoma cells, *Cell*, 17, 43, 1979.

117. **Mercer, W. E., Nelson, D., DeLeo, A. B., Old, L. J., and Baserga, R.,** Microinjection of monoclonal antibody to protein p53 inhibits serum-induced DNA synthesis in 3T3 cells, *Proc. Natl. Acad. Sci. U.S.A.,* 79, 6309, 1982.

118. **Eliyahu, D., Raz, A., Gruss, P., Givol, D., and Oren, M.,** Participation of p53 cellular tumour antigen in transformation of normal embryonic cells, *Nature,* 312, 646, 1984.

119. **Hinds, P., Finley, C., and Levine, A. J.,** Mutation is required to activate the p53 gene for cooperation with the *ras* oncogene and transformation, *J. Virol.,* 63, 739, 1989.

120. **Finlay, C. A., Hinds, P. W., and Levine, A. J.,** The p53 proto-oncogene can act as a suppressor of transformation, *Cell,* 57, 1083, 1989.

121. **Eliyahu, D., Michalovitz, D., Eliyahu, S., Pinhasi-Kimhi, O., and Oren, M.,** Wild-type p53 can inhibit oncogene-mediated focus formation, *Proc. Natl. Acad. Sci. U.S.A.,* 86, 8763, 1989.

122. **Lane, D. P. and Crawford, L. V.,** T antigen is bound to a host protein in SV40-transformed cells, *Nature,* 278, 261, 1979.

123. **Sarnow, P., Ho, Y. S., Williams, J., and Levine, A. J.,** Adenovirus E1b-58kd tumor antigen and SV40 large tumor antigen are physically associated with the same 54 kd cellular protein in transformed cells, *Cell,* 28, 387, 1982.

124. **Werness, B. A., Levine, A. J., and Howley, P. M.,** Association of human papillomavirus types 16 and 18 E6 proteins with p53, *Science,* 248, 76, 1990.

125. **Scheffner, M., Werness, B. A., Huibregtse, J. M., Levine, A. J., and Howley, P. M.,** The E6 oncoprotein encoded by human papillomavirus types 16 and 18 promotes the degradation of p53, *Cell,* 63, 1129, 1990.

126. **Burger, C. and Fanning, E.,** Specific DNA binding activity of T antigen subclasses varies among different SV40-transformed cell lines, *Virology,* 126, 19, 1983.

127. **Kraiss, S., Quaiser, A., Oren, M., and Montenarh, M.,** Oligomerization of onco-protein p53, *J. Virol.,* 62, 4737, 1988.

128. **Isobe, M., Emanuel, B. S., Givol, D., Oren, M., and Croce, C. M.,** Localization of gene for human p53 tumour antigen to band 17p13, *Nature,* 320, 84, 1986.

129. **Sturzbecher, H. W., Addison, C., and Jenkins, J. R.,** Characterization of mutant p53-hsp72/73 protein-protein complexes by transient expression in monkey COS cells, *Mol. Cell Biol.,* 8, 3740, 1988.

130. **Eliyahu, D., Goldfinger, N., Pinhasi-Kimhi, O., Shaulsky, G., Skurnik, Y., Arai, N., Rotter, V., and Oren, M.,** Meth A fibrosarcoma cells express two transforming mutant p53 species, *Oncogene,* 3, 313, 1988.

131. **Lavigueur, A., Maltby, V., Mock, D., Rossant, J., Pawson, T., and Bernstein, A.,** High incidence of lung, bone, and lymphoid tumors in transgenic mice overexpressing mutant alleles of the p53 oncogene, *Mol. Cell Biol.,* 9, 3982, 1989.

132. **Ehrhart, J. C., Duthu, A., Ullrich, S., Appella, E., and May, P.,** Specific interaction between a subset of the p53 protein family and heat shock proteins hsp72/hsc73 in a human osteosarcoma cell line, *Oncogene,* 3, 595, 1988.

133. **Romano, J. W., Ehrhart, J. C., Duthu, A., Kim, C. M., Appella, E., and May, P.,** Identification and characterization of a p53 gene mutation in a human osteosarcoma cell line, *Oncogene,* 4, 1483, 1990.

134. **Takahashi, T., Nau, M. M., Chiba, I., Birrer, M. J., Rosenberg, R. K., Vinocour, M., Levitt, M., Pass, H., Gazdar, A. F., and Minna, J. D.,** p53: a frequent target for genetic abnormalities in lung cancer, *Science,* 246, 491, 1989.

135. **Hollstein, M., Sidransky, D., Vogelstein, B., and Harris, C. C.,** p53 mutations in human cancers, *Science,* 253, 49, 1991.

136. **Soussi, T., Caron de Fromentel, C., Mechali, M., May, P., and Kress, M.,** Cloning and characterization of a cDNA from *Xenopus laeris* coding for a protein homologous to human and murine p53, *Oncogene,* 1, 71, 1987.

137. **Lehman, T. A., Bennett, W. P., Metcalf, R. A., et. al.,** p53 mutations, *ras* mutations, and p53-heat shock 70 protein complexes in human lung carcinoma cell lines, *Cancer Res.,* 51, 4090, 1991.

138. **Reddel, R. R. and Harris, C. C.,** unpublished observations.

Chapter 11

EVENTS OF TUMOR PROGRESSION ASSOCIATED WITH CARCINOGEN TREATMENT OF EPITHELIAL AND FIBROBLAST COMPARED WITH MUTAGENIC EVENTS

George E. Milo and Bruce C. Casto

TABLE OF CONTENTS

I. INTRODUCTION

Transformation of human cells *in vitro* by any of several different carcinogens induces a sequence of phenotypic changes which ultimately results in the expression of a neoplastic phenotype.[1-23] Such sequences of events are often seen in animal skin bioassays, but the major difference between human and animal systems is that the target organ, skin, is structurally different in the mouse compared to man, i.e., in mouse, the skin is two layers in thickness, whereas in man it is 7 to 11 layers in thickness. Also, the organization of the nuclear chromatin is different in mouse compared to human cells.[24] Lastly, carcinogen-initiated human cells appear to undergo promotion as a silent stage, not a papillomatous stage as seen in mice, when carcinogen-exposed skin is treated with a promoter prior to expression of a carcinoma.[21,25]

Several stages of expression of carcinogen-initiated cells progressing toward a malignant phenotype have been defined. Populations of spontaneous human sarcomas and squamous cell carcinomas have expressed altered surface antigens similar to those found in carcinogen-initiated fibroblasts and keratinocytes that exhibit growth in soft agar.[2,4,13,15] However, within heterogeneous spontaneous carcinoma tumors are cells that possess surface antigens that are found on normal cells and a high-molecular-weight keratin that is associated with differentiated keratinocytes.[26-28] Many of the similarities between *in vitro* transformed human epithelial cells and fibroblasts are presented here in addition to comparisons between carcinogen-treated normal cells and their normal cell counterparts or spontaneous human tumor cells.[26]

Many carcinogen-treated human skin fibroblast cells exhibit anchorage-independent growth after the cells are treated with different concentrations of carcinogens, but do not form progressively growing tumors. On several occasions, changes in cellular characteristics have been observed that indicate a subtle progression of initiated cells to vigorous malignancy; it appears that the progression of initiated cells is regulated by a complex, controlling cellular mechanism.[29] Molecular control at the genomic level that regulates the entrance of initiated cells into the later stages of progression may explain why *in vitro* transformed cells and spontaneous human sarcoma cells do not produce a vigorous malignant phenotype when injected into a surrogate host.

It is reasonable to assume that a series of cellular events that lead to a change in phenotype may be defined by localizing and characterizing specific controlling genes in the cell.[30] The uniqueness and definition, including DNA modification (adduct formation), mutation, and cell transformation (anchorage-independent growth), can best be studied in the same cell system using compounds representative of different classes of xenobiotics in different stages of metabolic activation.

II. MATERIALS AND METHODS

A. GROWTH MEDIA

Keratinocytes were cultured in Eagle's minimum essential medium (MEM) with Hank's balanced salt solution containing 25 mM HEPES buffer at pH 7.2. The growth medium was supplemented with 1× essential amino acids, 1× vitamins, 1.0 mM sodium pyruvate, 2.0 mM glutamine, 0.1 mM nonessential amino acids, and 50 μg/ml of gentimycin. This growth medium (GM-K) was supplemented with 10% fetal bovine serum (FBS; Hyclone Sterile Systems, lot 100418).

Human foreskin fibroblasts were cultured in Eagle's MEM-Hank's balanced salt solution (HBSS) containing 25 mM HEPES buffer at pH 7.2, supplemented with 1.0 mM sodium pyruvate, 2.0 mM glutamine, 1× nonessential amino acids (Eagle's MEM), and gentamycin (final concentration, 50 μg/ml). This GM-F was supplemented with 10% FBS (Hyclone, Logan, UT) that was selected for its ability to support cell growth and cell cycle synchronization of human fibroblasts.[2,4,10] FBS selected for these experiments optimally yielded a 40 to 50% colony-forming efficiency in culture at a low cell density of 40 cells per square centimeter and 85 to 90% radiolabeled nuclei (S-phase of cells at a high cell density of 10,000 cells per square centimeter).[31]

The blocking medium (BM) used to block cells in G_1 prior to their release into S was Dulbecco's Eagle's MEM modified by deleting arginine, glutamine, and $Fe_2(NO_2)_3$, substituting $MgCl_2$ for $MgSO_4$, and eliminating the phenol red or using only 30% of the phenol red in the Grand Island Biological Formulation. This BM was supplemented with 10% FBS, which had been dialyzed exhaustively against changes of the BM minus phenol red (hereafter referred to as d-FBS).

The release medium (RM) was GM supplemented with 0.5 U of insulin

(Sigma Chemical Co., St. Louis, MO) and 10% FBS. The amino acid-enriched selective medium (SM) was GM supplemented with $8 \times$ nonessential amino acids, $2 \times$ essential vitamin mixture (Eagle's MEM $100 \times$ concentrated), and 20% FBS. Cloning medium (CM) was composed of GM supplemented with $1 \times$ essential amino acids ($50 \times$ concentrated amino acid mixture for Eagle's MEM), $1 \times$ essential vitamin mixture, and 20% FBS. Soft-agar cloning medium (SA-CM) was composed of a 5-ml base of 2.0% agar in medium and an overlay of 2 ml of 0.30% agar in medium. The 2.0% agar base layer contained McCoy's 5-A and the same supplements as used in the CM. The overlay contained LoCal Dulbecco's modified Eagle's medium supplemented with 20% FBS. This medium was extensively modified in formulation, including the content of calcium, phenol red, amino acids, vitamins, and inorganic salts, to optimize the growth of transformed human cells and human sarcoma cells while restricting the growth of normal cells.[33]

B. SOLVATION OF CHEMICALS

Each of the chemicals — *N*-methyl-*N'*-nitro-*N*-nitrosoguanidine (MNNG), propane sultone (PS), β-propiolactone (PL), unsymmetrical dimethylhydrazine (UDMH), benzo(a)pyrene diolepoxide I (+) (BPDE-I), 1-nitrosopyrene (1-NOP), and methylazoxymethanol acetate (MAMA) — was dissolved in acetone (Spectrar Acetone, Mallinckrodt, Inc., Paris, KY). These solutions were used immediately or stock solutions were prepared at 100 μg/ml or greater and stored under argon in the dark at $-20°C$. Prior to use, all chemical test articles were diluted for use with GM.

C. CELL CULTURES

Neonatal foreskin and surgically removed sarcoma lesions were collected daily from local hospitals. Each tissue was placed in a 15-ml vial containing 10 ml of Eagle's MEM supplemented with 5% FBS. Tissue was washed three times with 10 ml of CM-20% FBS and minced into 1 × 1-mm sections, transferred to centrifuge tubes, and incubated at 37°C for 2 to 4 h with a 0.25% collagenase solution in an atmosphere of 4% CO_2 in air. After completion of digestion, cells were harvested by centrifugation at 650 × *g* for 10 min and washed twice. Each cell suspension was then seeded into a T-75 flask at a density sufficient to yield an absolute plating efficiency of 10,000 cells per square centimeter in 24 h. The residual cells in suspension were poured off after 24 h and the mixed cell cultures rinsed once and refed with GM; 72 h later, the foreskin fibroblasts or sarcoma cells were selectively removed by initially rinsing the mixed cultures once with GM without FBS and then adding 1 ml of 0.1% trypsin in GM to the cultures at 21°C. The removal of fibroblasts was monitored under an inverted microscope (10 ×). At any indication of ruffling of the edges around the epithelial cell colonies, the trypsinization process was stopped by pouring 10 ml of GM supplemented with 10% FBS over the remaining attached cells. The suspended fibroblasts were recovered and reseeded at 10,000 cells per square centimeter in 75-cm² flasks or 150-cm² wells. These cells were identified as the primary culture (P_1) and were found to be free of keratinocytes. The presence of keratinocytes in these cultures was determined using a fluorescent antibody detection system described elsewhere.[34]

Sarcoma cell suspensions were prepared directly for seeding in soft agar as described here. Sarcoma tissue was finely minced and then sieved into a dish containing Dulbecco's LoCal MEM. After performing a cell count, an aliquot of these cells was seeded into 2 ml of SA-CM.

For preparation of keratinocytes, neonatal foreskin tissue was obtained from routine circumcision. The tissue was placed in a 15-ml vial containing 10 ml of CM medium supplemented with 10% FBS. The tissue was then cut into 4 × 4-mm sections and transferred to a vial containing 5 ml of cold (12°C), sterile 0.02% ethylenebis(oxyethylenenitrilo)tetracetic acid (EGTA; Eastman Kodak, Rochester, NY) in HBSS minus the calcium and magnesium and containing 0.1% trypsin (HBSS; TL 13 BP, Millipore Corp., Freehold, NY) at pH 3.5. After 32 to 40 h, the layers from the stratum spinosum through the squames were separated from the stratum basalis-dermis by mechanical removal. Ophthalmic forceps were used to slip off the upper layers, leaving the stratum basalis attached to the dermis. The epithelial cells of the stratum basalis were removed from the dermis by scraping the dermis lightly with curved ophthalmic forceps.[26] The epithelial cells were vortexed lightly and recovered by centrifugation at 260 × g for 10 min. The pellet was resuspended in a modified suspension medium with a lower calcium concentration of 0.7 mM (SM-LC) and seeded at a cell density of 50,000 cells per square centimeter. The individual cultures were serially passaged 1:2 at an 80% confluency. The absolute attachment efficiency for the epithelial cells was 90%. The cultures were routinely refed on a 2 to 3-d regimen and incubated in a 95% relative humidity, 3% carbon dioxide-enriched air atmosphere.

Epithelial cultures were used for toxicity and transformation experiments prior to three passages (5 population doublings [PDs] at 1:2 splits). Squamous cell carcinoma (SCC) cell suspensions were treated to disperse the cells in a fashion similar to that described for the preparation of sarcoma tumor cells.

D. CHARACTERIZATION OF CYTOTOXICITY

In previous work, cytotoxicity on isogenic fibroblasts was determined to select a noncytotoxic dose to use for transforming keratinocytes.[35] The keratinocytes prepared for these early transformation experiments had been isolated from foreskin primary populations by selective trypsinization of fibroblasts, were 12 d in age, and contained colonies with growth in vertical strata as well as longitudinal growth over the plastic substratum. Using the concentrations determined from fibroblast cultures, the highest concentration of carcinogen used as a transforming dose on keratinocytes did not lead to detachment of the epithelial colonies from the plastic substratum. Presently, keratinocytes are prepared from the stratum basalis and seeded at a cell density of 3,200 cells per square centimeter. The carcinogen is added 48 h later for 6 h, removed, and the cultures are refed with SM-LC; 10 d later, the cultures are fixed, stained with hematoxylin and eosin, and counted. Colonies containing 50 cells or more are counted in treated and untreated cultures. The relative colony-forming efficiency (RCE) is then computed by counting the number of colonies in treated cultures and dividing this number by the number of colonies in untreated cultures. The RCE is expressed as percent toxicity, i.e., RCE [value] × 100. Once the toxic and nontoxic concentrations of the carcinogen of interest has been determined, the transformation experiments are initiated. Cytotoxicity determinations in fibroblast cultures prepared from the dermis were evaluated as described previously.[10]

E. TRANSFORMATION PROTOCOL FOR KERATINOCYTES

Proliferating cell cultures at 500,000 cells per 25-cm^2 flask were transferred from SM-LC to Dulbecco's modified growth medium minus arginine and glutamine and retained in this medium for 48 h, after which they were refed with GM containing 1.0 U of insulin per milliliter and 10% dialyzed FBS (RM). After 2 h, when 20% of the cells were in early S, the RM medium was removed and the carcinogens were added in GM containing 1 U of insulin per milliliter of medium and 10% undialyzed FBS. The carcinogens were in contact with the cells for 6 h. At this time, the medium with carcinogen was removed and the cultures washed twice with GM supplemented with 10% FBS and refed with SM-LC without FBS. These treated cultures were incubated at 37°C in an atmosphere of 3.5% CO_2 in air for 2 weeks or until the epithelial cell sheet reached confluency. These cultures were then washed with 0.02% EDTA (Sigma Chemical Co., St. Louis, MO) and serially passaged 1:2 using a solution of 0.1% trypsin in 0.02% EDTA. The treated cultures were serially passaged twice more when they reached a culture density of 70 to 80% confluency. At this time, the treated cultures were either passaged in soft agar, evaluated for the presence of SCC-associated antigenic determinants, or injected into a nude mouse for evaluation of the neoplastic potential.[26]

F. TRANSFORMATION PROTOCOL FOR FIBROBLASTS

The transformation protocol used in these studies also requires treatment of cells in the early S-phase of the cell cycle with each carcinogen. Cells were seeded at a cell density of 10,000 cells per square centimeter into the G_1 blocking medium that lowered the radiolabeling index to less than 1%.[10] It is important to keep the cells in a replicative mode, because if confluent cultures are permitted to remain in a contact-inhibited stage for times in excess of 12 h, the probability of successful transformation is reduced to zero. If rapidly growing cells are used for the transformation experiments ($\geq$22% radiolabeling index) and are not blocked successfully in late G_1, the numbers of cells passing synchronously through S is reduced to 60% or less. In either case, there is a reduction in the number of cells passing through the S-phase from 85 to 95% in blocking medium to about 50 to 60% without it.

Transformation with each of the carcinogens (PS, PL, and UDMH) was performed by treating cells in the early S-phase at a density of 10,000 cells per square centimeter with ED_{50} dosages of each compound, i.e., the concentration of chemical that caused a 50% decrease in the RCE, compared to the control when the cells were seeded at 1000 cells per square centimeter. Control cultures routinely produced 23% cloning efficiencies when the cultures were seeded at 1000 cells per 25-cm^2 well. RCE is defined as the number of treated colonies from cells counted relative to the number of colonies from untreated cells × 100. For example, an effective cytotoxic dosage of 50 would be that concentration (μg/ml) of chemical that reduced the RCE to 50% of the control. The chemical concentrations determined from cells seeded at 1000 cells per square centimeter were then used to treat cells at 10,000 cells per square centimeter at 37°C. After 48 h in blocking medium, the cells were refed with RM, and 8 to 12 h later were treated with carcinogen for 12 h. The carcinogen-medium was then removed, the cells were fed with RM supplemented with 0.5 U of insulin, and incubation was continued for an additional 12 h. These cells were serially passaged at a density of 5000 cells per square centimeter into an 8× amino acid GM supplemented with 20% FBS. These proliferating cell cultures were subsequently serially passaged at 1:10 split ratios into an 8× amino acid-enriched GM supplemented with 20% FBS for 16 PDs.

G. GROWTH IN SOFT AGAR OF EITHER CARCINOGEN-TREATED KERATINOCYTES, FIBROBLASTS, OR SPONTANEOUS TUMOR CELLS

Treated cells were seeded into 2 ml of soft agar (SA-CM) overlay at a seeding density of 250,000 cells in a 25-cm^2 well.[33] The bottom agar layer in the 25-cm^2 well was prepared by mixing 2× McCoy's (v/v) prewarmed to 37°C with 4% agar at ≤41°C. These plates were incubated at 37°C in 3% CO_2 for up to 4 d. The top agar layer was prepared in the following manner: 2× LoCal Dulbecco's modified Eagle's medium supplemented with 40% FBS and containing 500,000 cells was mixed with an equal volume of 0.6% agar at 37°C to make a final volume of 2 ml per 25-cm^2 well and a final seeking density of 500,000 cells per well. These 25-cm^2 wells were incubated at 37°C in 4% CO_2 at a high (75%) relative humidity. Each culture was evaluated within 24 h for clumping and doublets. To date, use of reduced calcium in the medium and extensive washing of the agar has reduced clumping to ≤0 to 5 cell clumps per well.[26] Cell populations prepared from either squamous carcinoma or sarcoma lesions were suspended in soft agar using a uniform cell suspension at a cell density of 100,000 cells per 25-cm^2 area.[26] Colonies were removed from soft agar 14 to 21 d later, seeded in culture flasks, and established as monolayer cell cultures.[23]

H. INDIRECT IMMUNOFLUORESCENCE STAINING WITH MONOCLONAL ANTIBODIES

Immunofluorescence staining and visualization of the cell surface antigen were performed by overlaying selected 2.5-cm² areas of unfixed cells or 8-μm unfixed tumor tissue sections with the monoclonal antibody (MoAb) for 45 min in Dulbecco's phosphate-buffered saline (PBS) at pH 7.2. Following a 45-min incubation at 37°C and three washings with PBS, 0.05 ml of FITC-conjugated antimouse IgG antiserum was added to the 2.5-cm² area of cells or tissue and incubation in 4% CO_2 was continued at 37°C for 45 min. FITC-conjugated antibody was rinsed off with 5 ml of PBS in a series of washes. Slides were then overlaid with 10% glycerol-PBS and read under a Zeiss epifluorescent microscope.[26,36]

I. CELLULAR INVASIVENESS

Cells from colonies of AI-positive populations of either carcinogen-treated cells or spontaneous tumors were isolated from soft agar, reestablished in culture, and 50,000 cells in 0.04 ml seeded onto skin from 9-d-old chick embryos.[13] After 72 h, the inoculated tissues were fixed in Bouin's solution and embedded in paraffin, and 5-μm sections were taken serially throughout the block. These sections were stained with hematoxylin and eosin, and evaluated.[13]

J. TUMORIGENICITY EVALUATION

Ten days prior to receiving a cell inoculum, male Balb-C nude (nu/nu) mice of 2 to 3 or 4 to 6 weeks of age were caged in a sterile plastic container. The mice were splenectomized and graft rejection was checked using modified human skin. Twice a week, the mice received 0.2 ml of ALS. The mice were then prepared (7 to 10 d after surgery) to receive tumor cells, provided there were no signs of graft rejection in the nude mouse colony

Chemically transformed cells, preparations of spontaneous human tumor cells, or mouse cells were suspended in Dulbecco's LoCal supplemented with 0.3% agar, and the suspension was injected subcutaneously into the subscapular area about 1.5 cm lateral to the mid-dorsal line. Blebs created by the injection regressed in 48 h. Cells from untreated cultures, suspended in the same manner as described for treated cells, also created blebs which regressed. Tumors were excised about 4 to 6 weeks later for identification of histological tumor type, and cells were reestablished in monolayer culture.[37] An alternate route for injection of the cells into the nude mouse was the cerebral route. Inocula containing 50,000 to 100,000 cells in 0.02 ml of medium were injected intracranially into 4 to 6- or 2 to 3-week-old suckling nude mice.[23]

K. KARYOLOGY

Human tumor cells recovered from tumors in nude mice and free of mouse fibroblasts were treated with 2×10^{-7} M colcemid for 4 h, collected, incubated for 15 min in 0.075 M KCl, and fixed.[37] The chromosome spreads were stained with Giemsa and evaluated.

L. TREATMENT OF CELLS WITH CARCINOGENS: MUTATION PROTOCOL

Mutagenesis was accomplished using B(a)P following metabolic activation by a supernatant fraction (S9) mix from rat livers. Human fibroblasts, plated at 1×10^6 cells in a 75-cm² flask 24 h earlier, were treated for either 4 or 16 h with activated B(a)P at concentrations ranging from 0 to 24 μg/ml.

The S9 fraction was prepared using 8- to 10-week-old rats (Sprague-Dawley, Madison, WI) inoculated i.p. with 500 mg/kg of Aroclor 1254. After 5 d, each 10 g of liver was rinsed in 9.15% KCl, homogenized in 20 ml of 0.25 M sucrose solution, and centrifuged at 9000 $\times$ g. The S9 was collected and stored at $-70°C$. S9 batches were prescreened for toxicity, with the least toxic batches being used in the experiments.[38] For treatments, S9 was used at a concentration of 0.5%, which exhibited an average cell mortality rate of 60% in method control experiments. The cofactors used caused no detectable cell mortality at experimental concentrations (glucose-6-phosphate, 2.3 mg/ml; NADP, 1.15 mg/ml).[38]

M. ESTIMATION OF MUTATION FREQUENCY

After treatment with B(a)P, each flask of cells was cultured in FBS-F10 medium for 7 d with two passages. After a 7-d expression, cells were plated in dishes to detect mutants and to estimate cloning efficiency. For estimation of mutation frequency, 1.5×10^4 cells were plated per 60-mm dish; 120 dishes were prepared for each dose. All the dishes in the mutation assay received 6-thioguanine (6TG) at a final concentration of $7.5 \times 10^{-5} M$ within 3 h. The selective medium was renewed once every week. For determination of cloning efficiency, 100 cells per 60-mm dish were plated and grown in FBS-F10 medium. Dishes for estimation of cloning efficiency were stained with 2% crystal violet at the end of the second week, and mutagenesis dishes at the end of the third week. Mutant colonies were counted and the mutation frequencies were determined according to the Poisson expectation.[39-43]

N. ISOLATION OF DNA

Human cells in culture flasks were harvested immediately after removal of carcinogen, washed, suspended in 10 mM Tris·HCl, pH 7.6, 5 mM MgCl₂, 0.32 M sucrose, and 1% Triton X-100, and homogenized gently. Nuclei were recovered after centrifugation at 4000 rpm for 20 min and suspended in 75 mM NaCl and 25 mM EDTA, pH 8, adjusted to 1% SDS. The solution was extracted with an equal volume of phenol followed by an equal volume of chloroform. DNA was precipitated with ethanol, washed, dried, and dissolved at about 1 mg/ml in sterile Tris-EDTA (10 mM Tris·HCL, pH 8.0 1 mM EDTA), and stored at $-20°C$.[30]

O. SYNTHESIS AND PURITY OF [γ-^{32}P]ATP

The synthesis of [γ-^{32}P]ATP was performed with a Promega Biolectin Gamma Prep-A system (Madison, WI). Upon demand, approximately 2 μCi of [γ-^{32}P]ATP in 100 ml total volume was synthesized. The purity of the [γ-^{32}P]ATP checked before a reaction was ≥97%, as determined by TLC on a PEI-cellulose plate.[44-46]

P. DETECTION OF DNA ADDUCTS AND NORMAL NUCLEOSIDES BY ^{32}P-POSTLABELING

DNA (1 μg) was digested to deoxyribonucleoside-3'-monophosphates (dNp) and 230 gg postlabeled with [γ-^{32}P]ATP to 5'-^{32}P-labeled deoxyribonucleoside-3',5'-bisphosphates (dpNp).

Q. IDENTIFICATION AND QUANTIFICATION OF NONRADIOACTIVE AROMATIC CARCINOGEN-DNA ADDUCTS

Determination and quantification of carcinogen-DNA adducts were conducted using the ^{32}P-postlabeling assay method.[44-46] Control or carcinogen-modified DNA (1 μg) was digested with 2 μg each of micrococcal endonuclease and spleen exonuclease in 10 μl of 20 mM sodium succinate and 10 mM CaCl$_2$, pH 6.0, at 38°C for 2 h. The resulting deoxyribonucleoside-3'-monophosphates were then converted to (5'-^{32}P)deoxyribonucleoside-3',5'-bisphosphates by T4 polynucleotide kinase-catalyzed (^{32}P) phosphate transfer from [γ-^{32}P]ATP as follows. A 10-μl aliquot of DNA digest was added to a solution prepared by mixing 1.5 μl of 0.1 M Bicine-NaOH, 0.1 M MgCl$_2$, 0.1 M dithiothreitol, and 10 mM spermidine at pH 9.0; 5.0 μl of [γ-^{32}P]-ATP; and 1.0 μl of T4 polynucleotide kinase (3.0 U/μl). The solution was incubated at 38°C for 2 h.

The ^{32}P-labeled nucleotides were resolved by anion-exchange, thin-layer chromatography on polyethyleneiminecellulose (PEIC) sheets, and about 50 to 60 μCi of labeled digest was chromatographed. Development was in 1 M LiCl (D$_1$) and 2.5 M ammonium formate, pH 3.5 (D$_2$), resulting in the removal of normal nucleotides and ^{32}Pi, while the adducts were retained at or close to the origin. ^{32}P adducts were then resolved by development in 3 M lithium formate and 7 M urea, pH 3.5 (D$_3$), followed by 0.6 M LiCl, 0.5 M Tris-HCl, and 7 M urea, pH 8.0 (D$_4$). Screen-enhanced autoradiography was at −80°C for 16 to 24 h. The amount of ^{32}P-labeled digest applied to the TLCs was 58.2 μCi for the detection of adduct radioactivity (4-D system) and 0.07 μCi for the assay of the normal nucleotides (1-D system) with a dilution factor of 832. Adduct spots and spots of normal nucleotides were cut from the chromatograms and counted with a scintillation counter for Cerenkov radiation.[30,44] The frequency of modified base was calculated according to:

$$\text{Relative adduct labeling}^{46} = \frac{\text{cpm in adduct nucleotides}}{\text{cpm in total nucleotides} \times \text{dilution factor}}$$

TABLE 1
Dose Response-Dependent Induction of Anchorage-Independent Growth

Carcinogen treatment	Concentration (μg/ml)	Percent reduction in RCE	Number of colonies formed in soft agar per 10^5 seeded cells
PS	25.0	90	0
	7.5	50	0
	0.5	0	68 + 13
PL	12.0	80	31 ± 13
	7.5	50	23 ± 1
	0.1	0	3 ± 2
MNNG	0.4	90	0
	0.03	50	163.0 ± 12.7
	0.01	25	148.0 ± 63.5
UDMH	167 (μM)	50	146.0 ± 12.0

Note: Legend to Table 1: Cells at a density of 10,000 cells per square centimeter at the onset of the S-phase were treated with increasing amounts of each carcinogen. Column 2 lists the microgram per milliliter concentration of each chemical. Column 3 lists the values expressed in percent of reduction in relative cloning efficiency (RCE) by the chemical. The number of colonies formed in soft agar per 10^5 seeded cells is recorded as a mean value ± 1 σ SD for an n of 8. This mean colony count ± SD is presented in column 4. Colonies were counted visually (50 cells or more per colony) under a stereomicroscope at 4.5× magnification.

III. RESULTS

A. CYTOTOXICITY

It was important to devise a method for measuring cytotoxicity on either human keratinocytes or fibroblasts *in vitro* that permitted an examination of the transformation potential of these compounds at different cytotoxic and noncytotoxic concentrations.[23] Transformation with each of the carcinogens — PL, PS, MNNG, and UDMH (Table 1) — was performed by treating fibroblast cells in the early S-phase (at a density of 10,000 cells per square centimeter) with selected concentrations of each compound. However, before these procedures were implemented, it was of paramount importance to evaluate the cytotoxic effects of the compounds that were to be tested for their carcinogenicity. Control culture cloning efficiencies were defined as the number of colonies counted relative to the number of cells seeded × 100. For treated cells, the relative cloning efficiency (RCE) is defined as the number of colonies from treated cultures divided by the number of colonies from untreated cultures × 100. For example, an effective cytotoxic dosage of 50 (ED_{50}) would be that concentration of chemical that reduces the RCE to 50% of the control value. It was important to use concentrations of chemicals with keratinocytes that did not lead to detachment of the epithelial colonies from the plastic substratum following treatment.[26,47] (To confirm that the cells in question in culture were either keratinocytes or fibroblasts, the epithelial cells

TABLE 2
Cytotoxicity, RCE (%), Adduct Modification, and Expression of AIG at Different Concentrations of Benzo(a)pyrene Diolepoxide-I, Benzo(a)pyrene, and 1-Nitrosopyrene

Compound	Concentration (μg/ml)	RCE (%)	Number of modified dG-nucleotides/10^8 nucleotides	Frequency AIG ± SD
BPDE-I	0	100	0	0
	1.0	100	0	0
	2.5	100	7.7	95.0 ± 8.7
	4.6	50	15.3	121.0 ± 6.2
	7.5	35	17.4	70.0 ± 12.0
	17.0	20	23.2	12.0 ± 3.2
	25.0	6	18.3	3.0 ± 2.0
B(a)P	0	100	—	0
	0.1	100	12.4	29.7 ± 14.5
	2.5	100	18.6	16.0 ± 12.0
	10.0	100	13.4	103.0 ± 41.0
	25.0	100	11.7	101.0 ± 2.8
1-NOP	0	100	0	0
	5.0	100	0.2	0
	7.5	100	0.7	5.0 ± 3.0
	10.0	92	7.0	6.0 ± 2.0
	18.0	80	11.0	15.0 ± 4.0
	30.0	60	24.0	23.0 ± 2.0
	95.0	50	51.0	41.0 ± 7.0
	105.0	21	60.0	10.0 ± 12.0

Note: Legend to Table 2: The compounds benzo(a)pyrene diolepoxide-I (BPDE-I), benzo(a)pyrene (B(a)P), and 1-nitrosopyrene (1-NOP) were evaluated for cytotoxicity (RCE %), modification of dG, and anchorage-independent growth (AIG). The RCE % (relative cloning efficiency) represents a mean value for an n of 8. The data in column 4 represent the number of modified dG nucleotides per 10^8 nucleotides for an n of 4. The frequency of anchorage-independent growth is expressed as the number of colonies former per 10^5 seeded cells with a diameter greater than 60 μm for an n of 12, ± 1 σ SD.

were identified by a MoAb specific for the normal proliferating epithelial cells and the presence of keratin.[26,28]) A typical cytotoxicity response for fibroblasts is presented in Table 1.[35] Other cytotoxic profiles were evaluated for B(a)P, BPDE-I, and 1-NOP (Table 2).

The toxic concentrations determined from cells seeded at 1000 cells per square centimeter were then used to treat cells seeded at high cell densities of either keratinocytes or fibroblasts in the transformation protocol.[23,26] Treatment was initiated as described in Section II. Following treatment with each carcinogen, the treatment medium was removed and incubation was continued.

B. CELLULAR TRANSFORMATION

Cell populations of either fibroblasts or keratinocytes treated with either PL or PS exhibited transient changes in morphology early after concluding treatment. Cells exhibited both altered cellular morphology and altered colony morphology. PL treatment resulted in the greatest changes in cellular morphology, while more subtle changes followed PS treatment. The changes in the cultures typically included a criss-cross (fibroblasts), piled-up orientation of cells on a lawn of normal cells with a more regular orientation; UDMH treatment did not cause changes in either cellular or colony morphology. When altered colony or cellular changes did occur, the changes were transient and did not persist for more than 1-2 PDs.

C. ANCHORAGE-INDEPENDENT GROWTH IN SOFT AGAR

The relationship between concentration and transformation was investigated using either S-phase cells blocked in G_1 or randomly proliferating fibroblasts at 1000 cells per square centimeter. The fibroblasts were treated with four concentrations of either PS or PL. All carcinogens increased the frequency of colony formation in agar with an increase in cytotoxic dose. This increase in agar colony formation response, while dosage dependent, was not linear. A correction for dead cells at the 75% cytotoxic dose resulted in a 10% increase in the frequency of soft-agar colonies. Moreover, the number of anchorage-dependent colonies observed following treatment at any one concentration was dependent on both the length of time the fibroblasts were treated with the chemical and the number of PDs that occurred following the chemical treatment. Limiting the MNNG treatment of fibroblasts to 3 h resulted in a frequency of 163 ± 12 colonies per 100,000 seeded cells (Table 1). Increasing the treatment time to 24 h decreased the number of colonies in soft agar to 2 or less per 100,000 viable seeded cells (data not shown). This low number represents a maximal level, since it is based on the number of colonies arising from the surviving fraction of carcinogen-treated cells. The treatment with other carcinogens — PS, UDMH, MAMA, or PL — yielded similar but not parallel results (Table 1). Twenty PDs after either PS or UDMH treatment, the cells exhibited at least 147 colonies per 100,000 seeded cells. The cells entered Phase III after PDs 51 to 53 and ceased to proliferate, and the transformed phenotype was then no longer detectable.

The transformation frequency ranged from 29.7 ± 14.5 to 101 ± 28 colonies per 10^5 seeded cells over a range of concentrations of B(a)P from 0.1 to 25.0 μg/ml (see Table 2). When the direct-acting metabolite BPDE-I was evaluated for its transformation capability, it was observed that as the concentration increased from 1.0 to 25.0 μg/ml, the expression of AIG decreased from 121 colonies to 3.0 ± 2.0 colonies per 10^5 seeded cells in soft agar. Another bulky, direct-acting carcinogen (1-NOP) treatment induced a similar response profile, i.e., at higher concentrations of the compound, there was a decrease in the frequency of AIG (see Table 2).

Transformation of keratinocytes was undertaken at different dosages of the compounds of interest. Cells treated with MNNG at a toxic concentration of 0.4 μ*M* (the ED_{50}) yielded 134 ± 9 colonies in soft agar (Table 3). At

TABLE 3
Expression of AI of Carcinogen-Treated Keratinocytes and Fibroblasts

Compound	Conc (μM)		Colonies per 10^5 seeded cells	
	Fibroblasts	Epithelial	Fibroblasts	Epithelial
MNNG	43.0	0.4	163 ± 12	134 ± 9
PS	3.7	7.5	109 ± 11	58 ± 9
PL	82.0	7.5	43 ± 12	23 ± 1
UDMH	167.0	18.3	146 ± 12	0

treatment concentrations greater than the ED_{50} in the keratinocyte transformation protocol, transformation still occurred, whereas when fibroblasts were treated with concentrations greater than the ED_{50}, either no transformants or only a low level of transformants were observed. When keratinocytes were treated with PS at concentrations of 7.5 μM, no transformants were observed; however, treatment with 0.75 μM yielded 23 ± 1 colonies. Treatment of keratinocytes with PL resulted in the formation of transformants at highly toxic concentrations ($\geq ED_{80}$) and at a nontoxic transforming dose of 0.1 $\mu g/ml$. After treatment of keratinocytes with UDMH at 18.3 μM, no colonies were observed in agar, whereas 146 colonies developed from treated fibroblasts even when the dermal fibroblasts and epidermal keratinocytes were obtained from the same tissue (see Table 3).

D. SURFACE CHARACTERIZATION OF CARCINOGEN-TRANSFORMED PHENOTYPE

Each of the carcinogen-transformed fibroblast or sarcoma cell cultures treated with either normal goat serum or rabbit anti-mouse IgG-antiserum were fluorescent negative when the conjugated chromophore, fluorescein isothiocyanate, was added to the incubation medium. When using fresh human sarcoma isolates, no cross-reactivity was seen by fluorescence with normal cells in the tissue surrounding the invasive lesion.[23] When the carcinogen-transformed fibroblasts were passaged through soft agar, they did react with the MoAb 345.134S and exhibited a positive, detectable fluorescence (Table 4). When either MNNG or PL carcinogen-transformed keratinocytes were grown *in vitro* after passage through soft agar, they both reacted with MoAb OSU 22-3, but not MoAb EP-16 or the polyclonal antibody against keratin. The MoAb EP-16, as reported before,[36] reacted strongly (3$^+$) with normal keratinocytes. Cells from squamous cell carcinomas (SCCs) that grew in soft agar also reacted strongly (3$^+$) with MoAb OSU 22-3, but no EP-16 or the polyclonal antibody against keratin. Sister cultures of cells that stained with the specific MoAb and grew in soft agar invaded the CES, but formed only localized tumors in nude mice approximately 0.8 to 1.2 cm in diameter.[26]

E. CELLULAR INVASIVENESS

Carcinogen-treated keratinocyte cells that were reestablished from agar

TABLE 4
Reactivity of the Anti-115K-GP Moab 345.134S with Chemically Transformed Human Skin Fibroblasts Passaged Through Soft Agar and Reestablished *In Vitro*

Chemically transformed cell line[a]	Number positive/number tested	Reaction with cells
PL	2/2	Positive fluorescence
MAMA[b]	8/8	Positive fluorescence[c]
MNNG	4/4	Positive fluorescence[c]

Note: Transformed keratinocytes that exhibit AIG reestablished in culture exhibit a reaction with OSU-EP-16 MoAB (see Section II). This MoAb is specific against the SCC cell surface antigen-associated tumor phenotype.

[a] Each of the reactions was evaluated at least twice for each chemical treatment.
[b] The MAMA treatments were evaluated at least eight times, and all lines were positive. In other lines, with other foreskin fibroblast populations where transformation experiments were performed with either PL or MNNG, >95% of the cultures exhibited a positive fluorescence in three separate experiments.
[c] From Milo, G. E., Casto, B., and Ferrone, S., *Mutat. Des.*, 199, 387,1987.

and following 2 PDs in culture, seeded onto the CEs, invaded the dermal layer of the CES in a fashion similar to transformed fibroblasts.[13] From the histological sections, cells in these lesions simulated the action of human tumor cells. The incidence of invasion of the CES was 100% for all carcinogen-treated cells that exhibited AIG.

F. LOCALIZED TUMORS

Cultures that demonstrated cellular invasiveness were injected subcutaneously into the subscapular area on a nude mouse, and 3 weeks later a representative 0.8- to 1.2-cm nodule was removed and submitted for histopathology.

Cell populations from spontaneous SCCs were injected subcutaneously into the subscapular area of the mice; out of 16 mice injected, 4 developed tumors. When the cells were injected into another group of mice by the intracranial route, a low incidence of tumor formation (one tumor in ten mice) was observed after 4 weeks, even though the tumor cell injections were repeated three times. The histological interpretation of the subcutaneous and intracranial tumors was that both were SCCs. When the chemically transformed keratinocytes were injected intracranially into the mouse, the transformed cells elicited hyperkinetic activity and progressive paralysis of the mouse approximately 4 weeks after injection. However, as was previously reported with transformed human epithelial cells,[26] we were unable to detect, after histopathological examination of the cranial tissue, the presence of transformed human epithelial cells using a hematoxylin-eosin stain or the specific

MoAb staining procedures described above. Mice receiving intracranial injections of normal cells lived a normal life-span without any evidence of neurological dysfunction or any overt pathology. The intracranially injected mice did not develop brain infections, whether injected with human tumor cells, *in vitro* transformed cells, or normal cells.

Mice receiving subcutaneously either the carcinogen-transformed fibroblasts or sarcoma tumor cells formed blebs that lasted approximately 24 h. After 4 weeks, 0.8- to 1.2-cm nodules had developed which would often regress 3 to 6 months later. When a part of the tumor was removed for histopathological examination at 4 weeks, a portion of it was reestablished in culture for karyological evaluation.

Unlike the results obtained with transformed keratinocytes, intracranial inoculation of either chemically transformed fibroblasts or human sarcoma cells resulted in the formation of intracranial tumors in nude mice.[13] Subcutaneously formed tumors from chemically transformed fibroblasts were diagnosed as undifferentiated mesenchymal tumors.[2,4,10,23] Localized tumors that formed from sarcoma cell lines were identical histologically to the tumor of origin. In all cases, tumors remained local and did not advance to the stage of a progressively growing tumor.

G. KARYOLOGICAL EVALUATION OF CARCINOGEN-TREATED KERATINOCYTES AND FIBROBLASTS

The subcutaneous tumors, when excised, were often encapsulated with a layer of mouse cells. To establish the human tumor cells in culture for karyological evaluation, it was necessary to grow the tumor cells *in vitro* in the presence of antisera prepared against mouse fibroblasts that would selectively kill the mouse cells; this antiserum did not require complement or activated killer cells for its activity. The concentration of antiserum to be used was estimated by dilution experiments to prevent the proliferation of normal mouse skin fibroblasts without interfering with the *in vitro* proliferation of tumor cells. Every 24 h the medium was replenished with the antiserum up until 8 d, after which the mouse skin fibroblast antiserum was deleted from the medium. The remaining live, attached cells were shown to be of human origin, and the karyotype was either diploid or pseudodiploid.[23,26]

H. MUTAGENIC ACTIVITY OF ACTIVATED BENZO(A)PYRENE

The induced mutation frequency of cells treated with human S9-activated B(a)P is presented in Table 5. In contrast to other experiments with other mutagens,[38,48] strict dose-dependent increases in mutatin responses did not appear[49] when compared to increases in the cytotoxicity of the compound of interest. Repeat experiments did not show any major statistically significant variation in results between experiments.[38] The 6TG-resistant, activated B(a)P induced mutants increased in numbers as the concentrations of B(a)P were increased (see Table 5). Following a 4-h treatment, the correlation between

TABLE 5
Adduct Formation at Different Concentrations of Benzopyrene[a]
Compared to Frequency of 6-Thioguanine Resistance

Treatment (h)	Conc (μg/ml)	RCE (%)	Number of modified dG nucleotides/ 10^8 nucleotides	Induced mutation frequency ($\times 10^{-6}$)
4	0	100	—	0
	3	111	0.24	3.29
	6	107	0.41	4.06
	12	86	3.00	9.34
	24	119	1.66	5.34
16	0	100	—	0
	3	71	1.27	4.74
	6	71	8.42	14.06
	12	42	7.46	14.48
	24	44	9.24	22.18

Note: The average number of mutant colonies (X) per dish was obtained according to $P_0 = e^{-x}$, where P_0 was the fraction of dishes bearing no mutants. Multiplication of the initial number of cells (1.5×10^4) per dish by the cloning efficiency of the cells gave the number of surviving cells per dish. Mutation frequencies were calculated by dividing (X) by the number of surviving cells per dish. Induced mutation frequencies were obtained by subtracting the background mutation frequency from the calculated mutation frequencies.

[a] Activated B(a)P.

induced mutation frequency and cytotoxicity of activated B(a)P was low. However, exposures of 16 h produced more consistent mutation frequencies as a function of increasing concentrations of S9-activated B(a)P than those seen following 4 h of treatment.

I. BENZO(A)PYRENE-DNA ADDUCT FORMATION

In both the transformation system and the HGPRT mutagenesis system, no adducts were detected in the untreated cells. In the S9-activated, B(a)P-treated cells, dG adducts ranged from 0.24 adducts per 10^8 nucleotides after an exposure of 4 h to 9.24 adducts per 10^8 nucleotides after an exposure of 16 h (see Table 5).

Following a 16-h exposure, increasing numbers of adducts were demonstrated as the concentration increased from 6 to 24 μg/ml; however, a relatively constant level of adducts, i.e., 24 adducts per 10^8 nucleotides, was produced above that concentration (data not shown) as the cytotoxic effect of S9-activated B(a)P continued to increase. The 4-h exposure gave increased numbers of adducts as the concentration of S9-activated B(a)P was increased up to 12 μg/ml. However, a decrease in adduct modification at concentrations greater than 24 μg/ml was observed. Both of these trends closely paralleled the induced mutation frequency response curves for those exposures (see Table 5).

When the number of B(a)P dG adducts per 10^8 nucleotides was compared to calculated mutation frequencies from the same treatments, a highly significant correlation resulted (slope $= 0.48 \pm 0.06$, significant at the 0.1 level). These results were unlike the B(a)P, BPDE-I, and 1-NOP dG-adduct results when compared with the transformation frequency (see Table 2). When BPDE-I and 1-NOP were used in the treatment regimen (see Table 2), which was roughly comparable to the 4-h treatment in the mutation assays (see Table 5), the slopes of the transformation curves flattened out at the higher concentrations of the compounds, and transformation decreased as cytotoxicity increased with those treatments. Continued adduct modification at increasing concentrations of the environmental carcinogens did not parallel the transformation frequency. Another major difference between the two systems was that in BPDE-I- and 1-NOP-treated cells, more cytotoxicity at higher concentrations of the chemicals was observed than with the S9-activated B(a)P-treated cells (see Table 5). At higher concentrations of the S9-activated B(a)P, an increased cytotoxic response (16-h treatment only) was observed without an apparent increase in dG modification or increase in the induced mutation frequency.

As shown in Table 2, a range of dG adducts was formed over a concentration range of 2.5 to 17 μg/ml for BPDE-I and 1-NOP treatment. Moreover, it was observed that adduct modification of DNA occurred at a zero cytotoxicity level for both BPDE-I and 1-NOP, which resulted in a tenfold greater transformation in BPDE-I-treated cells compared to B(a)P-treated cells. At the 50% cell survival point (4.6 μg/ml BPDE-I; 95 μg/ml 1-NOP), the number of adducts in the 1-NOP-treated cells was higher than in BPDE-I-treated cells, yet more transformants (121 $\pm$ 6.2 colonies per 10^5 seeded cells) formed in BPDE-I-treated cells than in 1-NOP-treated cells (41.0 $\pm$ 7.0 colonies per 10^5 seeded cells).

IV. DISCUSSION

Although there are many reports of transformation *in vitro* of human diploid fibroblasts using chemical or physical agents as carcinogens, there are only a few reports of *in vitro* transformation of epithelial cells. Comparison of the phenotypes of carcinogen-initiated fibroblast and epithelial cells with cells found in spontaneous SCC and sarcoma lesions should facilitate our understanding of the probable progressive stages of transformation. Progression is the process by which the initiated cells proceed sequentially from a normal phenotype toward a neoplastic phenotype. During this period, important cellular and possibly genetic changes have been identified, i.e., intracellular changes, morphological alteration of cells and colonies, cell surface modifications, anchorage-independent growth, cellular invasiveness, localized tumor formation, and formation of a progressively growing tumor. A recent reference to multistage carcinogenesis of human epithelial cells listed the possible involvement of four distinct stages of expression leading to the malignant transformation of epithelial cells. The apparent lack of occurrence of some of these stages in carcinogen-initiated cells may be due to certain critical stages (which usually are expressed as distinct events) existing as "silent" stages in the process, due to the events being telescoped into a more narrow window of time. This seeming absence of expression of these stages should not be confused with heterogeneity. Our interest in comparing the different stages of expression of two types of carcinogen initiated cells with cells from similar human tumors led us to the conclusion that human tumors contain multiple subpopulations in different stages of expression that are as unique in their biological and growth characteristics as cells transformed *in vitro* by carcinogens. For example, some of the human tumor cell populations have retained functions associated with differentiated cells; however, when these populations were placed into nude mice, some formed a progressively growing tumor in the surrogate host. Other tumor populations grew in soft agar and induced localized tumors that regressed after 4 to 6 weeks, but did not advance to the stage of a progressively growing tumor.

These observations imply that human spontaneous tumor lesions contain mixed subpopulations of tumor cells, some of which have progressed to fixed endpoints. Some populations are fixed in the AIG stage, while others are fixed in the preliminary or final stages of tumorigenicity. While some cells were permanently fixed in one stage of expression, others remained only temporarily in a particular stage and continued on to full neoplastic expression. The data obtained with transformed human cells *in vitro* suggest that these carcinogen-transformed cells develop characteristics similar to those of human tumor cells, including surface-associated antigen expression, AIG, cellular invasiveness, and tumorigenicity. Very seldom do any of these tumor cells or transformed cells exhibit an infinite life-span or form progressively growing, invasive tumors.

Comparison of the biological endpoints, cytotoxicity, transformation, and mutagenicity with adduct modification at different concentrations of B(a)P, BPDE-I, or 1-NOP suggested that the extent of modification of dG does not correlate as a linear relationship[11,50-52] with the biological endpoints. However, the results of this investigation do show a strong correlation between the frequency of mutations induced by B(a)P and the formation of dG-B(a)P adducts. This result indicates that adduct formation more closely parallels the biological endpoint of mutagenesis than the early events of transformation, i.e., expression of anchorage-independent growth. The interesting feature of these two different biological endpoints is that after initiation at a specific site, the initiated cells must progress through different stages to express transformation (but do not require further treatment with an environmentally supplied exogenous agent in order to express anchorage-independent growth), but the expression of mutagenesis is detected in a relatively short time. The nature of these two dissimilar biological endpoints would, by themselves, suggest that if the biological expression of each of these endpoints and adduct formation were direct correlations, the responses over a concentration range should exhibit similar response patterns in both systems.

The lack of a strong linear relationship between biological endpoint and the extent of modification of specific bases in the genomic DNA complicates the mechanistic interpretation of both the carcinogenesis and mutagenesis assays and the relationships, if such exist between mutagenesis and transformation.[53] The association between major or minor specific-base modification, leading to an expression of either cell transformation or mutagenesis, at best serves as an approximation of a relationship between endpoint and modification of specific target molecules.

In other mammalian systems,[54] there are incongruities when mutagenesis is correlated with transformation.[55] For example, Elmore et al.[56] also showed that increased mutation rates in carcinogen-treated human fibroblasts did not correlate with induced transformation frequencies of human fibroblasts.

Another anomaly was observed between the use of the activated metabolite BPDE-I, S9 mix-activated B(a)P, and parent B(a)P in treated fibroblasts. These data apparently suggest that the P450 complexes of each of the latter systems form different metabolites. Recently, we[35,57] reported that freshly isolated human foreskin cells *in vitro* can metabolize B(a)P without adding activated S9 mix.

The S9 mix used in these studies (see Table 5, 16 h) biotransformed the B(a)P; however, the distribution of intracellular vs. extracellular metabolites was different when compared to the endogenous cellular metabolism of B(a)P.[57] This may account for the greater amount of adduct modification when the cells without exogenous S9 mix metabolized the B(a)P. Exogenously supplied S9 mix for the B(a)P reaction produces a two- to fourfold increase in tetrols outside the cell compared to the intracellular amount.

Direct modification of DNA by a direct-acting carcinogen led to the formation of initiated cells that proceed into progressive stages that at 30 PDs exhibit anchorage-independent growth.[55,58]

It is our opinion that the use of activation systems to metabolize carcinogens can produce equivocal results. Moreover, the extent of metabolism by exogenously supplied S9 mix definitely has an effect on the distribution of intracellular, nonbound metabolites. As the concentration of the compounds increases, the extent of modification of specific dG molecules also increases, but we do not see a corresponding direct increase in the expression of transformation. Therefore, the increase in DNA modification does not directly correlate with a corresponding direct increase in toxicity or transformation over a broad concentration range.

In summary, the first change in the expression of a normal cell to a carcinogen-transformed phenotype following treatment with a carcinogen may be an altered cellular morphology (focus on a lawn of normal cells or a colony of altered cells when cloned at a low cell density). The presence of morphologically altered cells depends on the cell type and the carcinogen used for initiation; however, the altered morphology usually occurs only transiently between 2 and 10 PDs following treatment. After treatment, both epithelial and fibroblast-treated cell populations exhibit anchorage-independent growth following passage in culture. These observations suggest that the treated cells progress toward a malignant phenotype with passage. Treated cell populations that exhibit anchorage-independent growth exhibit cellular invasiveness on chick embryonic skin and subsequently form intracranial or subcutaneous localized tumors 0.8 to 1.2 cm in size in nude mice. These sequential, progressive developments in the carcinogen-transformed keratinocytes or fibroblasts are consistent with the concept of multistep development of cancer cells. Tumor cells isolated from spontaneous tumors exhibit low AIG, specific cell surface antigens, and growth in nude mice; however, the progressive development from a subcutaneous nodule to a progressively growing tumor in the surrogate host was not routinely expressed by either carcinogen-transformed epithelial or fibroblast cells isolated from human sarcomas or carcinomas.

ACKNOWLEDGMENTS

The research this laboratory has undertaken to study human cell carcinogenesis and the comparisons contained herein would not have been possible without the contributions from many collaborators too numerous to cite individually (see references); however, we extend our appreciation to these contributors for their scientific effort. We acknowledge further the partial support for this work from the National Institutes of Heath-National Cancer Institute (NIH-NCI) RO1 CA25907-07 (G.E.M.), Environmental Protection Agency (EPA) R813254 (G.E.M.), NIH-NCI P30 CA16058 (The Ohio State University Comprehensive Cancer Center), and EPA 68-02-4456 (B.C.C.).

REFERENCES

1. **Casto, B. C. and Hatch, G. G.**, Developments in neoplastic transformation, in *In Vitro Toxicity Testing*, Berky, M. and Sherrod, C., Eds., Franklin Institute, Philadelphia, 1977, 192.
2. **Milo, G. E. and DiPaolo, J. A.**, *In vitro* transformation of diploid human cells with chemical carcinogens, *In Vitro*, 13, 193, 1977.
3. **Kakunaga, T.**, Neoplastic transformation of human diploid fibroblast cells by chemical carcinogens, *Proc. Natl. Acad. Sci. U.S.A.*, 75, 1334, 1978.
4. **Milo, G. E. and DiPaolo, J. A.**, Neoplastic transformation of human diploid cells *in vitro* after chemical carcinogen treatment, *Nature*, 275, 130, 1978.
5. **Namba, M., Nishitami, K., and Kimoto, T.**, Carcinogenesis in tissue culture. Neoplastic transformation of a normal human diploid cell strain, WI-38, with Co-60, -rays, *Jpn. J. Exp. Med.*, 48, 303, 1978.

6. **Borek, C.**, X-ray-induced *in vitro* neoplastic transformation of human diploid cells, *Nature*, 283, 776, 1980.

7. **Milo, G. and DiPaolo, J.**, Presensitization of human cells with extrinsic signals to induced chemical carcinogenesis, *Int. J. Cancer*, 26, 805, 1980.

8. **Sutherland, B. M., Cimino, J., Delihas, N., Shih, A., and Oliver, R. P.**, Ultraviolet light-induced transformation of human cells to anchorage-independent growth, *Cancer Res.*, 40, 1934, 1980.

9. **Greiner, J., Evans, C., and DiPaolo, J.**, Carcinogen-induced anchorage-independent growth and *in vitro* lethality of human MRC-5 cells, *Carcinogenesis*, 2, 359,1981.

10. **Milo, G. E., Oldham, J. W., Zimmerman, R. J., Hatch, G. G., and Weisbrode, S. A.**, Characterization of human cells transformed by chemical and physical carcinogens *in vitro*, *In Vitro*, 17, 719, 1981.

11. **Silinskas, K. C., Kateley, S. A., Tower, J. E., Maher, V. M., and McCormick, J. J.**, Induction of anchorage independent growth in human fibroblasts by propane sultone, *Cancer Res.*, 41, 1620, 1981.

12. **Zimmerman, R. J. and Little, J. B.**, Starvation for arginine and glutamine sensitizes human diploid cells to the transforming effects of *N*-acetoxy-2-acetyl aminofluorene, *Carcinogenesis*, 2, 1303, 1981.

13. **Donahoe, J., Noyes, I., Milo, G. E., and Weisbrode, S.**, A comparison of expression of neoplastic potential of carcinogen-induced-transformed human fibroblasts in nude mice and chick embryonic skin, *In Vitro*, 18, 429, 1982.

14. **Dorman, H. B., Siegfried, J. M., and Kaufman, D. G.**, Alterations of human endometrial stromal cells produced by *N*-methyl-*N'*-nitro-*N*-nitrosoguanidine, *Cancer Res.*, 43, 3348, 1983.

15. **Kun, E., Kirsten, E., Milo, G. E., Kurian, P., and Kumari, H. L.**, Cell cycle dependent intervention by benzamide of carcinogen induced neoplastic transformation and *in vitro* poly(ADP-ribosyl)ation of nuclear protein in human fibroblasts, *Proc. Natl. Acad. Sci. U.S.A.*, 80, 7219, 1983.

16. **Zimmerman, R. J. and Little, J. B.**, Characterization of a quantitative assay for the *in vitro* transformation of normal human diploid fibroblasts to anchorage independence by chemical carcinogens, *Cancer Res.*, 43, 2176, 1983.

17. **Zimmerman, R. J. and Little, J.**, Characteristics of human diploid fibroblasts transformed *in vitro* by chemical carcinogens, *Cancer Res.*, 43, 2181, 1983.

18. **Farber, E.**, The multistep nature of cancer development, *Cancer Res.*, 44, 4217, 1984.

19. **Sutherland, B. M. and Bennett, P.**, Transformation of human cells by DNA transfection, *Cancer Res.*, 44, 2769, 1984.

20. **Steele, V. and Mass, M.**, A rat tracheal cell culture transformation system for assessment of environmental agents as carcinogens and promoters, *Environ. Int.*, 11, 323, 1985.

21. **Chang, S.**, *In vitro* transformation of human epithelial cells, *Biochim. Biophys. Acta*, 823, 161, 1986.

22. **Popescu, N. C., Amsbauch, S. C., Milo, G., and DiPaolo, J. A.**, Chromosome alterations associated with *in vitro* exposure of human fibroblasts to chemical or physical carcinogens, *Cancer Res.*, 46, 4720, 1986.

23. **Milo, G. E., Casto, B., and Ferrone, S.**, Comparison of features of carcinogen-transformed human cells *in vitro* with sarcoma-derived cells, *Mutat. Res.*, 199, 387, 1987.

24. **Hilwing, I. and Groop, A.**, Staining of constitutive heterochromatin in mammalian chromosomes with a new fluorochrome, *Exp. Cell Res.*, 75, 122, 1970.

25. **Poste, G., Tzeng, J., Doll, J., Greig, R., Rieman, D., and Zeidman, I.**, Evolution of tumor heterogenicity during progressive growth of individual lung metastasis, *Proc. Natl. Acad. Sci. U.S.A.*, 79, 6475, 1982.

26. **Milo, G. E., Yohn, J., Schuller, D., Noyes, I., and Lehman, T.**, Comparative stages of expression of human squamous carcinoma cells and carcinogen transformed keratinocytes, *J. Invest. Dermatol.*, 92, 848, 1989.

27. **Kumari, H.L., Shuler, C., and Milo, G. E.,** HNF transfection with chondrosarcoma DNA results in tumorigenicity and appearance of a sarcoma cell epitope, *Exp. Mol. Pathol.*, 1990.

28. **Milo, G. E., Shuler, C., Kurian, P., French, B. T., Mannix, D. G., Noyes, I., Hollering, J., Sital, N., Schuller, D. E., and Trewyn, R. W.,** Nontumorigenic SCC cell line converted to tumorigenicity by MMS without activation of H-*ras* or c-*myc*, *Proc. Natl. Acad. Sci. U.S.A.*, 87, 1268, 1990.

29. **Shuler, C., Kurian, P., French, B., Noyes, I., Sital, N., Hollering, J., Trewyn, R. W., Schuller, D., and Milo, G. E.,** Noncorrelative c-*myc* and *ras* oncogene expression in squamous cell carcinoma cells with tumorigenic potential, *Teratogen. Carcinogen. Mutagen.*, 10, 53, 1990.

30. **Kumari, L., Shuler, C., Lehman, T., Ferrone, S., and Milo, G. E.,** Development of a neoplastic phenotype following transfection of HNF cells with sarcoma DNA, *Carcinogenesis*, 10, 401, 1990.

31. **Milo, G. E., Malarkey, W., Powell, J., Blakeslee, J., and Yohn, D.,** Effects of steroid hormones in fetal bovine serum on plating and cloning of human cells *in vitro*, *In Vitro*, 11, 23, 1976.

32. **Huttner, J., Milo, G. E., Panganamala, R. V., and Cornwell, D. G.,** Fatty acids and the selective alteration of *in vitro* fibroblasts and guinea pig smooth muscle cells, *In Vitro*, 14, 854, 1978.

33. **Milo, G. E. and Casto, B. C.,** Conditions for transformation of human fibroblast cells: an overview, *Cancer Lett.*, 31, 1, 1986.

34. **Yohn, J., Lehman, T. A., Kurian, P., Ribovich, M., and Milo, G. E.,** Benzo[z]pyrene diol epoxide I modification of DNA in human skin xenografts, *J. Invest. Dermatol.*, 91, 363, 1988.

35. **Kurian, P., Nesnow, S., and Milo, G. E.,** Quantitative evaluation of the effects of putative human carcinogenesis and related chemicals on human foreskin fibroblasts, *Cell Biol. Toxicol.*, 6, 171, 1990.

36. **Hamburger, A. W., Reid, Y. A., Pelle, B., Milo, G. E., Noyes, I., Krakauer, H., and Fuhrer, J. P.,** Isolation and characterization of a monoclonal antibody specific for epithelial cells, *Cancer Res.*, 45, 783, 1985.

37. **Popescu, N. C., Amsbauch, S., Milo, G., and DiPaolo, J. A.,** Stable chromosome alterations associated with *in vitro* exposure of human fibroblasts to chemical or physical carcinogens, *Proc. Am. Assoc. Cancer Res. Abstr.*, 26, 30, 1985.

38. **Huang, S. L. and Waters, M. D.,** Two methods to induce 6-thioguanine resistance in human fibroblasts in the presence of rat-liver microsomes, *Mutat. Res.*, 121, 71, 1983.

39. **Albertini, R. J. and DeMars, R.,** Detection and quantitation of X-ray-induced mutation in cultured diploid human fibroblasts, *Mutat. Res.*, 18, 199, 1973.

40. **Maher, V. M. and Wessel, J. E.,** Mutation to azaguanine resistance induced in cultured diploid human fibroblasts by the carcinogen, *N*-acetoxy-2-acetylaminofluorene, *Mutat. Res.*, 28, 277, 1975.

41. **Jacobs, L. and DeMars, R.,** Chemical mutagenesis with diploid human fibroblasts, in *Handbook of Mutagenicity Test Procedures*, Kilby, B., Legator, M., Nichols, W., and Ramel, C., Eds., Elsevier/North-Holland, Amsterdam, 1984, 193.

42. **Jacobs, L. and DeMars, R.,** Quantification of chemical mutagenesis in diploid human fibroblasts; induction of azaguanine-resistant mutants by *N*-methyl-*N'*-nitrosoguanidine, *Mutat. Res.*, 53, 29, 1978.

43. **Huang, S. L. and Lieberman, M. W.,** Induction of 6-thioguanine resistance in human cells treated with *N*-acetoxy-2-acetylaminofluorene, *Mutat. Res.*, 57, 349, 1978.

44. **Randerath, K., Reddy, M. V., and Randerath, K.,** ^{32}P-labeling test for DNA damage, *Proc. Natl. Acad. Sci. U.S.A.*, 78, 6126, 1981.

45. **Gupta, R. C., Reddy, M. V., and Randerath, K.,** ^{32}P-postlabeling analysis of non-radioactive aromatic carcinogen-DNA adducts, *Carcinogenesis*, 9, 1081, 1982.

46. **Ribovich, M., Kurian, P., and Milo, G. E.**, Specific BPDE modification of replicating and parental DNA from early S phase human foreskin fibroblasts, *Carcinogenesis,* 7, 737, 1986.

47. **Lehman, T. A., Noyes, I., and Milo, G. E.**, Establishment and chemical transformation of human skin epithelial cells *in vitro, J. Tissue Cult. Methods,* 10, 197, 1986.

48. **Huang, S. L., Huang, S. M. S., Casperson, C., and Waters, M. D.**, Induction of 6-thioguanine resistance in synchronized human fibroblast cells treated with methylmethanesulfonate, *N*-acetoxy-2-acetyl-aminofluorene and *N*-methyl-*N'*-nitro-*N*-nitrosoguanidine, *Mutat. Res.,* 83, 251, 1981.

49. **Huang, S. L., Biddix, G., and Waters, M. D.**, Use of prolonged treatment and the fluctuation test to detect mutations in human fibroblasts treated with methyl methanesulfonate, *Mutat. Res.,* 105, 175, 1982.

50. **Maher, V. M. and Wessel, J. E.**, Mutation to azaguanine resistance induced in cultured diploid human fibroblasts by the carcinogen, *N*-acetoxy-2-acetylaminofluorene, *Mutat. Res.,* 28, 277, 1975.

51. **Shugart, L., Holland, J. M., and Rahn, R. O.**, Dosimetry of PAH skin carcinogenesis: covalent binding of benzo(a)pyrene to mouse epidermal DNA, *Carcinogenesis,* 4, 195, 1983.

52. **Irvin, T. R. and Wogan, G. N.**, Quantitation of aflatoxin B_1 adduction within the ribosomal RNA gene sequences of rat liver DNA, *Proc. Natl. Acad. Sci. U.S.A.,* 81, 664, 1984.

53. **Barrett, J. C. and Ts'o, P. O. P.**, Relationship between somatic mutation and neoplastic transformation, *Proc. Natl. Acad. Sci. U.S.A.,* 75, 3297, 1978.

54. **Huberman, E., Mager, R., and Sachs, L.**, Mutagenesis and transformation of normal cells by chemical carcinogens, *Nature,* 264, 360,1976.

55. **Barrett, J. C., Tsutsui, T., and Ts'o, P. O. P.**, Neoplastic transformation induced by a direct perturbation of DNA, *Nature,* 274, 229, 1978.

56. **Elmore, E., Kakunaga, T., and Barrett, J. C.**, Comparison of spontaneous mutation rates of normal and chemically transformed human skin fibroblasts, *Cancer Res.,* 43, 1650, 1983.

57. **Cunningham, M. J., Kurain, P., and Milo, G. E.**, Metabolism and binding of benzo[a]pyrene in randomly-proliferating, confluent and S-phase human skin fibroblasts, *Cell Biol. Toxicol.,* 5, 155, 1989.

58. **Barrett, J. C. and Ts'o, P. O. P.**, Evidence for the progressive nature of neoplastic transformation *in vitro, Proc. Natl. Acad. Sci. U.S.A.,* 75, 3761, 1978.

Chapter 12

PROGRESSION FROM PIGMENT CELL PATTERNS TO MELANOMAS IN PLATYFISH-SWORDTAIL HYBRIDS — MULTIPLE GENETIC CHANGES AND A THEME FOR TUMORIGENESIS

Juergen R. Vielkind

TABLE OF CONTENTS

I. INTRODUCTION

The idea that genetic factors play a causative role in the genesis of tumors is rather old and has been brought up by both the medical and scientific professions. In 1917, Norris,[1] a Scottish M.D., concluded that this disease is hereditary based on the clustering of melanoma in a family; in 1914, Boveri, a German zoologist, inspired by his observations on abnormalities in chromosome number and embryogenesis in the sea urchin,[2] published an article on the origin of tumors[3] in which he speculated that tumorous cell proliferation is caused by the predominance of chromosomes promoting cell proliferation or, alternatively, by the loss of chromosomes inhibiting unlimited growth. He further concluded that, because each cell has two chromosomes of each member, the depression of only one may go unnoticed.

Today, the genetic basis of tumorigenesis is undisputed. It is agreed that cancer cells contain changes in the complex genetic network that guarantee controlled cell division and progression of cells into the differentiated state. As a consequence of these genetic changes, cells fail to achieve or lose their differentiated state, leading to expression of undifferentiated characteristics such as cell growth. The changes are of two categories: they can be dominant (gain of function) and affect the gene class known as protooncogenes, or the changes can be of a recessive nature (loss of function), affecting the genes known as tumor suppressor genes (recessive oncogenes, anti-oncogenes, or growth suppressor genes). It is also clear from many studies that tumorigenesis is a multistep process; changes in only one of the two gene classes is not sufficient for cancer to occur, and a cooperative effect of genes of both classes has been suggested and some experimental evidence reported (see Reference 4; for an extensive review, see Reference 5).

Most of our current view on the genetic basis of human cancer has come indirectly from studies of the familial clustering of specific cancers and the accompanying changes in certain genes, a classic example being the retinoblastoma gene involved in human eye tumors (see Reference 6). However, there is one animal tumor model — hereditary melanoma in small tropical fish of the genus *Xiphophorus* known as platyfish and swordtails — in which the formation of a skin tumor, melanoma, can be delineated to various genetic factors. The melanomas in these fish are very similar to melanomas found in the mouse or in man.[7,8] The genetic principle of the melanomas in these fish was already recognized in the late 1920s (see References 9 to 11), and it appears that similar genetic principles underly other, i.e., carcinogen-induced tumors[12,13] in this model as well.

II. ORIGIN OF HERITABLE
MELANOMA:MACROMELANOPHORE PATTERNS

The melanomas in *Xiphophorus* fish have their origin in polymorphic pigment cell patterns. Several species exhibit these patterns, which represent

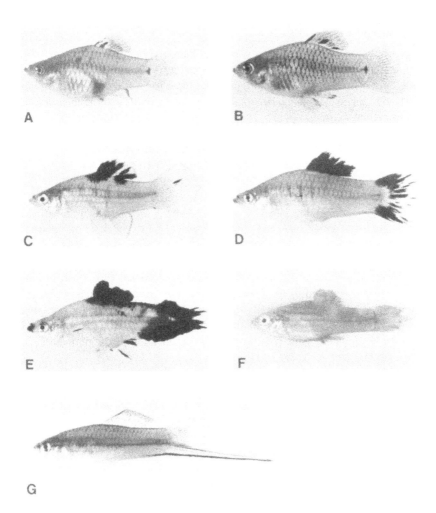

FIGURE 1. Parental *Xiphophorus* fish and backcross hybrids exhibiting heritable melanomas. (A) Platyfish, (*X. maculatus*) female exhibiting spotted dorsal pattern (black spots in the dorsal fin) due to *Sd* macromelanophore factor; (B) platyfish (*X. maculatus*) male exhibiting spotted-dorsal and stripe-sided pattern (zigzag bands on the flank) due to the *Sd* and *Sr* factors, respectively; (C) F₁ platyfish/swordtail hybrid exhibiting extended spotted-dorsal pattern; (D, E, F) backcross hybrids exhibiting (D) benign, (E) malignant, and (F) albinotic malignant dorsal fin melanoma; (G) swordtail (*X. helleri*) male.

spots or stripes on the fins or flank (Figure 1[A and B]); the patterns are composed of distinct, large, melanin-bearing pigment cells called macromelanophores.[14] The species *X. maculatus* (platyfish) exhibits the highest number of patterns, which are traditionally referred to as spot-sided, spotted-belly, black-banded or nigra, spotted-dorsal, and stripe-sided, terms reflecting the distribution and arrangement of the macromelanophores in the adult fish.[15]

These patterns are relatively invariably expressed in a given population; however, in the laboratory, significant changes in expression may be observed in interpopulation or interspecific hybrids. In these hybrids, the pattern may be suppressed or enhanced, or it may develop into a melanoma (reviewed in Reference 16). It is worth pointing out that these hybrids do not occur in the wild and can usually only be obtained in the laboratory through artificial insemination.

III. GENETIC FACTORS ASSOCIATED WITH MACROMELANOPHORE PATTERNS

In *X. maculatus,* the five patterns mentioned above are genetically determined by codominant, partially sex-linked allelic factors which are termed according to the pattern they code for, e.g., the *Sd* factor codes for the spotted-dorsal pattern (Figure 1[A and B]), the *Sr* factor for the stripe-sided pattern (Figure 1[B]), etc.[15] The identification of variants of these basic factors[9] and the observation of rare crossovers[17,18] within macromelanophore loci in laboratory strains has led to the conclusion that the macromelanophore factors comprise at least two functionally related genes: the macromelanophore gene (*M*), coding for the formation of the cell type (the macromelanophore) and a (linked) regulatory gene (*IR*), coding for the location of the pigment cells (the patterns) (see Reference 11).

IV. DEVELOPMENTAL ORIGIN OF MACROMELANOPHORE, THE MELANOMA CELL TYPE

The macromelanophore originates from a transient embryological structure, the neural crest, as was shown by transplantation experiments 35 years ago.[19] The neural crest in these fish is formed much as in other vertebrates and gives rise to segregating and migrating cells that appear in defined areas of the body, where they yield a variety of cell types; for example, cells that migrate under the ectoderm in the embryo finally reach the skin where they become melanoblasts and differentiate into melanocytes.[20,21] This is the pigment cell typically found in birds and mammals, including man. In amphibians and fish, however, the melanocyte is a transient stage and differentiation progresses into the melanophore, which is larger, shows more dendrites, and has a higher melanin content than the melanocyte. All *Xiphophorus* fish have melanophores which are evenly distributed in the skin and are termed micromelanophores (average, 100 to 200 μm in diameter) in order to distinguish them from the much larger macromelanophores (average, 400 to 500 μm in diameter), which are arranged into patterns and are only formed when the fish carries a macromelanophore factor.[14] Both cell types can be distinguished morphologically at the early melanocyte stage, i.e., when the melanoblasts develop into early melanocytes. This indicates that the two cell types are determined at the melanoblast stage or even before.[10]

V. GENETIC BASIS OF MELANOMA FORMATION

As indicated above, significant changes in pattern expression may be observed in interpopulation or interspecific hybrids. Mostly studied are the hybrids that arise from crosses of *X. maculatus* with another species, *X. helleri* (swordtails). The fact that each of the five patterns is enhanced to a different degree in the hybrids,[15] ranging from severe melanoma formation of the spot-sided pattern to only enhancement of the stripe-sided pattern, clearly indicates a genetic basis and suggests that a gene is encompassed within the macro-melanophore factors influencing melanoma formation. There are also differences in the potential of tumor formation among factors yielding the same pattern, e.g., some *Sd* factors do and some do not yield melanoma in certain hybrids, once again pointing to genetic determinants mediating melanoma formation. Similarly, the same macromelanophore factor can yield melanomas in *X. helleri* hybrids, but not in hybrids with another species such as *X. couchianus*.[9] Thus, the occurrence of a melanoma depends on the genetic composition of both the complex macromelanophore factor and the foreign gene pool into which it is introduced.

VI. GENETIC MECHANISMS OF MELANOMA FORMATION

A. ONCOGENIC ACTIVATION OF *Sd*-MACROMELANOPHORE FACTOR IN HYBRIDS

Although several species exhibit pigment cell patterns which may give rise to melanoma in hybrids, the patterns of the platyfish *X. maculatus* have been studied most, in particular the spotted-dorsal pattern (see Reference 11). This is because the changes in expression of this pattern are easily recognized in the dorsal fin over changes in expression of the other patterns appearing on the side of the fish's body. The change in spotted-dorsal pattern expression resulting in dorsal fin melanoma also best documents Mendelian inheritance of the melanoma phenotype. The crosses that result in melanoma formation in hybrids are illustrated in Figure 2(A and B), and photographs of the parental and *Sd*-carrying offspring fish are presented in Figure 1(A, C, D, E, and G). A female inbred platyfish homozygous for the *Sd* factor is crossed with a male from an inbred swordtail (*X. helleri*) strain that does not exhibit a macromelanophore pattern and is thus assumed not to carry genetic factors homologous to factors such as *Sd*. This cross gives rise to F$_1$ hybrids that exhibit an extended spotted-dorsal pattern, i.e., the number of macromelan-ophores is highly increased, resulting in a pattern with spreads over almost the entire dorsal fin. These F$_1$ fish can be considered to be hemizygous for the *Sd* factor and, consequently, when backcrossed to the swordtail parent, yield backcross fish which segregate 1:1 with respect to the *Sd;* the half of the offspring that inherit *Sd* exhibit an overgrowth of macromelanophore cells and express the pattern as a dorsal fin melanoma. It should be mentioned that

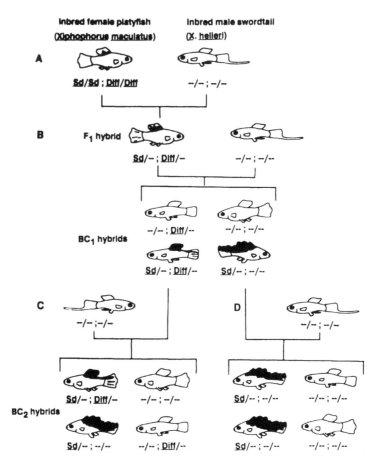

FIGURE 2.   Mendelian segregation of the X-chromosomal *Sd* factor and the autosomal tumor suppressor gene (*Diff*) and illustration of changes in the spotted-dorsal phenotype in F_1 and backcross platyfish/swordtail hybrids. (A) The cross between a platyfish (*X. maculatus*) female homozygous for *Sd* and *Diff* with a swordtail (*X. helleri*) male results in F_1 hybrids exhibiting an extended spotted-dorsal pattern; (B) the backcross of an F_1 to the swordtail results in the first backcross generation (BC_1) in which the *Sd* and *Diff* segregate, yielding 50% *Sd* offspring in which, due to the independent segregation of *Diff*, benign and malignant melanomas occur in a 1:1 ratio; (C, D) further backcrosses documenting that the benign melanoma phenotype is due to heterozygosity, and the malignant phenotype to the nullizygosity of *Diff*. In the crosses illustrated, the male is always *X. helleri;* reciprocal crosses lead to the same segregation patterns and phenotypes. Absent or diverged *X. helleri* allelic loci of *Sd* and *Diff* are represented by – –.

reciprocal crosses yield principally the same results. Obviously, the *Sd* factor that directs development of the macromelanophore cell pattern becomes deregulated when introduced into the gene pool of the swordtail which, because it does not form macromelanophores, does not possess regulatory mechanisms for the proper differentiation of this cell type. The formation of melanomas can be reverted to pattern formation by backcrossing hybrids carrying melanomas with the platyfish and thus reintroducing platyfish regulatory genes. Thus, the melanomas do not arise through mutation or rearrangement of the *Sd* factor during the crosses, but the *Sd* remains intact and is able to direct a normal phenotype in a proper genetic background.

B. LOSS OF TUMOR (MELANOMA) SUPPRESSOR GENE

The most interesting result of these crosses is the observation that, within the melanoma carriers, a second Mendelian segregation has occurred. Approximately half of the backcross hybrids which inherited the *Sd* exhibit a slow-growing melanoma which is operationally referred to as a benign melanoma because it does not affect the animal's life, while the other half exhibit a fast-growing malignant melanoma which is characterized by large nodular lesions — massive tissue destruction following invasion into the surrounding tissue — and by metastases.[10] This 1:1 segregation of melanoma types suggests an independently segregating autosomal gene which influences melanoma type. This is further demonstrated by backcrossing a benign and a malignant melanoma carrier with the parental swordtail (Figure 2[C and D]). As expected, in the first case a 1:1 segregation of benign:malignant melanomas in the *Sd*-inheriting offspring is observed, whereas in the second case only malignant melanomas are observed.

These segregation patterns, then, suggest that the benign phenotype is due to heterozygosity of a specific chromosome pair, i.e., one chromosome is derived from the platyfish and the other from the swordtail, while the malignant phenotype is due to homozygosity, i.e., both members of the chromosome pair are derived from the swordtail and both members of the homologous chromosomes from the platyfish have been lost as the result of the backcrosses. Direct proof for this interpretation comes from classical linkage analysis of isozyme markers that are polymorphic for the two species.[22] While in these analyses a linear linkage map for several enzyme markers was established, suggesting homology of the chromosomes from the two species, the map was nonlinear with regard to the gene influencing the melanoma phenotype, suggesting nonhomology for this particular region.[11] These data indicate that the swordtail chromosome may not contain a similar gene. Taken in sum, these data support the interpretation that a platyfish gene is homozygously lost and, as a consequence, a malignant melanoma can form. This gene, by definition, must be classified as a tumor or melanoma suppressor gene, although it is not recessive at the cellular level since it already allows an overgrowth of macromelanophores in the heterozygous state (see below).

C. MELANOMA SUPPRESSOR GENE HAS DIFFERENTIATION FUNCTION

It was already suggested in the late 1950s by Gordon,[23] one of the discoverers of the melanoma formation in these fish, that melanoma may be a consequence of improper pigment cell differentiation. This view was more

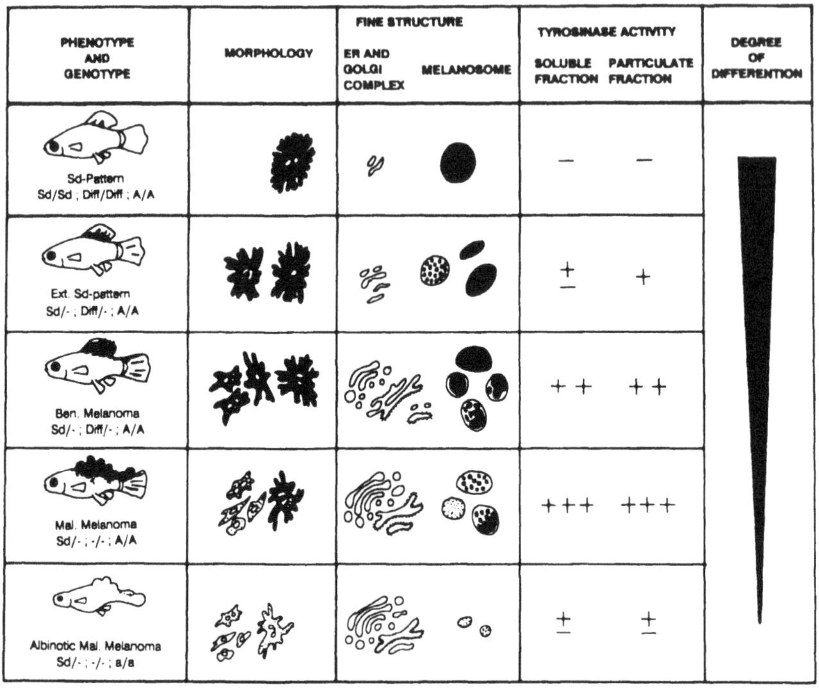

| PHENOTYPE AND GENOTYPE | MORPHOLOGY | FINE STRUCTURE | | TYROSINASE ACTIVITY | | DEGREE OF DIFFERENTION |
		ER AND GOLGI COMPLEX	MELANOSOME	SOLUBLE FRACTION	PARTICULATE FRACTION	
Sd-Pattern Sd/Sd ; Diff/Diff ; A/A				–	–	
Ext. Sd-pattern Sd/- , Diff/- ; A/A				±	+	
Ben. Melanoma Sd/- ; Diff/- ; A/A				+ +	+ +	
Mal. Melanoma Sd/- ; -/- ; A/A				+ + +	+ + +	
Albinotic Mal. Melanoma Sd/- ; -/- ; a/a				±	±	

FIGURE 3. Degree of macromelanophore differentiation in various *Diff* and *a* genotypes as judged by morphological, structural, and biochemical criteria. *Sd*, spotted-dorsal macromelano-phore factor; *Diff*, differentiation gene; *A*, wild type albino gene; *a*, recessive albino gene.

recently proven correct by studying morphological and biochemical markers of the macromelanophores in the normal patterns and in benign and malignant melanomas.[24-26] Each differentiation stage of the macromelanophore is well characterized by morphological criteria and the built-in biochemical markers, the melanin pigment and the enzyme necessary to synthesize the pigment. As shown in Figure 3, fully differentiated macromelanophores are found in the patterns, but the degree of differentiation is lower in the benign melanoma and even lower in the malignant melanoma; it is composed of mostly early melanocytes having a high proliferative capacity. Thus, it appears that the loss of the differentiation capability of the macromelanophore is paralleled by the loss of the platyfish melanoma suppressor gene. This further suggests that homozygosity of the melanoma suppressor gene in the platyfish is nec-essary for complete macromelanophore differentiation, and the gene has there-fore been termed differentiation gene (*Diff*).[25]

D. COMPLEX MACROMELANOPHORE LOCUS CONTAINS GENE PERMITTING OR PREVENTING MELANOMA FORMATION

The other basic macromelanophore factors — *Sp*, *Sb*, and *N* — are also enhanced to melanoma formation in hybrids, with *Sp* yielding the strongest and *N* yielding the weakest melanomas. Although not investigated as thoroughly, it appears that two groups of melanoma, i.e., benign and malignant, occur, suggesting that a *Diff*-type melanoma inheritance may also be involved. The *Sr* factor, however, does not give rise to melanomas, nor does a certain spotted-dorsal factor, *Sd**. Furthermore, after X-ray-induced mutation of the *Sr* factor[27] or after crossing over between the *Sr* factor and other factors, such as the *Sp* or *Sd* factor,[18] melanomas do occur in hybrids. These observations can be explained by assuming that the complex macromelanophore factors encompass a gene that, depending on its allelic state, either allows or hinders melanoma formation. Because, in the case of the changed *Sr* factor, the pattern phenotype was affected, it was proposed that this gene may actually be identical with the *IR* gene determining the pattern.[10] Thus, the loss of the *Diff* in hybrids only has a tumor-promoting effect if the macromelanophore factor is in the melanoma permissive state.[11]

VII. MODULATION OF MELANOMA PHENOTYPE

A. INFLUENCE OF GENETIC BACKGROUND

When using different laboratory swordtail strains, the severity of the melanoma in the backcross hybrids is strain specific (unpublished data, see also Reference 16). For example, when using one particular strain, a fast-growing melanoma is observed, also giving rise to secondary melanomas; using another strain, a malignant melanoma occurs, but no secondary tumors are observed; in yet a third case, a melanoma results that is almost of a benign type. Thus, while the loss of *Diff* seems to be the primary culprit, other factors have an influence on the tumor phenotype as well. Nothing is known about these factors, but one might speculate that they may be encoded by genes which influence proliferation and invasiveness or the metastasizing abilities of the melanoma cells. Genes that are connected to tumor metastasis have recently been identified in rat carcinoma and mouse melanoma cells.[28,29] Some, for example, are presumed to encode glycoproteins thought to be involved in the cell-cell and cell-matrix adhesion; overexpression in non-metastasizing tumor cells leads to metastatic behavior.[28]

B. MATERNAL EFFECT ON MELANOMA SEVERITY

In principle, reciprocal crosses also yield melanoma in backcross hybrids. However, the melanomas that arise from the crosses depicted in Figure 2(A and B) occur slightly earlier and are slightly more malignant than those that arise in reciprocal crosses in which the platyfish carrying the *Sd* is a male and the swordtail is a female, and similarly when the sexes are interchanged in the cross between the F_1 and swordtail. This suggests a maternal effect on the degree of malignancy of the melanoma (see Reference 30). An explanation for this may come from our studies on neural crest formation and the migration of its derivatives. As mentioned above, transplantation experiments have

FIGURE 4. Scanning electron micrograph of (A) a swordtail embryo at the stage of early truncal neural crest cell migration; (B) enlarged area shown as inset in (A) showing a few neural crest cells (arrows) migrating from the neural crest; and (C) equivalent area as shown in (B), but from a platyfish embryo at the same stage as the swordtail in (A), showing many more cells migrating since neural crest cell migration has begun earlier (arrows). M, mesencephalon; OpV, optic vesicle; OtV, otic vesicle; P, prosencephalon; Rh, rhombencephalon; S, somites; bar = 40 μm.

traced the origin of the macromelanophore to the neural crest. Unfortunately, other than those studies, nothing has been reported on neural crest formation and the fate of its derivatives since that time. We have investigated the formation of the neural crest and the segregation and migration behavior of its derived cells in embryos of the platyfish, the swordtail, and backcross hybrids. We found that the neural crest cells migrate earlier in the platyfish than in the swordtail (Figure 4).[20] Preliminary data indicate that the early migration behavior exists in approximately 50% of hybrids, of which 50% would exhibit the earlier appearance and slightly higher malignancy, i.e., in backcross hybrids resulting from the crosses shown in Figure 2, in which the macromelanophore factor is passed to the offspring through the female. Thus, the maternal effect of increasing melanoma severity may be directly or indirectly connected to those factors governing the early migration in both the platyfish and the melanoma hybrids. One can imagine that the early migrating cells in a swordtail background do not encounter the proper microenvironment during migration or at their final destination. Improper signals, however, could lead to higher proliferation and thus cause the more severe melanomas.

C. ALBINO GENE INCREASES MALIGNANCY OF MELANOMA

Following the same crossing scheme shown in Figure 2 but using a swordtail that is homozygous for a recessive albino (*a*) gene, backcross hybrids are obtained which carry albinotic (amelanotic) melanomas. These albinotic melanomas are even more malignant than their melanotic counterparts. They exhibit a higher rate of proliferation and lower degree of differentiation than cells of melanotic malignant melanomas.[24-26] The cells do contain some tyrosinase activity, but it must be stressed that melanin synthesis and the formation of melanomas are not causatively connected. The reason why the gene has such an effect is unknown.

VIII. CARCINOGEN-INDUCED NEOPLASMS IN BACKCROSS HYBRID FISH

The generation of melanomas in the backcross hybrids suggests a general principle in the genesis of tumors. As shown for the *Diff* gene-carrying chromosome, these hybrids carry a mixture of chromosome (gene) pairs that are homozygous for the swordtail or heterozygous for the platyfish and swordtail. Although the two species can be crossed and produce fertile offspring, some of their genes may have diverged sufficiently so that they are no longer able to act with or to regulate the genes of the other species. If some of these genes are involved in the biochemical pathways that regulate the growth and differentiation of various cell types or lineages, a mutation which disrupts the functional copy of such a gene is backcross hybrids could be the basis for the formation of a tumor. According to this hypothesis, treatment with carcinogens would be expected to result in the induction of various kinds of tumors in the backcross fish, but rarely in the F_1 and parental strains.[12,13] The results of these experiments support this hypothesis; the backcross hybrids are highly susceptible, and various kinds of neoplasms were observed in contrast to the F_1 and parentals, which rarely develop any neoplasm. One particularily interesting group for carcinogen-induced mutational changes are the backcross fish that are hemizygous for *Diff*. Only one mutational event is necessary to render this gene nonfunctional and lead to melanoma formation. Following this rationale, these fish have been used to study the effect of UV radiation.[31] Results of the carcinogen experiments allow two conclusions. First, these backcross fish represent a suitable and sensitive test system for the testing of potential carcinogenic substances,[32] due to the manipulated and thus susceptible genetic make-up. In other organisms, multiple genetic changes must accumulate, requiring presumably higher carcinogen concentrations and longer exposure times and also a greater number of animals to be treated. Second, because 20 out of the 24 chromosomes are marked by species-specific polymorphic isozyme markers,[33] this system allows the identification of chromosomes, and potentially genes, which hinder the formation of specific tumors and are necessary for proper differentiation of the cell type of a given tumor.

IX. CONCLUDING REMARKS AND FUTURE DIRECTIONS

Studies on the genetics of the heritable melanomas in platyfish/swordtail hybrids have clearly documented the complex relationship between genetic changes and the genesis of cancer. The formation of melanoma appears to parallel the progression of tumorigenesis. In individual fish, the normal macromelanophore pattern of the platyfish may be transformed into an extended pattern, a benign melanoma, or an invasive malignant melanoma. This "stepwise" progression is accompanied by a stepwise replacement of platyfish chromosomes (genes) by those of the patternless swordtail. Obviously, the swordtail cannot provide the genetic regulatory mechanisms necessary for the proper control of growth and differentiation of the macromelanophore. The accumulation of swordtail chromosomes is equivalent to the introduction of multiple genetic changes that lead to the genesis of the melanoma, and it is also the basis for the carcinogen susceptibility in these hybrid fish.

The progression of genetic changes leading to malignant melanoma can be summarized as follows:

1. The macromelanophore factor must be in a state permissive for melanoma development; if not, only an extended pattern is observed instead of the melanoma.
2. The loss of only one copy of the tumor suppressor gene *Diff* allows an expansion of the pattern.
3. The replacement of platyfish by swordtail chromosomes in combination with the loss of one *Diff* copy leads to a further expansion, i.e., a slowly growing benign melanoma.

4. The loss of the second *Diff* copy leads to a malignant melanoma.
5. Changes in other genes occur that allow malignant melanoma cells having lost both *Diff* copies to metastasize and form "full-fledged" melanomas; such genes were recognized in crosses with a particular swordtail strain in which the offpsring carried "malignant" melanomas but were unable to form any secondary melanomas. Very little is known about these genes, but they are presumed to play a role in the control of proliferation and homing of cells, and belong to the class of tumor suppressor genes.

One particular result in these multiple genetic changes needs comment. Since the *Diff* has a growth-promoting effect already in the heterozygous state, this gene cannot be considered as recessive at the cellular level. A similar situation for tumor suppressor genes has been found in colorectal cancer in man.[34] Therefore, it appears necessary to expand the recessive tumor suppressor model. For example, in the heterozygous state, reduced expression of the corresponding wild-type protein may alter growth regulation of the cell, resulting in clonal expansion of these cells. Similar to the platyfish/ swordtail model, in the colorectal cancer model, it is also possible to define a series of disease stages progressing from polyps to adenomas to the malignant carcinomas. This progression is characterized by an accumulation of mutations, including mutations of *ras* and losses in the 5q, 17p, and 18q chromosomal regions containing tumor suppressor genes such as *p53* and *DCC* (reviewed in Reference 34). Multiple genetic changes may also be involved in other cancers, but may be difficult to correlate with the progression of the disease because the various stages are hard to follow.

The tasks of the future are to analyze, at the molecular and cellular level, the genes and their physiological functions that, when altered, represent the various steps toward melanoma malignancy. Great efforts are underway in several laboratories to expand the gene map in *Xiphophorus*. Opportunities have developed to isolate the macromelanophore gene using a search-strategy assay because it was possible to induce macromelanophores in swordtails by injecting total genomic DNA[35] or DNA from genomic libraries (unpublished data) into the embryonic neural crest region. However, these fish are live bearers, and a sufficient number of embryos requires a large breeding colony and the sacrifice of many females. The recent development of transgenic fish systems[36,37] may provide an alternative and allow faster progression in the isolation of a macromelanophore-inducing genomic clone. Another opportunity has evolved as genes related to the *src*[38] and *erb* protoongogenes[39-41] have been found to be sex linked in *Xiphophorus*. In particular, two very similar *erb*-like genes have been found to be closely linked to the macromelanophore factors (Woolcock, Schmidt, and Vielkind, unpublished data). While their functions remain under speculation, one gene termed *Xmrk* has been considered the activated oncogene[41] which produces melanomas — without regard for the fact that melanomas occur only in hybrids; normal patterns are found in the platyfish. In addition, some confusion has been created by claiming that the gene is homologous to the tumor gene Tu,[41,42] a term that is very loosely defined.[43] Tu has also been claimed to cause not only the melanomas, but other neoplasms as well, and has been interchangeably used with the terms activated oncogene $Xmrk$, Tu, and macromelanophore gene.[39-43] Since it is quite clear that close linkage data are not sufficient to define function and that clearer definitions can be applied only after the function of the genes are better understood, it is preferable to use the term macromelanophore factor.

Expansion of the studies to the cellular level have become possible through the studies on the neural crest and the possibility of culturing neural crest cells.[44] These *in vitro* studies should facilitate the identification of extracellular signals important for migration and differentiation of neural crest-derived pigment cell precursors of the normal and melanoma genotype. Finally, it is noteworthy that heritable fish melanoma is a model for human familial melanoma, since both have a genetic basis and have their origin in neural crest-derived melanocytes. Human melanoma is a very serious problem not only because melanoma cells have a very high potential for metastasis, but also because the melanoma frequency is rising; by the year 2000, it is estimated that 1 in 90 persons in the U.S. will develop the disease.[45] Thus, it is necessary to identify the various genetic factors, most of them presumably in the tumor suppressor class, for better management of melanoma.

ACKNOWLEDGMENTS

The author gratefully acknowledges the excellent assistance of B. W. Woolcock and B. M. Schmidt during the writing of the manuscript and B. Sadaghiani for preparing the art and photographic work. Thanks are also due to M. Schartl for providing backcross embryos for the neural crest study. The Medical Research Council of Canada, The Cancer Society, Inc., and the National Institutes of Health (U.S.) are acknowledged for grant support. Special thanks are due to the MRC for supporting the author through a 5-year scholarship.

REFERENCES

1. **Norris, W.**, Case of fungoid disease, *Edinburgh Med. Surg. J.*, 16, 562, 1920.
2. **Boveri, T., Zellenstudien. V.** Ueber die Abhaengigkeit der Kerngroesse und Zellenzahl der Seeigellarven von der Chromosomenzahl der Ausgangszellen, *Z. Naturwiss.*, 39, 1, 1905.
3. **Boveri, T.**, Zur Frage der Enstehung maligner Tumoren, Fischer, Jena, 1914; transl. by Boveri, M., *The Origin of Malignant Tumors*, Williams and Wilkins, Baltimore, 1929.

4. **Weinberg, R. A.,** Oncogenes, antioncogenes, and the molecular basis of multistep carcinogenesis, *Cancer Res.,* 49, 3713, 1989.

5. Reviews by several authors in *Cell,* 64, 235, 1991.

6. **Hansen, M. F. and Cavenee, W. K.,** Retinoblastoma and the progression of tumour genetics, *Trends Genet.,* 4, 125, 1988.

7. **Grand, C. G., Gordon, M., and Cameron, G.,** Neoplasm studies. VIII. Cell types in tissue culture of fish melanotic tumors compared with mammalian melanomas, *Cancer Res.,* 1, 660, 1941.

8. **Grand, C. G. and Cameron, G.,** Tissue culture studies of pigmented melanomas: fish, mouse, and human, in *The Biology of Melanomas,* Vol. 4, New York Academy of Science, 1948, 171.

9. **Kallman, K. D.,** The platyfish, *Xiphophorus maculatus,* in *Handbook of Genetics,* Vol. 4, King, R. C. Ed., Plenum Press, New York, 1975, chap. 6.

10. **Vielkind, J. and Vielkind, U.,** Melanoma formation in fish of the genus *Xiphophorus:* a genetically-based disorder in the determination and differentiation of a specific pigment cell, *Can. J. Genet. Cytol.,* 24, 133, 1982.

11. **Vielkind, J. R., Kallman, K. D., and Morizot, D. C.,** Genetics of melanomas in *Xiphophorus* fishes, *J. Aquatic Anim. Health,* 1, 69, 1989.

12. **Schwab, M., Haas, J., Abdo, S., Ahuja, M. R., Kollinger, G., Anders, A., and Anders, F.,** Genetic basis of susceptibility for the induction of neoplasms by *N*-methyl-*N*-nitrosourea (MNU) and X-rays in the platyfish/swordtail tumor system, *Experientia,* 34, 780, 1978.

13. **Schwab, M., Abdo, S., Ahuja, M. R., Kollinger, G., Anders, A., Anders, F., and Freese, K.,** Genetics of susceptibility in the platyfish/swordtail tumor system to develop fibrosarcoma and rhabdomyosarcoma following treatment with *N*-methyl-*N*-nitrosourea (MNU), *Z. Krebsforsch.,* 91, 301, 1978.

14. **Gordon, M.,** The genetics of a viviparous top-minnow *Platypoecilus;* the inheritance of two kinds of melanophores, *Genetics,* 12, 253, 1927.

15. **Gordon, M.,** Effects of five primary genes on the site of melanomas in fishes and the influence of two color genes on their pigmentation, in *The Biology of Melanomas,* Vol. 4, New York Academy of Science, 1948, 216.

16. **Atz, J. W.,** Effects of hybridization on pigmentation in fishes of the genus *Xiphophorus, Zoologica (NY),* 47, 153, 1962.

17. **MacIntyre, P. M.,** Crossing over within the macromelanophore gene in the platyfish, *Xiphophorus maculatus, Am. Nat.,* 95, 323, 1961.

18. **Kallman, K. D. and Schreibman, M. P.,** The origin and possible genetic control of new, stable pigment patterns in the Poeciliid fish *Xiphophorus maculatus, J. Exp. Zool.,* 176, 147, 1970.

19. **Humm, D. G. and Young, R. S.,** The embryological origin of pigment cells in platyfish-swordtail hybrids, *Zoologica (NY),* 41, 1, 1956.

20. **Sadaghiani, B. and Vielkind, J. R.,** Neural crest development in *Xiphophorus* fishes: scanning and light microscopic studies, *Development,* 105, 487, 1989.

21. **Sadaghiani, B. and Vielkind, J. R.,** Distribution and migration pathways of HNK-1-immunoreactive neural crest cells in teleost fish embryos, *Development,* 110, 197, 1990.

22. **Morizot, D. C. and Siciliano, M. J.,** Linkage group V of platyfishes and swordtails of the genus *Xiphophorus* (Poeciliidae): linkage of loci for malate dehydrogenase-2 and esterase-1 and esterase-4 with a gene controlling the severity of hybrid melanomas, *J. Natl. Cancer Inst.,* 71, 809, 1983.

23. **Gordon, M.,** The melanoma cell as an incompletely differentiated pigment cell, in *Pigment Cell Biology,* Gordon, M., Ed., Academic Press, New York, 1959, 215.

24. **Vielkind, J. R., Vielkind, U., and Anders, F.,** Melanotic and amelanotic melanomas in xiphophorin fish, *Cancer Res.,* 31, 868, 1971.

25. **Vielkind, U.,** Genetic control of cell differentiation in platyfish-swordtail melanomas, *J. Exp. Zool.,* 196, 197, 1976.

26., Vielkind, U., Schlage, W., and Anders, F., Melanogenesis in genetically determined pigment cell tumors of platyfish and platyfish-swordtail hybrids: correlation between tyrosinase activity and degree of malignancy, *Z. Krebsforsch.*, 90, 285, 1977.

27. Anders, A., Anders, F., and Klinke, K., Regulation of gene expression in the Gordon-Kosswig melanoma system. I. The distribution of the controlling genes in the genome of the xiphophorin fish, *Platypoecilus maculatus* and *Platypoecilus variatus*, in *Genetics and Mutagenesis of Fish*, Schroeder, J. H., Ed., Springer-Verlag, New York, 1973, 33.

28. Guenthert, U., Hofmann, M., Rudy, W., Reber, S., Zoeller, M., Haussmann, I., Matzku, S., Wenzel, A., Ponta, H., and Herrlich, P., A new variant of glycoprotein CD44 confers metastatic potential to rat carcinoma cells, *Cell*, 65, 13, 1991.

29. Leone, A., Flatow, U., King, C. R., Sandeen, M. A., Margulies, I. M. K., Liotta, L. A., and Steeg, P. S., Reduced tumor incidence, metastatic potential and cytokine responsiveness of *nm23*-transfected melanoma cells, *Cell*, 65, 25, 1991.

30. Siciliano, M. J. and Perlmutter, A., Maternal effect on development of melanoma in hybrid fish of the genus *Xiphophorus*, *J. Natl. Cancer Inst.*, 49, 415, 1972.

31. Setlow, R. B., Woodhead, A. D., and Grist, E., Animal model for ultraviolet radiation-induced melanoma:platyfish-swordtail hybrid, *Proc. Natl. Acad. Sci. U.S.A.*, 86, 8922, 1989.

32. Vielkind, J. R., The oncofish: *Xiphophorus* hereditary melanoma and implications for carcinogenicity test model, in Proceedings of the Seventeenth Annual Aquatic Toxicity Workshop: Nov. 5-7, 1990, Vancouver, BC, Chapman, P., Bishay, F., Power, E., Hall, K., Harding, L., McLeay, D., Nassichuk, M., and Knapp, W., Eds., *Can. Tech. Fish. Aquat. Sci.*, No. 1774 (Vol. 2), 579, 1991.

33. Morizot, D. C., Slaugenhaupt, S. A., Kallman, K. D., and Cakravarti, A., Genetic linkage map of fish of the genus *Xiphophorus* (Teleostei: Poeciliidae), *Genetics*, 127, 399, 1991.

34. Fearon, E. R. and Vogelstein, B., A genetic model for colorectal tumorigenesis, *Cell*, 61, 759, 1990.

35. Vielkind, J. R., Haas-Andela, H., Vielkind, U., and Anders, F., The induction of a specific pigment cell type by total genomic DNA injected into the neural crest region of fish embryos of the genus *Xiphophorus*, *Mol. Gen. Genet.*, 185, 379, 1982.

36. Chong, S. S. C. and Vielkind, J. R., Expression and fate of CAT reporter gene microinjected into fertilized medaka (*Oryzias latipes*) eggs in the form of plasmid DNA, recombinant phage particles and its DNA, *Theor. Appl. Genet.*, 78, 369, 1989.

37. Stuart, G. W., Vielkind, J. R., McMurray, J. V., and Westerfield, M., Stable lines of transgenic zebrafish exhibit reproducible patterns of transgene expression, *Development*, 109, 577, 1990.

38. Vielkind, J. R. and Dippel, E., Oncogene-related sequences in xiphophorin fish prone to hereditary melanoma formation, *Can. J. Genet. Cytol.*, 26, 607, 1984.

39. Schartl, M., A sex chromosomal restriction-fragment-length marker linked to melanoma-determining *Tu* loci in *Xiphophorus*, *Genetics*, 119, 679, 1988.

40. Zechel, C., Schleenbecker, U., Anders, A., and Anders, F., v-*erb*B related sequences in *Xiphophorus* that map to the melanoma determining Mendelian loci and overexpress in a melanoma cell line, *Oncogene*, 3, 605, 1988.

41. Wittbrodt, J., Adam, D., Malitschek, B., Maeueler, W., Raulf, F., Telling, A., Robertson, S. M., and Schartl, M., Novel putative kinase receptors encoded by the melanoma-inducing *Tu* locus in *Xiphophorus*, *Nature*, 341, 415, 1989.

42. Zechel, C., Schleenbecker, U., Anders, A., Pfuetz, M., and Anders, F., Search for genes critical for the early and/or late events in carcinogenesis: studies in *Xiphophorus* (Pisces, Teleostei), in *Modern Trends in Human Leukemia VIII*, Neth, R. D. et al., Eds., Springer-Verlag, New York, 1989, 366.

43. **Anders, A. and Anders, F.,** Etiology of cancer as studied in the platyfish-swordtail system, *Biochim. Biophys. Acta,* 516, 61, 1978.
44. **Sadaghiani, B. and Vielkind, J. R.,** Explanted fish neural tubes give rise to differentiating neural crest cells, *Dev. Growth Differ.,* 32, 513, 1990.
45. **Rigel, D. S., Kopf, A. W., and Friedman, R. J.,** The rate of malignant melanoma in the United States: are we making an impact?, *J. Am. Acad. Dermatol.,* 17, 1050, 1987.

INDEX

Milton Keynes UK
Ingram Content Group UK Ltd.
UKHW031143141024
449569UK00024B/1101